T0393057

Computational Intelligent Security in Wireless Communications

Wireless network security research is multidisciplinary in nature, including data analysis, economics, mathematics, forensics, information technology, and computer science. This text covers cutting-edge research in computational intelligence systems from diverse fields on the complex subject of wireless communications security.

It discusses important topics including computational intelligence in wireless networks and communications, artificial intelligence and wireless communications security, security risk scenarios in communications, security/resilience metrics and their measurements, data analytics of cybercrimes, modeling of wireless communications security risks, advances in cyber threats and computer crimes, adaptive and learning techniques for secure estimation and control, decision support systems, fault tolerance and diagnosis, cloud forensics and information systems, and intelligent information retrieval.

The book –

- Discusses computational algorithms for system modeling and optimization from a security perspective.
- Focuses on error prediction and fault diagnosis through intelligent information retrieval via wireless technologies.
- Explores a group of practical research problems where security experts can help develop new data-driven methodologies.
- Covers application on artificial intelligence and wireless communications security risk perspectives.

The text is primarily written for senior undergraduate students, graduate students, and researchers in the fields of electrical engineering, electronics and communication engineering, and computer engineering.

The text comprehensively discusses a wide range of wireless communications techniques with emerging computational intelligent trends, to help readers understand the role of wireless technologies in applications touching various spheres of human life with the help of hesitant fuzzy set-based computational modeling. It will be a valuable resource for senior undergraduate students, graduate students, and researchers in the fields of electrical engineering, electronics and communication engineering, and computer engineering.

Wireless Communications and Networking Technologies: Classifications, Advancement and Applications

Series Editor:
D.K. Lobiyal, R.S. Rao and Vishal Jain

The series addresses different algorithms, architecture, standards and protocols, tools and methodologies which could be beneficial in implementing next generation mobile network for the communication. Aimed at senior undergraduate students, graduate students, academic researchers and professionals, the proposed series will focus on the fundamentals and advances of wireless communication and networking, and their such as mobile ad-hoc network (MANET), wireless sensor network (WSN), wireless mess network (WMN), vehicular ad-hoc networks (VANET), vehicular cloud network (VCN), vehicular sensor network (VSN) reliable cooperative network (RCN), mobile opportunistic network (MON), delay tolerant networks (DTN), flying ad-hoc network (FANET) and wireless body sensor network (WBSN).

Cloud Computing Enabled Big-Data Analytics in Wireless Ad-hoc Networks
Sanjoy Das, Ram Shringar Rao, Indrani Das, Vishal Jain and Nanhay Singh

Smart Cities
Concepts, Practices, and Applications
Krishna Kumar, Gaurav Saini, Duc Manh Nguyen, Narendra Kumar and Rachna Shah

Wireless Communication
Advancements and Challenges
Prashant Ranjan, Ram Shringar Rao, Krishna Kumar and Pankaj Sharma

Wireless Communication with Artificial Intelligence
Emerging Trends and Applications
Anuj Singal, Sandeep Kumar, Sajjan Singh and Ashish Kr. Luhach

Computational Intelligent Security in Wireless Communications
Suhel Ahmad Khan, Rajeev Kumar, Omprakash Kaiwartya, Mohammad Faisal and Raees Ahmad Khan

For more information about this series, please visit: https://www.routledge.com/Wireless%20Communications%20and%20Networking%20Technologies/book-series/WCANT

Computational Intelligent Security in Wireless Communications

Edited by
Suhel Ahmad Khan, Rajeev Kumar,
Omprakash Kaiwartya, Mohammad Faisal,
and Raees Ahmad Khan

CRC Press is an imprint of the
Taylor & Francis Group, an informa business

First edition published 2023
by CRC Press
6000 Broken Sound Parkway NW, Suite 300, Boca Raton, FL 33487-2742

and by CRC Press
4 Park Square, Milton Park, Abingdon, Oxon, OX14 4RN

CRC Press is an imprint of Taylor & Francis Group, LLC

ISBN: 9781032081663 (hbk)
ISBN: 9781032347028 (pbk)
ISBN: 9781003323426 (ebk)

DOI: 10.1201/9781003323426

Typeset in Times
by Deanta Global Publishing Services, Chennai, India

Contents

Preface.......vii
Editors.......ix
Acknowledgment.......xi
Contributor List.......xiii

Chapter 1 An investigation on Cooperative Communication Techniques in Mobile Ad-Hoc Networks.......1

Prasannavenkatesan Theerthagiri

Chapter 2 IoE-Based Genetic Algorithms and Their Requisition.......25

Neeraj Kumar Rathore, and Shubhangi Pande

Chapter 3 A Framework for Hybrid WBSN-VANET-based Health Monitoring Systems.......51

Pawan Singh, Ram Shringar Raw, and Dac-Nhuong Le

Chapter 4 Managing IoT – Cloud-based Security: Needs and Importance.......63

Sarita Shukla, Vanshita Gupta, Abhishek Kumar Pandey, Rajat Sharma, Yogesh Pal, Bineet Kumar Gupta, and Alka Agrawal

Chapter 5 Predictive Maintenance in Industry 4.0.......79

Manoj Devare

Chapter 6 Fast and Efficient Lightweight Block Ciphers Involving 2d-Key Vectors for Resource-Poor Settings.......99

Shirisha Kakarla, Geeta Kakarla, D. Narsinga Rao, and M. Raghavender Sharma

Chapter 7 Sentiment Analysis of Scraped Consumer Reviews (SASCR) Using Parallel and Distributed Analytics Approaches on Big Data in Cloud Environment.......121

Mahboob Alam, Mohd. Amjad, and Mohd. Amjad

Chapter 8 The UAV-Assisted Wireless Ad hoc Network.......131

Mohd Asim Sayeed, Raj Shree, and Mohd Waris Khan

Chapter 9 Integrating Cybernetics into Healthcare Systems: Security Perspective 161

Saquib Ali, Jalaluddin Khan, Jian Ping Li, Masood Ahmad, Kanika Sharma, Amal Krishna Sarkar, Alka Agrawal, and Ranjit Rajak

Chapter 10 Threats and Countermeasures in Digital Crime and Cyberterrorism 173

Mohit Kumar, Ram Shringar Raw, and Bharti Nagpal

Chapter 11 Cryptography Techniques for Information Security: A Review 191

Ganesh Chandra, Satya Bhushan Verma, and Abhay Kumar Yadav

Chapter 12 A Critical Analysis of Cyber Threats and Their Global Impact 201

Syed Adnan Afaq, Mohd. Shahid Husain, Almustapha Bello, and Halima Sadia

Chapter 13 A Cybersecurity Perspective of Machine Learning Algorithms 221

Adil Hussain Seh, Hagos Yirgaw, Masood Ahmad, Mohd Faizan, Nitish Pathak, Majid Zaman, and Alka Agrawal

Chapter 14 Statistical Trend in Cyber Attacks and Security Measures 241

Shirisha Kakarla, Deekonda Narsinga Rao, Geeta Kakarla, and Srilatha Gorla

Index 259

Preface

The widespread use of wireless technology in our daily lives has resulted in the increased demand for these devices. While the widespread use of wireless communications systems provides undeniable benefits to consumers, the communication exchanges are vulnerable to adversarial assaults due to the open broadcast nature of the wireless signals.

Wireless communications systems, unlike their wired equivalents, have major security risks from the physical layer to the application layer, which makes them less versatile than their wired counterparts. Security measures should be available to the user in order to secure wireless communications from harmful attacks. Wireless communications infrastructure and services require regular upgradation to manage the rapidly increasing demands to improve wireless communications security to fight against cybercriminal activities, especially because more and more people are using wireless networks (e.g., cellular networks and Wi-Fi) for online banking and personal emails, owing to the widespread use of smartphones. Wireless communications makes transmission of data more valuable than wired communication. Wireless communications have more vulnerable, secure, passive eavesdropping for data interception and active jamming. It needs authenticity, availability, confidentiality, and integrity requirements. To ensure the requirements, we need to design the wireless communications system to be secure and easy, to gain the users' satisfaction.

Further, due to the rapid expansion of modern and developing information technology such as social media, artificial intelligence, big data, Internet of Things (IoT), and smart devices in the past several decades, cyber threats and computer crimes have escalated in recent decades. Organizations due to actual and suspected cyber threats correlated with such developments have slowed the implementation of big data and the cloud. Secure communication will protect the cyber risks and it is an emerging area of ongoing research, as there is generally no clear view of how to model cyber risk and therefore how to price it. For companies, the value of cyber/wireless communications protection is rising. Security of wireless communications implies the ability to develop and assess a typology of cyber offenses and cyber threats in order to address them.

Wireless communications security research is multidisciplinary in nature, including researchers from data analysis, economics, mathematics, forensics, information systems, information technology, and computer science. The proposed book delivers an ideal platform to gather leading-edge work from diverse fields on the complex subject of Computational Intelligence Security in Wireless Communications.

Editors

Suhel Ahmad Khan is currently working as Assistant Professor in the Department of Computer Science, Indira Gandhi National Tribal University (A Central University), Amarkantak, Madhya Pradesh, India. He has 10 years of teaching and research experience. His areas of interest are Software Engineering, Software Security, Security Testing, Cyber Security, and Network Security. He has completed one major research project with PI funded by UGC, New Delhi, India. He has published numerous papers in international journals and conferences including IEEE, Elsevier, IGI Global, Springer, etc. He is an active member of various professional bodies such as IAENG, ISOC-USA, IACSIT, and UACEE.

Dr. Rajeev Kumar is currently working as Associate Professor in the Department of Computer Science and Engineering, Babu Banarsi Das University, Lucknow, Uttar Pradesh, India. He is a young and energetic researcher and has worked on two major projects (with PI) funded by University Grants Commission, New Delhi, India and Council of Science & Technology, Uttar Pradesh (CST-UP), India. He has more than five years of research and teaching experience. He has published numerous papers in international journals and conferences including IEEE, Elsevier, IGI Global, Springer, etc. His research interests are in the different areas of Security Engineering and Computational Techniques.

Omprakash Kaiwartya is currently working at the School of Science and Technology, Nottingham Trent University (NTU), Nottingham, UK, as Senior Lecturer and Course Leader for MSc Engineering (Electronics, Cybernetics and Communications). Previously, he was a research associate (equivalent to Senior Lecturer) in the Department of Computer and Information Science at Northumbria University, Newcastle, UK, where, he was involved in the gLINK, European Union project. Prior to this, he was a post-doctoral fellow (equivalent to Lecturer) in the Faculty of Computing, University of Technology (UTM), Johor, Malaysia. Before moving to Malaysia, he completed his BSc in Computer Science from Guru Ghansidas Central University, Bilaspur, Chhattisgarh, India, and combined MSc and Ph.D. from the School of Computer and Systems Science, Jawaharlal Nehru University (JNU), New Delhi, India. Overall, he has authored/co-authored over 100 international publications including journal articles, conference proceedings, book chapters, and books. His research interest focuses on IoT-centric smart environment for diverse domain areas including transport, healthcare, and industrial production. His recent scientific contributions are in Internet of Connected Vehicles (IoV), EMobility, Electronic Vehicles Charging Management (EV), Internet of Healthcare Things (IoHT), Smart use case implementation of Sensor Networks, and Next Generation Wireless Communication Technologies (6G and Beyond). Furthermore, he is Fellow of Higher Education Academy (FHEA), IEEE Senior Member and BCS Professional Member. He has served as a TPC member and reviewer in 100+

international conferences and workshops including IEEE Globecom, IEEE ICC, IEEE CCNC, IEEE ICNC, IEEE VTC, IEEE INFOCOM, ACM CoNEXT, ACM MobiHoc, ACM SAC, and many more. Furthermore, he has been reviewing papers for 30+ international journals including IEEE Magazines on Wireless Communications, Networks, Communications, IEEE Communications Letters, IEEE Sensors Letters, IEEE Transactions on Industrial Informatics, Vehicular Technologies, Intelligent Transportation Systems, Big Data, and Mobile Computing. Moreover, he has been an editorial member of various special issues with top-ranked journals in Communication Society and serving as Associate Editor of IET Intelligent Transport Systems, IEEE Internet of Things Journal, Springer, EURASIP Journal on Wireless Communication and Networking, MDPI Electronics, Ad-Hoc and Sensor Wireless Networks, and KSII Transactions on Internet and Information Systems.

Mohammad Faisal is currently working as Associate Professor and Head of the Department in the Department of Computer Application, Integral University, Lucknow, Uttar Pradesh, India. He has more than 15 years of teaching and research experience. His areas of interest are Software Engineering, Requirement Volatility, Distributed Operating System, Cyber Security, and Mobile Computing. He has published a book "Requirement Risk Management: A Practioner's Approach" published by Lambert Academic Publication, Germany, ISBN: 978-3-659-15494-2. He has published quality research papers in journals and national and international conferences of repute. He is contributing his knowledge and experience as a member of the Editorial Board/Advisory committee and TPC in various international journals/conferences of repute. He is an active member of various professional bodies such as IAENG, CSTA, ISOC-USA, EASST, HPC, ISTE, IAENG, and UACEE.

Raees Ahmad Khan (Member, IEEE, ACM, CSI, etc.) is currently working as Professor and Head of the Department in the Department of Information Technology, Dean of School for Information Science and Technology, Babasaheb Bhimrao Ambedkar University (A Central University), Vidya Vihar, Raebareli Road, Lucknow, India. He has more than 20 years of teaching and research experience. He has published more than 300 research publications with good impact factors in reputed international journals and conferences including IEEE, Springer, Elsevier, Inderscience, Hindawi, IGI Global, etc. He has published a number of national and international books (authored and edited) (including Chinese language). His research interests are in the different areas of Security Engineering and Computational Techniques.

Acknowledgment

The authors wish to express their sincere thanks to all those who participated in developing and evaluating the work contained in this book. The authors are especially grateful to CRC publishing and their representatives for administering and monitoring the process of the development of the manuscript, and for exercising such care and expertise to see the work of this book through to publication.

We are also thankful to all who participated in the review process of the book's chapters. The authors would also like to thank Mr. Gauravjeet Singh Reen, Senior Commissioning Editor-Engineering, for communicating, correcting errors, and careful reading of the various materials during the process of developing the book. Thanks also go to Dr. Ram Shringar Raw for his reading and administering the material in this book and for communicating with various parties during the process of finalizing the material covered by the book.

Dr. Suhel Ahmad Khan

Dr. Rajeev Kumar

Dr. Omprakash Kaiwartya

Dr. Mohammad Faisal

Prof. Raees Ahmad Khan

Contributor List

Syed Adnan Afaq
Integral University, Lucknow, Uttar Pradesh, India

Alka Agrawal
Babasaheb Bhimrao Ambedkar University, Lucknow, Uttar Pradesh, India

Masood Ahmad
Babasaheb Bhimrao Ambedkar University, Lucknow, Uttar Pradesh, India

Mahboob Alam
Jamia Millia Islamia University, New Delhi, India

Saquib Ali
Department of BCA, Azad Degree College, University of Lucknow, Lucknow, Uttar Pradesh, India

Mohd. Amjad
Jamia Millia Islamia University, New Delhi, India

Ganesh Chandra
Goel Institute of Technology and Management, Lucknow, Uttar Pradesh, India

Manoj Devare
Amity University, Bhatan Pada, Maharashtra, India

Mohd. Faizan
Babasaheb Bhimrao Ambedkar University, Lucknow, Uttar Pradesh, India

Srilatha Gorla
Ministry of Electronics and Information Technology, Hyderabad, Telangana, India

Bineet Kumar Gupta
Shri Ramswaroop Memorial University, Barabanki, Uttar Pradesh, India

Vanshita Gupta
Kamla Nehru Institute of Technology, Sultanpur, Uttar Pradesh, India

Mohd. Shahid Husain
University of Technology and Applied Sciences, CAS-Ibri Campus, Oman

Geeta Kakarla
Department of CSE, Sreenidhi Institute of Science and Technology, Hyderabad, Telangana, India

Shirisha Kakarla
Department of CSE, Sreenidhi Institute of Science and Technology, Hyderabad, Telangana, India

Jalaluddin Khan
University of Electronic Science and Technology of China, Chengdu, China

Mohd. Waris Khan
Integral University, Lucknow, Uttar Pradesh, India

Mohit Kumar
NSUT East Campus Formerly Ambedkar Institute of Advanced Communication Technologies and Research, New Delhi, India

Dac-Nhuong Le
Faculty of Information Technology, Hai Phong University, Hai Phong, Vietnam

Jian Ping Li
University of Electronic Science and Technology of China, Chengdu, China

Bharti Nagpal
NSUT East Campus Formerly Ambedkar Institute of Advanced Communication Technologies and Research, New Delhi, India

Mohd. Naseem
Baba Ghulam Shah Badshah University, Rajouri, Jammu and Kashmir, India

Yogesh Pal
Shri Ramswaroop Memorial University, Barabanki, Uttar Pradesh, India

Shubhangi Pande
Shri G. S. Institute of Technology and Science (SGSITS), Indore, Madhya Pradesh, India

Abhishek Kumar Pandey
Department of Computer Science, M. L. K. PG. College, Balrampur, India

Nitish Pathak
Guru Gobind Singh Indraprastha University, New Delhi, India

Ranjit Rajak
Dr. Harisingh Gour Central University, Sagar, Madhya Pradesh, India

D. Narsinga Rao
Directorate of Economics and Statistics (DES), Govt. of Telangana State, India

Neeraj Kumar Rathore
Indira Gandhi National Tribal University (A Central University), Amarkantak, Madhya Pradesh, India

Ram Shringar Raw
NSUT East Campus Formerly Ambedkar Institute of Advanced Communication Technologies and Research, New Delhi, India

Halima Sadia
Integral University, Lucknow, Uttar Pradesh, India

Amal Krishna Sarkar
Babasaheb Bhimrao Ambedkar University, Lucknow, Uttar Pradesh, India

Mohd. Asim Sayeed
Babasaheb Bhimrao Ambedkar University, Lucknow, Uttar Pradesh, India

Adil Hussain Seh
Babasaheb Bhimrao Ambedkar University, Lucknow, Uttar Pradesh, India

Kanika Sharma
Mangalmay Institute of Engineering and Technology, Greater Noida, Uttar Pradesh, India

M. Raghavender Sharma
Osmania University, Hyderabad, Telangana, India

Rajat Sharma
Shri Ramswaroop Memorial University, Barabanki, Uttar Pradesh, India

Raj Shree
Babasaheb Bhimrao Ambedkar University, Lucknow, Uttar Pradesh, India

Sarita Shukla
Shri Ramswaroop Memorial University, Barabanki, Uttar Pradesh, India

Pawan Singh
Indira Gandhi National Tribal University (A Central University), Amarkantak, Madhya Pradesh, India

Prasannavenkatesan Theerthagiri
GITAM University, Bengaluru, Karnataka, India

Satya Bhushan Verma
Goel Institute of Technology and Management, Lucknow, Uttar Pradesh, India

Abhay Kumar Yadav
Shri Ramswaroop Memorial University, Barabanki, Uttar Pradesh, India

Hagos Yirgaw
Adigrat University, Adigrat, Ethiopia

Majid Zaman
University of Kashmir, Srinagar, Jammu and Kashmir, India

1 An investigation on Cooperative Communication Techniques in Mobile Ad-Hoc Networks

Prasannavenkatesan Theerthagiri

CONTENTS

1.1 Introduction 1
1.2 Cooperation Techniques 3
1.2.1 Crediting Mechanisms 3
1.2.1.1 Incentive-based Approach 4
1.2.1.2 Reputation Schemes 5
1.2.1.3 Hybrid system 7
1.2.1.4 Trust-based schemes 8
1.2.2 Acknowledgment-based Mechanisms 8
1.2.2.1 End-to-End ACK Method 8
1.2.2.2 TWO ACK Method 9
1.2.2.3 Cryptographic-based signature 9
1.2.3 Punishment-based mechanism 10
1.2.3.1 Game-Theoretic Approach 10
1.2.3.2 Non-cooperative Game-Theoretic Approach 12
1.3 Discussion and Evaluation of research findings 14
1.4 Conclusion 20
References 21

1.1 INTRODUCTION

In mobile ad-hoc networks (MANETs), the terminals (nodes) such as mobile phones, gaming devices, laptops, tablets, and PDA communicate through cooperation. The wireless broadcasting mechanism is utilized for the communication of mobile nodes. In recent days, as many applications utilize MANETs specifically, cooperation is an essential issue in this kind of application, including discovery, military battlefield, event monitoring systems, and more civilian applications. In cooperation, any

DOI: 10.1201/9781003323426-1

information is processed, communicated, and forwarded by every node. The node cooperation with the other nodes is randomly generated, and the routes are discovered and used by MANET routing protocols [1]. As there is no centralized administration, cooperative communication plays a dynamic role in MANETs. The nodes in the network act as a router and host to perform all of its network operations. The MANET nodes are self-organizing and should cooperate well with the intermediate node to provide effective communication. In this concern, the neighbor nodes play a vital role in forwarding the packets to reach the destination [2]. If the source node's packet needs to reach the destination, which is within the source node's wireless communication (coverage) range, then the rate of successful data transmission is reasonable. If the destination is beyond the coverage area, they need to reach via the cooperative intermediate neighbor nodes intended to forward the packet to the destination [3, 4]. Figure 1.1 shows the cooperation of nodes in the MANET. Considering the example shown in Figure 1.1, node 1 needs to forward the packet to the destination node 6; thus, it needs the well cooperating neighbor nodes. As shown in Figure 1.1, nodes 4 and 5 are the well cooperating nodes so that the relaying packets can reach the destination node.

The best network performance can be achieved if the entire node in the network involves relaying the packets. It cannot be achieved in the real world because of MANET nodes' nature, such as resource constraints and dynamic topology. The behaviors of preserving their energy to survive in the network are called selfishness, and malicious activities can easily break this dynamic cooperation. The malicious activities are generated by the malicious node to either break the actual operation MANETs or disturb the whole MANET system. The selfish activity also shows the behavior of a disturbing network, but in the way of non-cooperation [5]. This kind of non-helping tendency in the forwarding of the packet is called selfish nodes. Selfishness is challenging to observe and qualify, unlike with the malicious nodes. Causes for selfishness are the heavy load in the network, which leads to dropping any incoming packet beyond the specific limit, and depletion of unwanted energy is

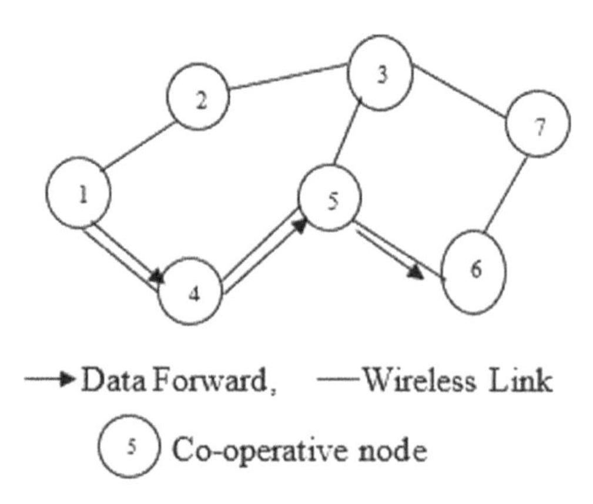

FIGURE 1.1 Cooperation.

sustained. It seems to behave like malicious nodes instead of selfishness [6]. These selfish behaving nodes are detected and avoided or stimulated to cooperate with other nodes in the network through cooperation mechanisms. One of the widely accepted models for stimulating cooperation is the reputation-based model. In this model, the reputation is collected directly from the node or indirectly from the neighbors' collection. By observing and monitoring each node's cooperation, the reputation is evaluated. The direct information from the node is more trusted than the neighbor's indirect information [7].

1.2 COOPERATION TECHNIQUES

Many researchers have devoted their work to the development of several cooperative techniques and have proposed many algorithms. These cooperative techniques are commonly categorized into three major modules: crediting mechanisms, acknowledgment-based mechanisms, and punishing mechanisms, based on the strategy utilized to enable cooperativeness. Figure 1.2 shows the different classification of cooperation techniques.

1.2.1 Crediting Mechanisms

The credits are used on a node for its cooperativeness with other nodes. The increase in the credits of a node shows that it helps in cooperatively forwarding the packets. The non-cooperation decreases the credit of a node in the network. A node earns the credits by forwarding the packets to others. Thus, it becomes the motivation for another node to behave cooperatively. The nodes which are not having credits will not participate in the cooperation process and not forward the packets. The crediting mechanism is further classified into four methods: incentive-based, reputation-based, hybrid-based, and trust-based approaches, as shown in Figure 1.3.

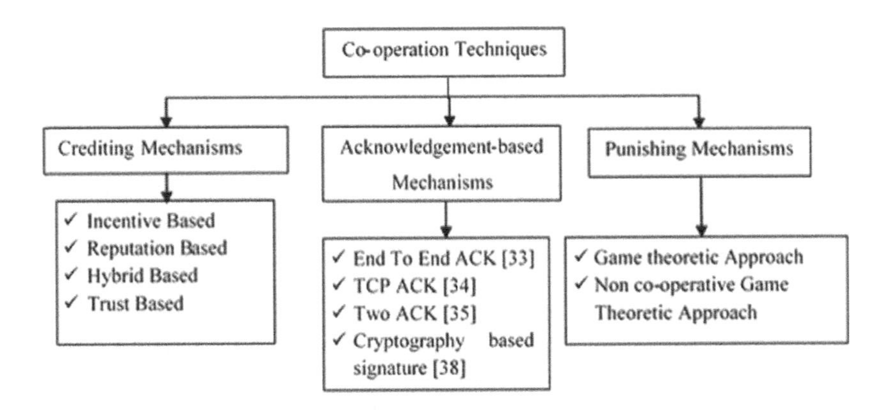

FIGURE 1.2 Cooperation Techniques.

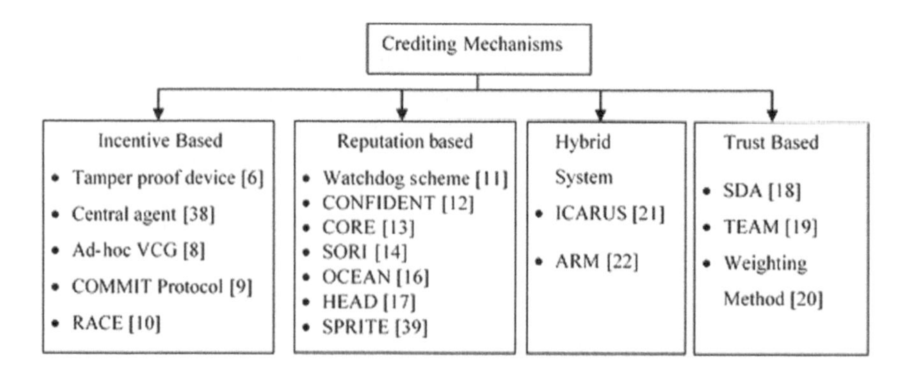

FIGURE 1.3 Crediting Mechanisms.

1.2.1.1 Incentive-based Approach

Incentive-based methods are intended to motivate the nodes in the network to cooperate with all intermediate neighbor nodes. The incentives are given to the nodes in the form of credits or awards, based on the observation and behaviors of the particular node. Many researchers analyzed the nodes' cooperation based on the incentive method, such as tamper-proof devices, central agent methods, ad-hoc Vickrey, Clarke, and Groves (VCG), COMMIT Protocol, and Report-based pAyment sChemE (RACE).

1.2.1.1.1 Tamper-proof device

Buttyan and Hubaux (2001) had proposed a tamper-proof device for assigning the credits. This device is installed on each node [8]. Credits are given to nodes based on the node's network services such as forwarding, sending packets. The authors also define two models as the Packet Purse Model (PPM) and the Packet Trade Model (PTM). In the PPM, the source node is responsible for paying the credit to the nodes based on their behavior along the path to the destination. The opposite method is used on the PTM; here, the destination node pays the credit to the node. The author introduces the credit count to avoid overloading of the packet by the source nodes. Counter increases/decreases based on cooperation. This scheme has many drawbacks as follows: first, the tamper-proof device should be installed at each node, which is impossible, and the device needs to be protected from external attack. It is not produced in the real world. Only the central node achieves more credits than the other nodes. Even for forwarding a node's packets, it should have enough credits. It affects the Packet Delivery Ratio (PDR), throughput of the network.

1.2.1.1.2 Central Agent

Zhong et al. (2003) proposed a credit-based scheme where the central agent is used for paying the credits to a node [9]. The central agent verifies the issuance of the credit paid to a node and then confirms the credit based on the reports by each forwarding node. The nodes have to keep track of their actions and then claim their credit from the central agent. After receiving the claims, the agent gives credit to the nodes; the agent gives credit to the participating nodes using the source node and the cooperating node. Because of the central node, it reduces the burden of each node credit assignment and tamper-proof device. However, it has the drawback of a communication bottleneck between the nodes and the central agent. The authors Anderegg et al. (2003) had introduced the ad-hoc VCG routing protocol; in this protocol, every node should generally publish their available energy in the route discovery process. The source node then chooses the best cost-effective energy path for data packet forwarding and assigning the credit to those participating intermediating nodes. It is an effective way of choosing a truthful energy-efficient route. However, it is not necessary for all the nodes to give genuine energy values. There may be the occurrence of collision between the nodes having similar energy values [10].

1.2.1.1.3 COMMIT Protocol

The COMMIT Protocol was proposed by Eldebenz et al. (2005). Rather than depending on all intermediate participating nodes, the source node only involves in the route discovery process. The source node should announce the maximum total credits that it offers for data forwarding. On receiving this information, all of the intermediate nodes agree or reject based on credits. When the offered credits are found, the destination will send a path to the sender. Even though, it is easier to control the sender, for the nodes with tedious communication overhead, it is not easier to achieve the offered credits [11]. A report-based payment scheme had been proposed by the authors Mohmoud et al. (2003). They had proposed the lightweight payment-reporting scheme. The credit reports are submitted to the central agent. The report contains only less important information. When extra information is needed, the central agent requests proof on the node. These proofs were stored only temporarily because of the storage overhead concern [12].

1.2.1.2 Reputation Schemes

The reputation is another type of scheme in monitoring node cooperation to anticipatory work with the intermediate neighbor nodes. In the reputation scheme, each node should monitor, observe, and collect the other nodes' behaviors and reports in the network. This information was used to evaluate the reputation value of the observed nodes. Based on these values, the node's selfishness level is determined. If the reputation value is less than a certain verge value, it shows the particular node's selfishness. This node needs to be avoided in the routing process. The important issue in evaluating reputation values is based on the node behavior to determine the passive/active or negative/positive acknowledgments. When the node's behaviors are examined, the reputation system takes further evaluation.

1.2.1.2.1 Trusting/voting system

In the reputation scheme, the trusting/voting technique is used by the monitoring nodes to determine the other node's cooperation or non-cooperation in the network service, such as forwarding the packets and routing, and the opinion on the particular node is examined from the other nodes, to detect the selfish node. The watchdog mechanism had been proposed by Marti et al. (2000), the rating scheme for computing the reputation in which the monitoring node's activities are involved [13]. The rate is assigned to the node by the information gathered from other nodes, and the node, which has a low rating, is avoided from the routing path. This type of rating of the nodes is called the path rating. Here, the watchdog node listens, the next neighbor hops transmission to know whether it helps in forwarding the relay or drops the packets. When the packets are dropped by such a node, then the counter value will be increased, that is, the misbehaving value for that particular node. When the rate reaches a certain threshold value, the watchdog node makes it as the misbehaving node, then the particular node will be excluded from the routing path established.

1.2.1.2.2 CONFIDENT Method

Authors Buchesser et al. (2002) had improved the path rating mechanism by a voting mechanism called Cooperation of Node: Fairness in Dynamic Ad-hoc Network (CONFIDENT). When the neighbor node detects misbehavior in the network, this information is forwarded to the reputation system. It gathers information from all observing nodes and enough proof for such misbehavior. The modified Bayesian approach gives less importance to the past observations. Here, assigning the rate to different functions is based on their behavior in the network. If this rating exceeds the threshold values, then the particular node is punished by not forwarding any packet to it. In confident schemes, each node uses four components for detecting and isolating the selfish nodes. (1) The monitor is to listen to the nodes with deviating behavior from the transmission in the network. Deviating node's information (ALARM) was alerted to the next component. (2) The Trust Manager analyzes these ALARM messages and decides whether to route with the node or not. (3) The Reputation System provides the reputation value based on the behaviors in the network. (4) The Path Manager finally avoids the nodes from the routing path, which are determined as the malicious non-cooperative nodes. However, it requires periodic packet exchanges as overhead [14].

1.2.1.2.3 CORE Method

In Collaborative Reputation Mechanism (CORE), Michiardi et al. (2002) had calculated the reputation values in three ways: the reputation value calculated by directly observing node behavior is called subjective reputation. The reputation value calculated based on the information provided by other nodes is called indirect reputation. The functional reputation is a combination of both. During the routing discovery process (in the route request (RREQ), route reply (RREP)), the reputation values are updated on each node's reputation table. The nodes that did not relay the packet were punished. It is important to the confident concept that the problem in confidence is false node voting. Here, it is conquered by restricting the negative rating dissemination in the network. The CORE only allows the good rating node [15]. Even when the node has a false rating as acting good, it is actually a bad node.

1.2.1.2.4 SORI Method

On improving these two CONFIDENT, CORE, instead of globally broadcasting the two-reputation information, is shared only on the intermediate neighbor nodes in the Secure and Objective Reputation-based Incentive Scheme for Ad-hoc Network (SORI) system. In SORI authors He et al. (2004) aim to reduce selfishness by encouraging the nodes to participate in the relaying of packets [16]. The reputation values are assigned to nodes by the neighbor nodes using the one-way hash chain-based algorithm. In this way, the penalties are given to the selfish nodes refusing in the packet relaying.

1.2.1.2.5 OCEAN Method

The Observation-based Cooperative Enforcement in Ad-hoc Networks (OCEAN) was proposed by Bansal et al. (2003). In the OCEAN, the observation of the node's activities is not yet shared globally like in SORI. Here, each node maintains the rating information of the neighbor node [17]. Based on the cooperation of the nodes in packet forwarding, the rating value can be increased or decreased. When the node's value is less than the threshold value, then such a node is listed in the misbehaving node list. The services of such nodes are also avoided. The feature of this scheme is that the nodes that are in the misbehaving list are inactive for some period. After that, it allows the network to behave cooperatively, reducing the chance of false detection. This scheme betters in all previous attempts at detecting cooperativeness between the nodes in the network. Even though it was affected by false nodes, it will be reported as misbehaving node by other nodes. The attacks are also possible for those nodes in the misbehaving list. Guo et al. (2007) developed the Hybrid mechanism to Enforce Nodes Cooperation (HEAD). The HEAD scheme uses the alerting messages instead of broadcasting the faulty misbehavior lost in the route discovery process [18]. S. Zhong et al. (2003) had developed a simple cheat-proof, credit-based system (SPRITE) for effective communication [19]. In this strategy, cooperation among the nodes is encouraged by assigning the reward in terms of credits called the cheat-proof system. Tamper-proof hardware is not used in any node; instead, it uses the receipts. The receipts from the forwarded or routed message are validated, and the credits are assigned to each node in the system. The central credit clearance service (CCCS) is to manage all the credits assigned to the nodes.

1.2.1.3 Hybrid system

The hybrid system utilizes the advantages of both credit and reputation schemes to stimulate the MANET's cooperation.

1.2.1.3.1 ICARUS Method

Charilas et al. (2012) had presented the ICARUS: Hybrid incentive mechanism for the cooperation stimulation scheme to control the credit exchanges between the participating nodes by utilizing the reputation schemes [20]. In the ICARUS, credit account service (ICAS) is used by the central agent for assigning credits to the nodes to determine the selfish non-cooperative nodes. However, in ad-hoc networks, the reliance on a central agent is difficult because of its dynamic nature. The account-based hierarchical Reputation Management (ARM) system uses the credit and reputation scheme proposed by Shen et al. (2008) but does not use any central agent for the detection [21]. In this scheme, each node maintains the credits for the peer nodes based on the reputation value. The nodes act as reputation managers for relaying the packets. However, in heavy traffic, the relaying task on nodes could not be a good way.

1.2.1.4 Trust-based schemes

The trust-based mechanism uses trust features for determining node behaviors. Based on the trust values, the good or bad trust on the particular node is to take. Several researchers worked on computing and evaluating trust features and node's trustworthiness in determining cooperative nodes.

1.2.1.4.1 SDA Method

The authors Z. K. Chong et al. (2013) adopted the Separation of Detection Authority (SDA) for detecting the selfish nodes in the network and improving the trustworthiness of the nodes [22]. The improvement to the node behavior's trustworthiness is done by using three components, such as the reporting node, agent node, and central authority. The reporting node finds out the misbehaving non-cooperative node and generates reports to forward it to the central authority. The central authority investigates the reports by using the agent nodes. The agent nodes are the neighbor monitoring nodes. Finally, all the agents submit the reports about the suspicious node to the central authority. However, here, communication overload is tight between the three entities and may degrade network performance.

1.2.1.4.2 TEAM Method

In the Trust-based Exclusion Access Control Mechanism, authors L. H. G. Ferraz et al. (2014) used the two modules for computing node's trustworthiness as local and global. The nodes observe and gather the one-hop neighbor's information from the local module. Then, this collected behavioral information is forwarded to this global module. The global module consists of nodes that evaluate the evidence on such trusts by using the voting mechanisms [23]. Then the non-cooperative nodes are defined in the access to the network. Here, the main advantage is the low overhead of messages that have been used for detecting and excluding such misbehaving nodes. However, the trust features and the friendship mechanisms need to be strengthened in the modules.

1.2.1.4.3 Weighting Method

The authors Yu et al. (2009) had calculated the average of the observed node's performance to detect the node's trustworthiness. The link quality information is also measured to improve the accuracy of the trust value [24]. However, the routing anomalies, interference problems, need to be avoided in these trust-based techniques.

1.2.2 Acknowledgment-based Mechanisms

1.2.2.1 End-to-End ACK Method

End-to-End ACK is an acknowledgment-based mechanism proposed by Conti et al. (2003) in which when the destination node receives the packet from the source node, it has the responsibility to send back an ACK to the source node [25]. The source node waits for some time to receive an ACK from the destination; if it does not receive ACK within the time, it assumes that the packet did not reach the destination. Here, authors use the reliability index for maintaining the performance and reliability of packets. When the index exceeds threshold values, then the corresponding route needs to be excluded. The reliability index is updated based on ACK from the destination node to the source node. However, every node's reliability index is visible to other nodes; it provides the attackers with chances to making an attack [26]. Refaei et al. (2005) had developed TCP acknowledgment, an ACK-based mechanism, but the neighbor node's activity was only used to measure the RI (Reliability Index). The authors [27] used the same ACK scheme and maintained the RI for neighbor nodes. When the source node receives ACK from the destination node, the RI of their neighbor nodes increases. It decreases when the source node does not receive ACK, or when some node retransmits the same packet.

1.2.2.2 TWO ACK Method

Instead of using many nodes in the network for end-to-end ACK, in TWO ACK authors, Balakrishnan et al. (2005) used two hops for ACK. This scheme works by sending ACK by a two-hop query from the sender node [28]. The two-hop away node is required to send an ACK back to the sender node on receiving the packet. The RI is calculated by the waiting time. The sender node waits for some time to receive the ACK from the two-hop away node. The RI is increased or decreased in threshold time values. When the RI is decreased to a low, the node and the link are avoided from routing and are marked as misbehaving. However, there may be the occurrence of a potential false neighbor on the RI. It is not yet analyzed in this paper. It is needed to gather the evidence on such occurrences from the neighbor node. Also, the congestion causes misbehavior when the network traffic is high [29]. Figure 1.4 illustrates the working of the TWO ACK scheme.

1.2.2.3 Cryptographic-based signature

Authors H. Yang et al. (2002) adopted the cryptographic-based signature with the Ad-hoc On-demand Distance Vector (AODV) to protect data forwarding and routing

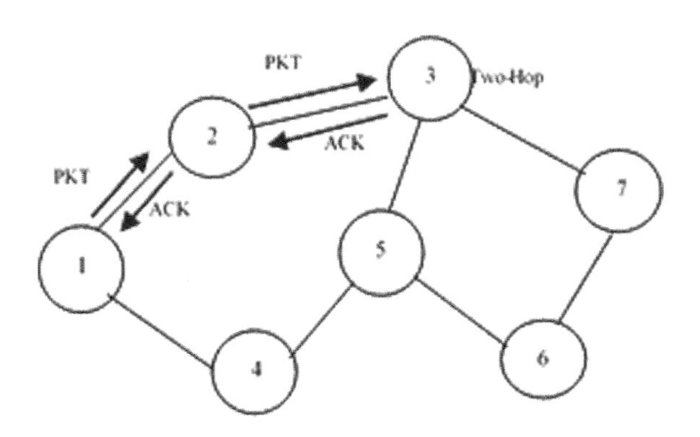

FIGURE 1.4 Two ACK Scheme.

in the network. This technique uses the cryptographically signed tokens to provide such protection against misbehaviors [30]. This token has an expiry period, which depends on the token holder's behavior in the network. It means that the token holder relays the packet as receive means, it will have a better expiry. It is also required to renew the token before its expiration. The renewing is done by gathering the k-number of different signals from its neighbor nodes. The neighbor node monitors every other node for detecting the misbehavior about whether the data packet has been dropped uncertainly. Based on such values, the nodes renew the request for the token signature that is needed to be granted. It has a better solution in advance to the watchdog. However, only depending on k-neighbor nodes for detecting misbehavior causes the drawback. Even though the node is not misbehaving, the low k-valued nodes are declared a suspicious node. The high k-value means that it has high connectivity, which is not possible in MANETs.

1.2.3 Punishment-based mechanism

In the punishment-based mechanism, the penalties are given to the node, which behaves selfishly. Moreover, they are isolated from the routing path and avoided in any services because of the penalty. The punishing mechanism uses game-theoretic approaches and non-cooperative game-theoretic approaches depicted in Figure 1.5.

1.2.3.1 Game-Theoretic Approach

1.2.3.1.1 DECADE Method

The selfish nodes, which will not participate in the cooperative forwarding of packets, are detected by using the Dynamic Source Routing (DSR) protocol in the distributed emergent cooperation through adaptive evolution in the MANET (DECADE)

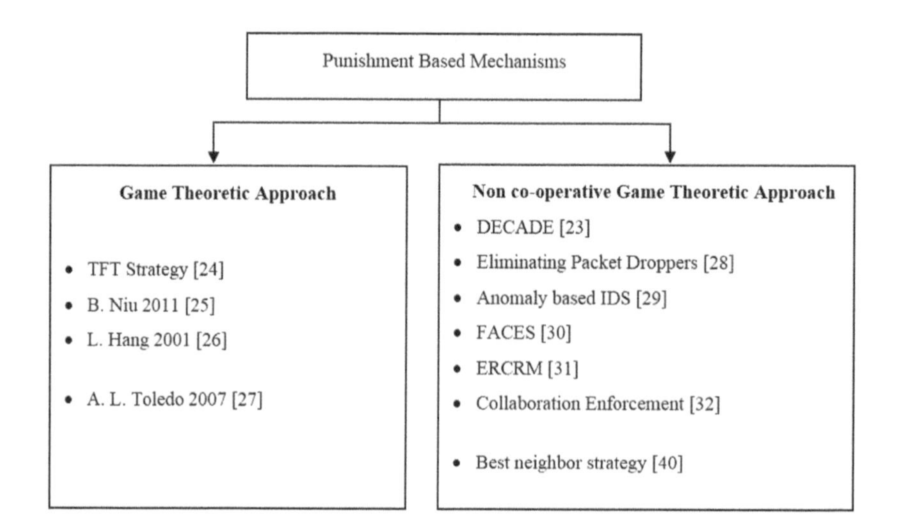

FIGURE 1.5 Punishing Mechanisms.

technique proposed by M. Majia et al. (2012). In this method, the author uses this scheme for each node to enable cooperation [31]. It uses the non-cooperative game theory in which each node in the network is encouraged to maximize the successful packet delivery, thus isolating the selfish node. By introducing the sociality parameter for each node in the network, the node's interactions are also to be improved. If any node forwards the packet to others, which they receive from neighbors, they will be rewarded with "X." However, cooperating intermediate node's energy will be reduced to the amount of "Y;" thus, "X-Y" reward will be given. Considering the example as node "A" sends a packet to node "B;" here, the rewarding and punishing the node are done by three scenarios. (1) If both nodes cooperatively transmit the packet without any deviations, then the reward is "X-2Y." (2) If any node does not forward the packets cooperatively to others, then the reward is "X-Y." (3) If both nodes are unwilling to cooperate in forwarding the packets, punishment such as "–Y" will be assigned to that node for that selfishness. The DECADE uses the classical cellular algorithm; it is the feedback-based algorithm. It works based on the intermediate node's feedback about the packet, whether to forward or drop to detect the cooperating nodes and isolate the selfish nodes. The sociality parameter was included in the DECADE algorithm, which improves the general trust model's performance by encouraging the nodes to choose the best path by availing the wider information from the intermediate nodes and providing the best performance when the network environment changes, i.e., on mobility. However, in the DECADE mechanism, too many algorithms and parameters such as availability and sociality are utilized to perform better on detecting and isolating the selfish nodes. It may degrade the overall network performance, and the computations used are more complex. The authors Niu et al. (2011) had proposed the approach, which uses three aspects for cooperation stimulation among the nodes. (1) Using infinite reported game concepts to determine the optimal cooperativeness. (2) The worst behavior tit-for-tat strategy for punishing the node cooperatively. (3) The realistic estimation mechanism for node behavior monitoring, which is maybe imperfect. The monitoring results in punishing nodes that act selfishly [32]. However, the monitoring results may be affected by external factors like noise and interference.

1.2.3.1.2 Tit-for-tat (TFT) strategy:

The game-theoretical strategy called the tit-for-tat (TFT) strategy was used to detect the selfish node by S. Ng et al. (2010). All nodes in the network will be cooperating in the first stage of the game. Based on the opponent's behavior, further actions are taken in the preceding stage. It works based on the forwarding game and repeated game. The connectivity among the node, routing path, and network property for the cooperation is also determined for detecting the node's willingness to forward the packets to others [33]. Huang et al. (2001) had proposed the resolving technique for dropping of route request (RREQ) control packets, which are received during the route discovery process. It is a monitoring approach utilized during the transmission for detecting the selfish node. Authors utilized the concept that the number of RREQ packets should be equal to the number of nodes in the network. When it is unequal, it is assumed that there are some nodes acting as selfish nodes [34]. To reduce false detection, this gaming approach was used. However, it uses indirect communication and complex distributed monitoring schemes for detecting the selfish nodes. Toledo et al. (2007) had improved the non-cooperative behaviors by concentrating on layers such as the network layer and the medium access control (MAC) layer. The selfishness occurs in the network layer by refusing the route discovery process, delaying or dropping the forwarding of packets by idle state nodes [35]. In the MAC layer, the selfishness can be monitored as the uneven channel access, false signal state.

1.2.3.2 Non-cooperative Game-Theoretic Approach

1.2.3.2.1 Eliminating Packet Droppers scheme

The authors Djenouri et al. (2009) had proposed the Eliminating Packet Droppers scheme. It includes five modules to detect and eliminate the misbehaving nodes, such as monitors, detectors, isolators, investigators, and witnesses. These modules have the responsibility to detect these misbehaving nodes [36]. The monitor controls the relaying of data packets. The detector is responsible for detecting the misbehaving node, based on the monitor module response. Isolator collaborates with the witness module, which means it gets enough evidence before any node is isolated as the misbehaving node. The investigator module investigates such suspected node's accusations for whether it has enough experience for the accusation. Finally, the witness module helps the isolator in isolating the misbehaving node. Isolating means that the data packets cannot be forwarded to other nodes. Here, the randomized two-hop acknowledgment algorithm is utilized for better performance. The randomized two-hop acknowledgment algorithm has less overhead compared to the two-hop acknowledgment algorithm. However, using the randomized ACK, the possibility of accurately detecting the misbehavior is not possible. The authors also made some assumptions for detecting the misbehavior. The isolated nodes are permanently excluded from the network, and they are not provided the option to rejoin the network after some occurrences if it is falsely isolated.

1.2.3.2.2 Cross-layer Anomaly-based Intrusion Detection System (IDS)

The authors L.S. Casado et al. (2015) had developed the IDS system model for detecting dropping attacks in the network. Here, authors concentrate on the mobility, collisions, and packet errors, which will deeply affect the communication between the nodes in the network [37]. The RTS/CTS data-forwarding model is intensely examined under the MANET conditions. The enhanced windowing method is used to detect mobility and channel errors; this windowing helps to detect packet dropping under these conditions. When the RTS is sent to a node, which is not replied because of mobility, it is also categorized as the misbehaving node, even though it uses temporal time-based windowing. By using the temporal window, the node can be considered illegitimately as malicious, because it is not able to detect the node's mobility in a particular temporal windowing period. To overcome this, the authors use event-based windowing instead of a time-based one, thus avoiding the previously mentioned issues. However, using the time-based windowing, each node exclusively needs to monitor and collect information of all other nodes and carry out all features to determine the network's misbehaving actions. In addition, it is not feasible to trust each node's monitoring and the implementation of the detection techniques.

1.2.3.2.3 FACES Method

The authors S. K. Dhurandher et al. (2011) had developed the Friend-based Ad-hoc Routing using Challenges to Establish Security in MANET system (FACES) algorithm to isolate the malicious nodes from the network. It uses the concept of sharing the friend lists and sending the challenges to each node. The friend list is a list of nodes having a friendship with other nodes obtained from the previous transmission and it will be used for routing, instead of proving the list of trustful nodes to the source node [38]. To improve the friendship with other nodes, the periodic process called "Share Your Friend" is initiated into the network. By doing so, the friends for each node are updated. The friends mean that the nodes that fully cooperate in the relaying process of the network. The FACES works in four stages: in the "Challenge Your neighbor," stage nodes are challenged to provide the neighbor node's authentication details, and nodes which provide these details are listed in the "friend list". The "Question mark list" contains the nodes, which have not completed the first stage. The nodes in the question mark list are not used for the relaying packets. The nodes, which did not perform well in the friend list, are also moved to the question mark list. In "Rate Friends," the nodes are rated based on their performance in the network activities such as relaying packets, cooperation, etc. Based on the involvement in the relaying process, the rating is given from 0 to 10. The authors have used the DSR protocol for routing by checking whether the node is in the friend list and not in the question mark list. Quality of the route is checked by every node, and the source node is encrypted by the public key cryptography to protect against eavesdropping and man in the middle attack. However, maintaining too many lists such as the question mark list, friend list, unauthorized list on each node may cause network overhead. Moreover, it gradually decreases the network throughput on larger networks. In highly dynamic battery-constrained mobile nodes, preserving these lists also consumes some battery power.

1.2.3.2.4 ERCRM Method

In the Exponential Reliability coefficient-based Reputation Mechanism for isolating selfish nodes in MANET (ERCRM) method, the authors J. Sengathir et al. (2015) estimate the energy measures on nodes and manipulate the reliability coefficient for isolating selfish nodes from the routing path in the MANET [39]. By using the exponential failure rate on nodes in networks through the moving average method, it is highly utilized on the nodes for calculating the reliability coefficient. The moving average method works by the most recent mobile nodes' most recent past behavior in relaying the packets to the neighbor nodes. The authors categorize the selfish nodes into three types as, type I, II, and III. In type I, the selfish node participates in the route discovery and maintenance process but does not participate and refers to relay packets to other nodes. In type II, it participates neither in the route discovery and maintenance process nor in relaying packets. In type III, the selfish nodes dynamically change their behavior by forwarding the packets and dropping the packets. The protocol used for the routing process itself removes the type II selfish nodes. The type III selfish nodes are removed from the network by using the estimated energy metric. When this value falls below the threshold value, the mobile node is isolated as the selfish node. By using both the energy metric and reliability coefficient values, the type I, selfish nodes are isolated from the routing path in the network. However, the estimated energy metric and reliability coefficient are computed for every node in the network; it increases the network overhead gradually. Moreover, the nodes that are not in the routing path and maintaining these values make the energy of nodes decrease. The authors use the predefined estimated energy metric of 0.45 Joules and the reliability coefficient of 0.4 for determining the MANET's selfish nodes. It should be dynamically changed under different conditions for better results in isolating the selfish nodes.

1.2.3.2.5 Collaboration enforcement in MANET

Authors N. Jiang et al. (2007) developed the novel approach instead of using reputation mechanisms. It has the scalability issues that cause bad accusations on the neighbor nodes; it can lead to the misjudgment of determining malicious nodes [40]. To overcome these issues, the author introduced a one-hop neighbor observation and rerouting the packets. Even though in this technique, each node needs to maintain the list of observations and actions of all nodes in the network. Moreover, the DSR protocol maintains the list of route caches; it also adds the routing overhead to the network. The Route Redirect (RRDIR) concept used for dynamic rerouting has another route discovery process; it may also increase the throughput of the network and the overall performance of the network. In addition, the number of the node used for the simulation is 50; it is very small as the scalability concern. For the simulation analysis, the selfish node's percentage increases by up to 20% only. As the scalability and robust concern, the nodes' selfishness should be evaluated up to 60%.

1.2.3.2.6 Best neighbor strategy

K. Komathy et al. (2007) used the best neighbor strategy to maintain scalability and robustness among neighbor nodes while enforcing cooperation among selfish nodes [41]. In this strategy, each node should play a packet forwarding to its neighbors. Recording and updating these values are done in each round. Trust values are obtained by this way of interaction with the individual neighbor nodes.

1.3 DISCUSSION AND EVALUATION OF RESEARCH FINDINGS

In this section, the discussion about the various categories of cooperation techniques and their challenges in establishing cooperation are to be presented. The summary and the comparison of different cooperative solutions for the MANET are tabulated in Table 1.1. The cooperation in the wireless and Ad-hoc network faces several issues, and those were considered in this summary. Moreover, it is not feasible to assure the network similarity, and its stability and scalability create uneven consequences for cooperative schemes. Mobile Ad-hoc networks contemporarily have several constrictions such as mobile resources, storage capacity, battery, processor

TABLE 1.1
Summary and Comparison of the Cooperation Techniques

S. No.	Authors	Cooperation Strategy	Cooperative Approach Type	Key Concepts	Features
1.	L. Buttyan and J. Hubaux [8]	Tamper-proof device	Incentive-based credit mechanism	Assigning credits based on cooperation ability	Packet Purse Model and Packet Trade Model were adopted
2.	S. Zhong et al. [19]	Central agent	Incentive-based crediting mechanism	Paying credits by central agents and it verifies paid nodes	Central agents reduce the burden of other nodes in the credit assignment
3.	L. Anderegg and S. Elderbenz [10]	Ad-hoc VCG	Incentive-based crediting mechanism	Nodes publish remaining energy for routing	Best available energy path discovered based on their values
4.	S. Elderbenz et al. [11]	COMMIT Protocol	Incentive-based crediting mechanism	Source takes the responsibility of credit assignment	Source nodes reveal maximum total credit to participating nodes
5.	M. Mohmoud and X. S. Shen [12]	RACE	Incentive-based crediting mechanism	Central agents manage reports from nodes	Central agents store the data temporarily for future evidence
6.	S. Marti et al. [13]	Watchdog scheme	Reputation-based crediting mechanism	A rating scheme for reputation computation	Neighbor node activities were monitored, and low rated nodes avoided
7.	S. Buchegger and J. Y. Boudec [14]	CONFIDENT	Reputation-based crediting mechanism	Misbehaving nodes and their reputations are collected	Lower reputation nodes from other neighbor nodes were identified
8.	P. Michiardi and R. Molva [15]	CORE	Reputation-based crediting mechanism	False reputation values were restricted	Good rating nodes only allowed for the routing

(Continued)

TABLE 1.1 (CONTINUED)
Summary and Comparison of the Cooperation Techniques

S. No.	Authors	Cooperation Strategy	Cooperative Approach Type	Key Concepts	Features
9.	Q. He et al. [16]	SORI	Reputation-based crediting mechanism	One-way hash chain algorithm for packet relay	Reputation values shared only on neighbor nodes
10.	S. Bensal and M. Baker [18]	OCEAN	Reputation-based crediting mechanism	Rating value, the misbehaving list is used for cooperation	Rating increased or decreased based on node cooperation and reduced false detection
11.	J. Guo et al [22]	HEAD	Reputation-based crediting mechanism	Enforces cooperation by the alert-based route discovery process	Alerting messages are used for faulty nodes instead of broadcasting
12.	D. Charles et al [21]	ICARUS	Hybrid-based crediting mechanism	Controls credits exchanged between nodes	The central agent uses credit control service for credit assignment to nodes
13.	H. Shen and Z. Li [31]	ARM	Hybrid-based crediting mechanism	No central agents were used for the for-credit assignment	Each node maintains credits for peer nodes based on the reputation value of nodes
14.	Z. K. Chong et al [23]	SDA	Trust-based crediting mechanism	Reporting node, agent node, the central authority for trust analysis	Central authority investigates the trust reports from all agents for trustworthiness
15.	L. H. G. Ferraz et al. [24]	TEAM	Trust-based crediting mechanism	The local and global model is used to calculate the trust of node	The global model observes and gathers the one-hop neighbor values to evaluate trust
16.	M. Yu et al [20]	Weighting Method	Trust-based crediting mechanism	Link quality and average weights are calculated	Accurate trust values are calculated by averages for trusted node detection

(Continued)

TABLE 1.1 (CONTINUED)
Summary and Comparison of the Cooperation Techniques

S. No.	Authors	Cooperation Strategy	Cooperative Approach Type	Key Concepts	Features
17.	S. K. Ng and W. K. G. Seah [33]	TFT Strategy	Game theory-based punishment mechanism	Forwarding game and repeated game are adopted	The game theory defines the routing path, network topology for forwarding the packets
18.	B. Niu et al [34]	Game Theory	Game theory-based punishment mechanism	Stimulation of nodes for the cooperation among nodes by game theory mechanism	The infinite repeated game, TFT worst behavior strategy, and realistic estimation mechanism are developed for cooperation
19.	L. Huang et al [35]	Game Theory	Game theory-based punishment mechanism	Resolves the packet dropping problems in the network	The concept of route request packets should equal to the number of nodes adopted
20.	A. L. Toledo and X. Wang [36]	Game Theory	Game theory-based punishment mechanism	Development of the network layer and the Medium Access Layer (MAC)	The MAC layer designed to determine the cooperativeness of nodes using channel access, fake signal state
21.	M. Mejia et al [32]	DECADE	Non-cooperative Game-Theoretic Approach	Sociality parameter and encouraging the nodes for cooperation	Rewarding and punishment are given to nodes based on packet delivery and cooperation in three ways

(Continued)

TABLE 1.1 (CONTINUED)
Summary and Comparison of the Cooperation Techniques

S. No.	Authors	Cooperation Strategy	Cooperative Approach Type	Key Concepts	Features
22.	M. Conti et al [26]	End-to-End	Acknowledgment based	The receiver node and source node are responsible for acknowledgment (ACK)	The reliability index is updated based on acknowledgment and cooperative nodes determined
23.	M. Refaei et al [28]	TCP ACK	Acknowledgment based	Neighbor node activities were utilized for the reliability index	The neighbor nodes determine the reliability of nodes and neighbor works by ACK of other nodes
24.	K. Balakrishnan et al. [29]	TWO ACK	Acknowledgment based	Two nodes were adopted for the acknowledgment instead of the end-to-end acknowledgment process	Sending ACK by two hops away from the sender node. Corresponding two hops required to send acknowledgment
25.	H. Yang et al [19]	Cryptography-Signatures	Acknowledgment based	Cryptographic Token renewal, K-neighbors for detecting misbehaving nodes	Cryptographically signed tokens on nodes protect nodes on cooperation
26.	D. Djenouri and N. Badache [37]	Eliminating packet Droppers	Non-cooperative Game-Theoretic Approach	Monitor, detector, isolator, witness, investigator, randomized ACK modules adopted	Based on the evidence, the misbehaving nodes are isolated by using each stage of modules

(Continued)

TABLE 1.1 (CONTINUED)
Summary and Comparison of the Cooperation Techniques

S. No.	Authors	Cooperation Strategy	Cooperative Approach Type	Key Concepts	Features
27.	L.S. Casaclo et al [38]	Anomaly-based IDS	Non-cooperative Game-Theoretic Approach	To detect the packing dropping attack IDS system model developed and RTS/CTS model adopted	Mobility and collision packet error, the channel error event-based windowing method is adapted to detect misbehaviors
28.	S. Zhong et al [41]	SPRITE	Credit based	Credit proof system, receipts of nodes, assigning reward in terms of credits	Receipts of nodes are validated and credits assigned. CCCS encourages cooperation
29.	S. K. Dhurandher et al. [39]	FACES	Non-cooperative Game-Theoretic Approach	Friend list, Sharing friend list, share your friend, Challenging the neighbor nodes	Misbehaving nodes are listed in the Question mark list. A node in this list was not used for relaying
30.	J. Sengathir and R. Manoharan [40]	ERCRM	Non-cooperative Game-Theoretic Approach	Estimated energy node, Reliability coefficient (RC), Exponential failure rate	Based on the energy of nodes, the RC calculated and selfish nodes are isolated in routing
31.	N. Jiang et al [25]	SRACEM	Non-cooperative Game-Theoretic Approach	Scalability issues, one-hop neighbor observation, DSR protocol	Each node maintains the observation of all other nodes. It finds selfish nodes
32.	K. Komathy and P. Narayana samy [42]	Neighbor Strategy	Non-cooperative Game theory-based	Trust values are updated for each neighbor node for cooperation	Scalability, stability, and robustness of nodes are evaluated

power, limited bandwidth, network connections, and latency, which make the cooperation strategy deployment for the mobile environments as a challenging task. Most significantly, the cooperation mechanisms aim to solve such problems. The cooperation systems necessitate solving the network latency issues and low bandwidth for network efficiency and effectiveness. In addition, issues in synchronization, security, and trustfulness require extra add-ons to the systems.

Generally, the cooperation systems typically undertake the selfish node discovery to determine the superlative node to cooperate in the network and to accomplish the finest routing path. Nonetheless, this one could be toughest in the presence of unpredictable non-cooperative nodes. The cooperation strategy should also be capable of incorporating the node heterogeneity. Thus, cooperation schemes need to adjust their operating environments and control node communication to attain the best cooperation. For mobile environments, fault-tolerance could be often required to enhance the fault/disconnections in communication. Consequently, identifying the cause of disconnection, recovering the original message, and retransmitting the original is most essential. For the effective and efficient establishment and organization of cooperation mechanisms, all the aspects mentioned above are the key issues and should be considered. Moreover, the best cooperation-based approaches have an optimal QoS delivery and quality in the network performance.

In this paper, the major classifications of the cooperative schemes in MANETs are discussed. As mentioned previously, it is generally divided into three types of approaches as credit-based approaches, acknowledgment-based mechanisms, and punishment-based approaches. Based on the study performed in this paper, the reputation mechanisms have further concerns compared to acknowledgment- and credit-based methods. There is more ease of possibility of communication with the non-cooperative nodes to gain more reputation in reputation schemes. The efficient node reputation computation through the network is also another challenging issue. Likewise, the dependence on the wireless broadcast approach is another weakness in these schemes. Better reliability is achieved in the credit-based approaches to these issues.

However, the dependence of tamper-proof hardware and node cheating behaviors should be avoided for this scheme, and it may add more complexity to the cooperative system. Most of the credit-based systems attempted to avoid tamper-proof hardware even though it has many difficulties for the security mechanisms. The effective cooperation-based approaches should aim to provide better efficiency and performance for the mobile devices in terms of device battery, storage, and network. In many network architectures, the cooperation between the applications is much needed, supported by mobile devices. Generally, in industry services, the agents must cooperate and share information over mobile devices and healthcare services. In contrast, mobile health (m-health) applications share medical information with patients and physicians [42]. Likewise, in healthcare scenarios, the cooperation among applications is more challenging, which requires a comprehensive study of these approaches.

1.4 CONCLUSION

The cooperation between nodes is much needed for communication establishment in dynamically unstable network infrastructures, and many solutions (cooperation approaches) have been proposed to address the issues and challenges. The mobile nodes should cooperate with each other for relaying data and accomplishing all networking functions. In this paper, the elaborated deep literature analysis from various research developments and their limitations and features of the network is performed. In addition, the cooperation stimulation strategies such as cooperative game theory-based approaches, and non-cooperative game theory-based approaches, punishment mechanisms, incentive-based methods, and hybrid methods are focused on the MANETs. The challenges, merits, weaknesses of approaches, and their significance in cooperation establishment are also included in this paper. Furthermore, the open issues in the construction and design of cooperative solutions for MANETs are also surveyed.

REFERENCES

1. X. Wang, J. Li, and S. Member, "Improving the network lifetime of MANETs through cooperative MAC protocol design," *IEEE transactions on parallel and distributed systems* vol. 26, no. 4, pp. 1010–1020, 2015.
2. B. Karaoglu, W. Heinzelman, and S. Member, "Cooperative load balancing and dynamic channel allocation for cluster-based mobile ad hoc networks," *IEEE transactions on mobile computing* vol. 14, no. 5, pp. 951–963, 2015.
3. T. Prasannavenkatesan, "FUCEM: Futuristic cooperation evaluation model using Markov process for evaluating node reliability and link stability in mobile ad hoc network", *Wireless Networks*, Springer, (Article in Press), 2020, IF: 2.405, ISSN: 1572-8196, DOI: https://doi.org/10.1007/s11276-020-02326-y.
4. P. Theerthagiri, "COFEE: Context-aware futuristic energy estimation model for sensor nodes using Markov model and auto-regression", *International Journal of Communication System*, p. e4248, 2019. IF: 1.278, ISSN: 1099-1131, DOI: https://doi.org/10.1002/dac.4248.
5. T. Prasannavenkatesan, K. Udhayakumar, and R. Ramkumar "Security attacks and detection techniques for MANET," *Discovery Journal*, vol. 15, no. 42, pp. 89–93, Ghaziabad, March 2014.
6. J. N. Al-karaki, and A. E. Kamal, "Stimulating node cooperation in mobile ad hoc networks," *Wireless Personal Communications*, vol. 44, pp. 1–15.
7. N. Samian, Z. Ahmad, W. K. G. Seah, and A. Abdullah, "Cooperation stimulation mechanisms for wireless multihop networks: A survey," *Journal of Network and Computer Applications*, vol. 54, pp. 88–106, 2015.
8. L. Buttyan, and J. P. Hubaux, "Nuglets: A virtual currency to stimulate cooperation in self- organized mobile ad hoc networks," *Technical Report DSC/2001/001.* Swiss Federal Institute of Technology, Lausanne, Switzerland; 2001.
9. B. M. C. Silva, J. Rodrigues, N. Kumar, and G. Han, "Cooperative strategies for challenged networks and applications: A survey," *IEEE Systems Journal*, vol. 11, pp. 1–12, 2015.
10. L. Anderegg, S. Eidenbenz, "Ad-hoc-VCG: A truthful and cost-efficient routing protocol for mobile Ad-hoc networks with selfish agents," in: Proceedings of the 9th International Conference on Mobile Computing and Networking (MobiCom). San Diego, CA; 14–19 September 2003. pp. 245–259.
11. S. Eidenbenz, G. Resta, P. Santi, "COMMIT: a sender-centric truthful and energy-efficient routing protocol for Ad-hoc networks with selfish nodes," in Proceedings of 19th IEEE International, Parallel and Distributed Processing Symposium (IPDPS). Denver, Colorado, USA; 3–8 April 2005. pp. 239–49, 2005.

12. M. Mahmoud, X. Shen, "A secure payment scheme with low communication and processing overhead for multihop wireless networks," *IEEE Trans Parallel Distrib Syst*, vol. 24, no. 2, pp. 209–24, 2013.
13. S. Marti, T. J. Giuli, K. Lai, M. Baker, "Mitigating routing misbehavior in mobile ad-hoc networks," In: Proceedings of the 6th Annual International Conference on Mobile Computing and Networking. Boston, MA; August 2000, pp. 255–65, 2000.
14. S. Buchegger, J. Y. L. Boudec, "Performance analysis of the CONFIDANT protocol," in Proceedings of the 3rd ACM International Symposium on Mobile Ad-Hoc Networking and Computing (MOBIHOC). Lausanne, Switzerland; 9–11 June 2002, pp. 226–236, 2002.
15. P. Michiardi, R. Molva, "CORE: A collaborative reputation mechanism to enforce node cooperation in mobile Ad-hoc networks," in Proceedings of the 6th Joint Working Conference on Communications and Multimedia Security. Netherlands; 26–27 September 2002, pp. 107–12, 2002.
16. Q. He, D. Wu, P. Khosla, "SORI: A secure and objective reputation-based incentive scheme for Ad-hoc networks," in Proceedings of the IEEE Wireless Communications and Networking Conference (WCNC). NewOrleans, LA; 21–25 March 2004, pp. 825–30, 2004.
17. S. Bansal, M. Baker, "Observation-based cooperation enforcement in ad hoc networks," Technical Report cs.NI/0307012. Computer Science Department, Stanford University, USA; 2003.
18. J. Guo, H. Liu, J. Dong, X. Yang, "HEAD: A hybrid mechanism to enforce node cooperation in mobile Ad-hoc networks," *Tsinghua Science and Technology*, vol. 12, no. 1, pp. 202–207, 2007.
19. S. Zhong, Y. R. Yang, J. Chen. "Sprite: A simple, cheat proof, a credit-based system for mobile Ad-hoc networks," in: Proceedings of the 22nd IEEE International Conference on Information Communications (INFOCOM). San Francisco; 1–3 April 2003, pp. 1987–1997, 2003.
20. D. E. Charilas, K. D. Georgilakis, A. D. Panagopoulos, "ICARUS: hybrid inCentive mechanism for cooperation stimulation in ad-hoc networks," *Ad-hoc Networks*, vol. 10, no.6, pp. 976–989, 2012.
21. H. Shen, Z. Li, "ARM: an account-based hierarchical reputation management system for wireless ad-hoc networks," in 28th International Conference on Distributed Computing Systems Workshops (ICDCS'08). Beijing, China; 17–20 June 2008, pp. 370–375.
22. Z. K. Chong, S. W. Tan, B. M. Goi, B. C. K. Ng, "Outwitting smart selfish nodes in wireless mesh networks", *International Journal of Communication System*, vol. 26 no. 9, 2013, ISSN: 1163–1175.
23. L. H. G. Ferraz, P. B. Velloso, O. C. M. B. Duarte, "An accurate and precise malicious node exclusion mechanism for Ad-hoc networks", *Ad-hoc Networks*, vol. 19, pp. 142–155, 2014.
24. M. Yu, M. Zhou, W. Su. "A secure routing protocol against Byzantine attacks for manets in adversarial environments," *IEEE Trans Veh Technol*, vol. 58, no. 1, pp. 449–460, 2009.
25. M. Conti, E. Gregori, G. Maselli, "Towards reliable forwarding for ad-hoc networks," in: Proceedings of the Personal Wireless Communications (PWC). Venice, Italy; 23–25 September 2003, pp. 790–804.
26. T. Prasannavenkatesan, R. Raja, P. Ganeshkumar, "PDA-misbehaving node detection & prevention for MANETs," in IEEE Explore and Proceedings of International Conference on Communication and Signal Processing (ICCSP). Melmaruvathur, pp. 1808–1812, April 2014.

27. M. T. Refaei, V. Srivastava, L. Da Silva, M. Eltoweissy, “A reputation-based mechanism for isolating selfish nodes in ad-hoc networks,” in Proceedings of the Second Annual International Conference on Mobile and Ubiquitous Systems: Networking and Services (MobiQuitous). SanDiego, CA; 17–21 July 2005, pp. 3–11, 2005.
28. K. Balakrishnan, J. Deng, P. K. Varshney, “TWOACK: Preventing selfishness in mobile ad hoc networks,” in Proceedings of the IEEE Wireless Communications and Networking Conference (WCNC). New Orleans, LA; 13–17 March 2005, pp. 2137–2142, 2005.
29. T. Prasannavenkatesan, P. Rajakumar, A. Pitchaikkannu, “An effective intrusion detection system for MANETs,” in Proceedings on International Journal of Computer Applications (IJCA) and ICACEA. Ghaziabad, vol. 3, pp. 29–34, March 2014.
30. H. Yang, J. Shu, X. Meng, S. Lu, “SCAN: Self-organized network-layer security in mobile ad hoc networks." *IEEE Journal on Selected Areas in Communications* vol. 24, no. 2, 261-273, 2006.
31. M. Mejia, N. Pena, J. L. Munoz, O. Esparza, and M. Alzate, “Ad hoc networks DECADE: Distributed emergent cooperation through adaptive evolution in mobile ad hoc networks,” *Ad Hoc Networks*, vol. 10, no. 7, pp. 1379–1398, 2012.
32. S. K. Ng, W. K. G. Seah, “Game-theoretic approach for improving cooperation in wireless multihop networks,” *IEEE Transactions on Systems, Man, and Cybernetics, Part B*, 40(3), 559–574.
33. B. Niu, H. V. Zhao, J. Hai, “A cooperation stimulation strategy in wireless multicast networks,” *IEEE Transactions on Signal Processing*, vol. 59, no. 5, pp. 2355–2369, 2011.
34. L. Huang, L. Li, L. Liu, H. Zhang, L. Tang, “Stimulating cooperation in route discovery of ad hoc networks,” in Proceedings of the 3rd ACM Workshop on QoS and Security for Wireless and Mobile Networks (Q2SWinet'07). Chania, Crete Island, Greece; 22–26 October 2007, pp. 39–46, 2007.
35. A. L. Toledo, X. Wang, “Robust detection of selfish misbehavior in wireless networks,” *IEEE Journal on Selected Areas in Communications*, vol. 25, no. 6, pp. 1124–1134, 2007.
36. D. Djenouri and N. Badache, “Ad hoc networks on eliminating packet droppers in MANET: A modular solution,” *Ad Hoc Networks*, vol. 7, no. 6, pp. 1243–1258, 2009.
37. L. Sanchez-casado, G. Macia-Fernandez, P. García-Teodoro, and R. Magán-carrión, “A model of data forwarding in MANETs for lightweight detection of malicious packet dropping,” *Computer Networks*, vol. 87, pp. 44–58, 2015.
38. S. K. Dhurandher, M. S. Obaidat, K. Verma, and P. Gupta, “FACES: Friend-based ad hoc routing using challenges to establish security in MANETs systems,” *IEEE Systems Journal*, vol. 5, no. 2, pp. 176–188, 2011.
39. J. Sengathir and R. Manoharan, “Exponential reliability coefficient based reputation mechanism for isolating selfish nodes in MANETs,” *Egyptian Informatics Journal*, vol. 16, no. 2, pp. 231–241, 2015.
40. N. Jiang, K. A. Hua, and D. Liu, “A scalable and robust approach to collaboration enforcement in mobile ad-hoc networks,” *Journal of Communications and Networks*, vol. 9, no. 1, pp. 56–66, 2007.
41. K. Komathy and P. Narayanasamy, “Best neighbor strategy to enforce cooperation among selfish nodes in wireless ad hoc network,” *Computer Communications*, vol. 30, pp. 3721–3735, 2007.
42. G. F. Marias, P. Georgiadis, D. Flitzanis, and K. Mandalas, “Cooperation enforcement schemes for MANETs : A survey,” *Wireless Communications and Mobile Computing*, vol. 6, pp. 319–332, 2006.

2 IoE-Based Genetic Algorithms and Their Requisition

Neeraj Kumar Rathore, and Shubhangi Pande

CONTENTS

2.1 Introduction 25
2.2 Need of IoE-Based Genetic Algorithms 27
2.3 Basic terminologies related to IoE-based GAs 27
  2.3.1 Chromosomes 28
  2.3.2 Population 28
  2.3.3 Genes 28
  2.3.4 Allele 29
  2.3.5 Genotype and Phenotype 29
2.4 General Genetic Algorithm 29
2.5 Operators in GAs 31
  2.5.1 Encoding 31
2.6 Stopping Condition for the Genetic Algorithm 40
2.7 Constraints in the Genetic Algorithm 41
2.8 Similarity and Comparison between IoE and IoT 42
  2.8.1 Similarity between IoE and IoT 42
  2.8.2 IoE vs IoT: What's the Difference? 43
2.9 Problem-Solving Using the GA 43
2.10 Advantages 44
2.11 IoE-based GA Application 45
2.12 Limitations 45
2.13 Conclusion 46
References 47

2.1 INTRODUCTION

The IoE technology is aimed primarily at transforming information gathered into actions, making decisions on a data basis, and providing new capacity and enriched experiences (Figure 2.1). A catechistic searching approach known as the genetic algorithm has arisen from evolutionary notions of natural selection and genetics. Here, a heuristics mean a technique designed to solve a problem quickly, which is guaranteed to give the best solution, but the optimal solution may or may not be

DOI: 10.1201/9781003323426-2

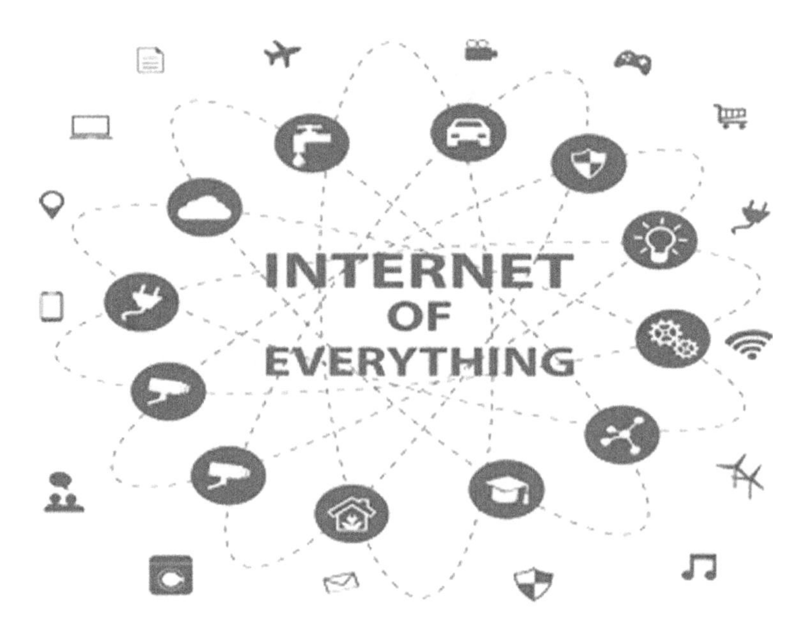

FIGURE 2.1 IoE concept.

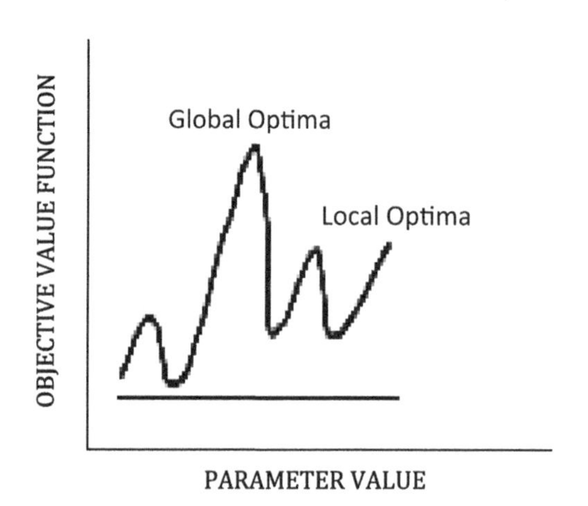

FIGURE 2.2 Position global optima and local optima.

obtained (Figure 2.2). The IoE-based genetic algorithm (GA) focused on optimization, which means the process of making something better. In biology, we looked at genes, chromosomes, and how two parents generate new offspring and pass on their greatest traits to the next generation, which then tries to come up with the best solution. Here, genes travel from next to next generation. The IoE-based GA is based on this idea, which is used in artificial intelligence (AI) and machine learning (ML), where we have a lot of solutions and have to find optimal solutions among them [1].

In general terms, if we have *N* number of solutions for resolving a specific problem, all the solutions cannot yield an optimal solution for that problem but IoE-based GAs assist in finding the optimal solution for all kinds of problems, whether it is constrained or unconstrained. It emerges over time so that a reform solution can be achieved and searching can be made refined [1,2].

Charles Darwin's theory:

- Within any population, there exist natural variations. Some individuals exhibit more favorable variation than others.
- The struggle for survival eliminates the unfit individuals. The fit individuals processing favorable variations survive and reproduce. This is referred to as "natural selection" (or survival of the fittest).
- The individuals having favorable variations pass on these variations to their progeny (offspring) from generation to generation.

IoE-based genetic algorithms now have the capability to promote a "good enough" solution "quick enough." It is known that evolution is a successful and strong biological system of adaptation strategy. Motivation to adopt IoE-based GAs comes from the following given points [1–3]:

- Solving difficult problems: The IoE-based GA is a powerful tool to produce valuable near-optimal solutions in less time.
- Breakdown of gradient-based methods: In most of the real-world situations, where a very complex problem undergoes an implicit propensity to become local optima-trapped.
- Obtaining a good solution fast: Few laborious problems such as the traveling salesman problem (TSP) possess practical applications such as path finding and VLSI design.

2.2 NEED OF IOE-BASED GENETIC ALGORITHMS

- They are more robust than conventional algorithms.
- If the inputs are modified to some extent or within the presence of noise, the IoE-based GA does not break easily, unlike in older AI systems.
- The IoE-based GA may exhibit compelling benefits over further usual optimization approaches (e.g., linear programming, depth first search (DFS), breadth first search (BFS), heuristics, etc.) when searching in a giant state-space, multimodal state-space, or n-dimensional surface.

2.3 BASIC TERMINOLOGIES RELATED TO IOE-BASED GAS

In a biological sense, the smallest unit of life is a human/animal cell. **The cell nucleus** can be demonstrated as the center of the cell. The information related to genetics is stored in the cell nucleus. In the nucleus, there is a nuclear envelope, within this, there is a nucleolus and a set of chromosomes. **Chromosomes** hold on to all the genetic data. Every chromosome is built up of DNA (deoxyribonucleic acid). **Genes** can be stated as the division of several parts of chromosomes. **Alleles** can be demonstrated as the plausibility of gene aggregation for one property, and distinct alleles constitute a gene. Take into account a gene that determines eye color, and the numerous allele possibilities are brown, black, green, and blue. A **gene pool** is defined as the collection of all possible alleles found in a population. The gene pool can regulate all the probable divergence for upcoming generations [3–15].

A **genotype** can be stated as the complete aggregation of genes for a selected individual. The substantial aspect of decoding a genotype can be signified as a **phenotype**. Reproduction recombines genotypes and the choice is always done on phenotypes [5].

2.3.1 Chromosomes

To figure out a specific problem, in general, if we have *N* number of solutions. So, one such solution to a given problem is a chromosome (Figure 2.3).

2.3.2 Population

It is a group of individuals (chromosomes). This is a subset of all total solutions to the given problem. The two key elements of the population exploited in IoE-based GAs are:

(a) Initial population generation: In utmost cases, it is selected randomly. Typically, for provoking the initial population, a sort of catechistic searching approach can be used.
(b) Population size: It is predicated on the problem's intricacy. For population size, random initialization is commonly used. Every bit is initialized to either 0 or 1 in binary-encoded chromosomes [15–37].

2.3.3 Genes

The sequence of genes constitutes a chromosome. We can also say that one element position of the chromosome (Figure 2.4).

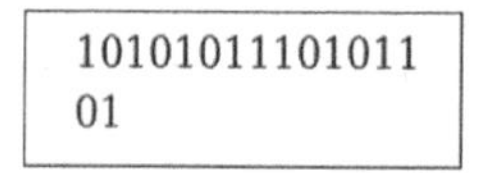

FIGURE 2.3 Representation of chromosomes.

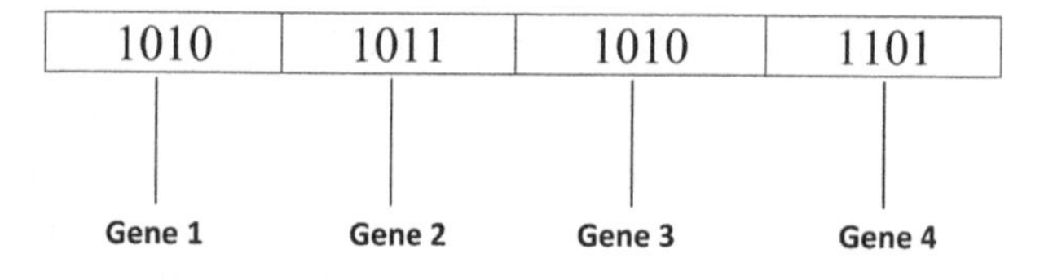

FIGURE 2.4 Representation of genes.

2.3.4 Allele

For a particular chromosome, a gene takes a value that is called an allele (Figure 2.5).

2.3.5 Genotype and Phenotype

The population in the actual real-world solution space is referred to as the phenotype, i.e., row solution having no proper representation in writing.

But, when we give input in the algorithm, proper representation is required, for that we use the genotype. The population in the computational space is termed genotype.

Decoding can be defined as the process of transforming a solution from the genotype to phenotype space. The approach of encoding entails redesigning a solution from the phenotype to genotype space (Figure 2.6) [37–51].

2.4 GENERAL GENETIC ALGORITHM

The general procedure of the IoE-based genetic algorithm [52–71] is shown in Figure 2.7:

Step 1: Initial Population

Initial population means randomly selecting solutions among all the possible solutions. Its important thing is diversity, which means selected solutions do not focus on particular points; they should focus on diverse points. As much as our input is diverse, the algorithm will properly work.

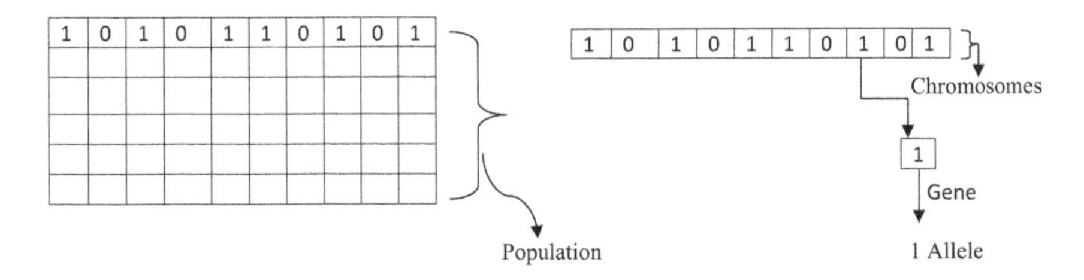

FIGURE 2.5 Representation of allele, gene, chromosome, and population concept.

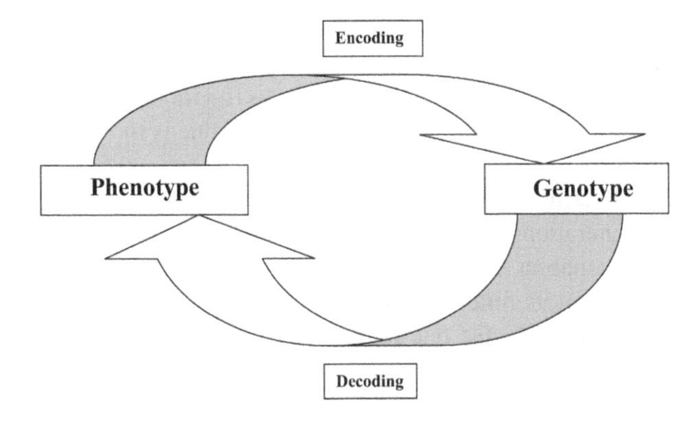

FIGURE 2.6 Representation of encoding and decoding.

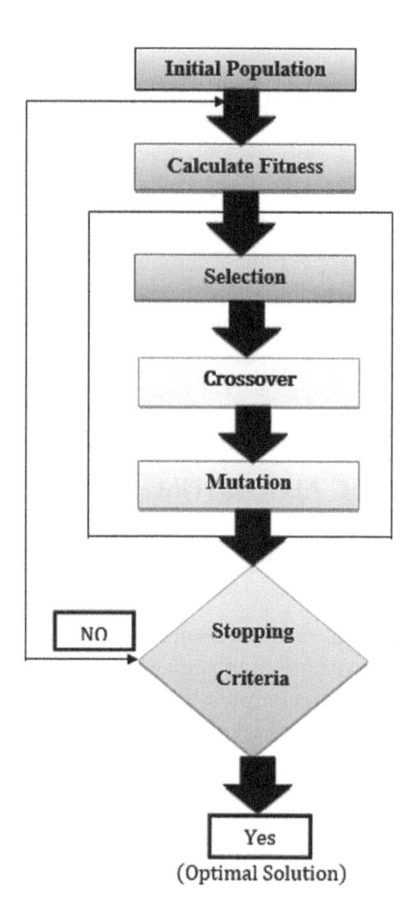

FIGURE 2.7 Flowchart for GA.

Step 2: Calculate the Fitness Function

Each individual in the original population is assigned a fitness function. It is a function you want to optimize or we can say that a function that takes solution as input and produces a more relevant (stable) output.

Step 3: Selection

Selection is directly proportional to the fitness function, which means select that promising solution whose fitness function value is more.

Step 4: Crossover

In biological terms, combining two parents to generate children for the upcoming generations is a crossover. In other terms, we can say that in the crossover, two-parent solutions are taken into account and produce a new population so that its fitness value is more. Crossover can be performed by various techniques like the one-point crossover, multipoint crossover, etc.

Step 5: Mutation

The new population (strings) generated by the crossover process are subjected to mutation. To obtain a new solution, a modest random change in the chromosome is made. Mutation can be of different forms like flipping, interchanging, reversing, etc. Mutation maintains the diversity of the population and protects the algorithm from becoming stuck in a local minimum. It includes a bit of riffling from 1 to 0 and back.

Step 6: Stopping criteria

Stopping criteria means the convergent point where the fitness value is the highest. It means, all the newly generated populations give optimal solutions. If there is a condition where optimal solutions are not produced, then the whole process is repeated until we reach the convergent state or endpoint.

2.5 OPERATORS IN GAs

Encoding, selection, crossover, and mutation are the core operators utilized in IoE-based genetic algorithms. These operators along with their types [13–26] are discussed below:

2.5.1 Encoding

Encoding is the process of representing particular genes. Bits, numbers, trees, arrays, lists, and other structures can all be used to encode data [19–39].

- Binary encoding: It is the most often used coding method. Each chromosome is represented by a binary string (bits 0 and 1). Each bit in a string indicates a small set of solution attributes. Here, the string length depends on accuracy (Figure 2.8).
- Octal encoding: In this type of encoding scheme, octal numbers (0–7) constitute a string (Figure 2.9).
- Hexadecimal Encoding: In this scheme, strings are encoded using hexadecimal numerals (0–9, A–F) (Figure 2.10).
- Permutation encoding: It is also known as real number encoding. Every chromosome is represented by a string of integer/real values that represent a number in a sequence in this encoding. Only ordering problems benefit from this type of encoding (Figure 2.11).

Chromosome 1	1 0 1 0 0 0 0 1 0 0 0 0
Chromosome 2	0 1 1 0 1 1 0 1 1 0 10

FIGURE 2.8 Binary encoding.

Chromosome 1	02466237
Chromosome 2	16315661

FIGURE 2.9 Octal encoding.

Chromosome 1	8BE9
Chromosome 2	4FAC

FIGURE 2.10 Hexadecimal encoding.

Chromosome 1	4 5 3 2 6 1 7 9 8
Chromosome 2	8 9 7 1 6 2 3 4 5

FIGURE 2.11 The permutation encoding scheme.

Chromosome A	4.3234 0.1234 5.5446 1.9293 6.5454
Chromosome B	ABDJEIFJDHDIERTUGNMOPLDGT
Chromosome C	(right), (back), (forward), (left), (right)

FIGURE 2.12 The value encoding scheme.

- Value encoding: Each chromosome in this sort of encoding is made up of a string of values. Here, values that are associated with the problem may be in the mode of real numbers, numbers, or characters for few sophisticated objects. This type of scheme serves as a decent choice for specific issues wherever a different variety of encoding would be troublesome (Figure 2.12).
- Tree encoding: Genetic programming employed this type of encoding scheme for emerging program expressions. Here, each chromosome is represented in the form of a tree of a few objects and these objects can be functions and commands used in programming languages.

a. ***Selection***

In this technique, two parents are selected from the population for crossing. The consequent step once agreeing on encoding is to find out a way to select individuals within a population that may turn out descendants for the consequent generation and the way through which several descendants each individual will generate. This choice aims to intensify in hops competent individuals within the population that their descendants are more fit. Chromosomes are chosen as reproductive parents from the initial population. The main focus is on the way to select chromosomes. The strongest ones live for producing new descendants, as per Darwin's theory of evolution.

In regards to the evaluation function, the selection process randomly selects chromosomes within the population. Higher the fitness level, the greater the chance of being chosen as an individual. The extent to which the most appropriate individuals are preferred is termed as the selection pressure. If the selection pressure is high, it means more preference is granted to the best individuals.

The IoE-based genetic algorithm convergence rate is primarily dictated by the selection pressure level, and increased selection pressures lead to increased convergence rates. Under a range of selection scheme pressures, genetic algorithms are capable of classifying optimal or nearly optimal solutions. IoE-based genetic algorithms can take an unnecessarily long time to search out optimization solutions if the convergence rate is slow because of too low selection pressure. Prematurely converging to an incorrect solution can result in an increased IoE-based genetic algorithm shift, happening because of too high selection pressure. To prevent premature convergence, in conjunction with the selection constraints, the distinction of populations should also be retained by the selection scheme [14,15].

Proportionate and ordinal selection systems are the two most common types of selection systems. Individuals can be eliminated in the first scenario depending on their fitness value in comparison to the fitness of others in the population. Individuals are picked in ordinal selection systems based on their rank in the population rather than their raw fitness. This shows the independence of selection pressure on population fitness distribution and is focused exclusively on population ranking.

A scaling function may also be used to reconstruct the population's fitness range, to regulate the selection pressure. As an illustration, if the entire solution has its fitness within the range [888,999], the chance of choosing a prominent individual as compared to that employing a proportionate-based approach would not be significant. Once every individual's fitness range is equal to that of [0,1], the chance of choosing a better individual rather than a poor one would be essential.

Selection needs to be stabilized with mutation variability and crossover variability. Sub-optimal and extremely fit individuals can seize the population if too strong selection is carried out which decreases the heterogeneity required for amendment and growth; too slow progression is resulted if too slow selection is carried out. There are numerous strategies of selection, which [29–49] are given below:

(a) Roulette wheel selection: One of the most frequent approaches for selecting IoE-based genetic algorithms is to use a roulette wheel. Here, the proportionate selection operator is used. This method is often a comparatively robust selection strategy, as fit individuals do not seem to be expected to be picked for, however, are far more possible to be. In roulette selection theory, a linear search is performed with the slots weighted proportionately to the individual's fitness rating, using a roulette wheel, by determining a desired value based on the random proportion of the population's fitness. The population is shuffled through until it reaches the target value.

This methodology can be explained below in a stepwise fashion:

i. Denote T as the sum of the total population value inside the population.
ii. The following process is repeated N number of times:
   - Select any random integer number "r" that comes within the range 0 to T.

- Continue looping through the population and accumulating the predicted value until this sum is larger than or equal to "*r*." That individual is selected whose sum of expected value is greater than this limit.

The main benefit of this method is that it is simple to apply, but it has the disadvantage of being noisy.

(b) Random selection: As the name indicates, the random selection of a parent from the population is involved. On average, random selection is a little more tedious as compared to the roulette wheel selection method in the sense of interruption of genetic codes.

(c) Rank selection: The roulette wheel would be having an issue when the fitness value is very different, as in this case, the chromosomes whose fitness value is very low have fewer chances of selection. The rank selection approach involves ranking the population and determining the fitness of each chromosome based on the ranking. Fitness 1 indicates the worst and fitness N means the best chromosome. This leads to gradual convergence but avoids convergence too quickly. It retains diversity and, thus, results in a favorable search. Capable parents are chosen in turn.

(d) Tournament selection: In this selection strategy, based on the fitness value, a tournament is carried out between N individuals in a N-way tournament selection. So, in this way, only the fittest individual is selected and passed to the successor generation. We get our final selection of individuals by performing many such tournaments and these passes to the next generation. The competition winner is determined by who has the highest fitness value. The selection pressure is given by the gap in the fitness value, whose advantage is that it pushes the IoE-based genetic algorithm for generating the fittest successor gene. This technique is known to be more powerful and able to identify an optimal solution.

(e) Boltzmann selection: Here, the selection rate is monitored by a frequently fluctuating temperature. Here, the selection pressure is negatively correlated to the temperature. As a result of the high initial temperature, the selection pressure is low in comparison to their temperature relationship. Gradually, as the temperature is dropped, the selection pressure increases. As a result, the search space becomes contracted in conjunction with the preservation of population divergence. Boltzmann probability for individual selection is determined as:

$$P = \exp \frac{\left[fmax - f\left(Xi\right)\right]}{T} \tag{2.1}$$

Here, $T = T_0\left(1-\alpha\right)^k$ and $k = (1 + 100*g/G)$; g = current generation number, and G = maximum value of g.

In this selection strategy, the probability is very high of choosing the best string formatting. Also, this method incorporates a very small execution time. One of the drawbacks of this method is that during the mutation stage some information may have vanished. But this can be avoided by elitism.

Elitism is a method that consists of the inclusion of the first prime chromosome or some prime chromosomes are transcribed to the new population.

(f) Stochastic Universal Sampling: This type of sampling method produces zero bias and minimal distribution. Here, the individual is mapped to coterminous fragments of the line, and like roulette wheel selection, each individual's fitness is proportional to the size of its fragments. As much as individuals are to be selected, in this method evenly spaced pointers are put along a line. Let the *N* pointer represent the number of individuals to be chosen, the 1/*N* pointer represents the distance between the pointers, and the range [0, 1/*N* pointer] represents the position of the first pointer, which can be a randomly generated value.

The stochastic universal sampling concept can easily be understood from Figure 2.13. For a group of six people, the distance between the pointers is 1/6 = 0.167. As a sample, one of the random numbers to be chosen within the range [0, 0.125] is 0.1.

Individuals after the selection procedure are assigned to the mating population as follows: 1, 2, 3, 4, 6, and 8.

When opposed to the roulette wheel selection, stochastic universal sampling ensures a more accurate selection of offspring.

b. ***Crossover***

IoE-based GA's basic technique for producing the best solutions/offspring is to crossover the parent genes. Different crossover strategies are designed to get the optimal solution in minimal generations as early as possible. Crossover operator selection has more impact on IoE-based GA's efficiency. A selection of suitable breeding operators may prevent premature convergence in the GA [51–71].

- Single point crossover

  A crossover point on the string is selected for the parent organism. In the string of species, all the data which are beyond that stage are exchanged between the two-parent species. Strings are labeled with positional bias.
- Two-Point Crossover

  This type of crossover technique is a special case by *N*-point. On the individual chromosomes (strings), the genetic material is exchanged at the points where random points are chosen.

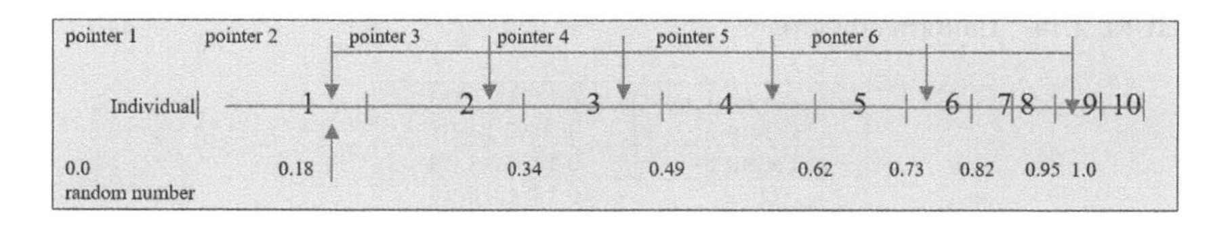

FIGURE 2.13 Stochastic universal sampling.

- Multipoint Crossover

  This crossover has two ways to go: an even number of cross-sites and an odd number of cross-sites. The even number of cross-sites entails randomly selected cross-sites around a circle, and information is exchanged along those cross-sites. Because of the odd number of cross-sites, a distinct cross point is frequently assumed at the start of the series.
- Uniform Crossover

  The chromosomes are not divided into fragments in a uniform fusion; rather, they handle every gene on an individual basis. To check whether every chromosome is enclosed within the offspring or not, we tend to primarily flip a coin for every chromosome. We might tend to skew the coin to one parent, too, to possess a lot of genetic material from that parent within the infant (Figure 2.14).
- Three-Parent Crossover

  Three parents are selected at random in this crossover technique. Every bit of the first parent is compared to the bit of the second parent. If all of the parents are identical, the offspring inherits the bit; otherwise, the offspring inherits the third parent's part (Figure 2.15).
- Crossover with Reduced Surrogate

  The crossover is constrained by the reduced surrogate operator to create new individuals whenever viable. This is accomplished by restricting the location of the crossover point so that crossover points occur only where the gene value fluctuates.
- Shuffle Crossover

  The shuffle crossover has to do with a standard crossover. A single location is picked on the crossover. However, until the variables are switched, they are jumbled at random in both parents. After recombination, the variables in the offspring are reset. Because the variables are reassigned at random each time the crossover is done, positional bias is eliminated.

Parent 1	1 0 1 0 1 0 1 0
Parent 2	0 1 0 1 0 1 0 1

Child 1	1 1 0 0 1 0 01
Child 2	0 0 1 1 0 1 10

FIGURE 2.14 Uniform crossover.

Parent 1	0 1 0 1 0 0 0 1
Parent 2	1 0 0 0 1 0 01
Parent 3	0 1 0 1 0 1 0 1
Child	0 1 0 1 0 0 0 1

FIGURE 2.15 Three-parent crossover.

- Precedence Preservative Crossover

  Blanton and Wainwright (1993) created PPX autonomously for vehicle routing problems, while Bierwirth et al. (1994) developed it for scheduling problems (2008). No new previous relationships are generated in motion as a result of the operator passing priority operation relationships supplied in two parental permutations for one offspring at the same rate. The PPX for an issue with six operations A–F is illustrated below. This is how the operator works (Figure 2.16):
  - The length vector sigma, sub i = 1 to mi, which represents that the number is randomly filled with operations involved in the problem place elements {1, 2}.
  - The sequence in which the operations are performed is determined by this vector, which is obtained sequentially from parent 1 and parent 2.
  - The operations "append" and "delete" can also be considered permutations for parents and offspring as lists.
  - It can be started by first creating an empty offspring.
  - According to the parent order indicated in the vector, one of two parents has the leftmost operation picked.
  - Once an operation has been identified, all parents uninstall it.
  - The preferred operation is eventually appended to the offspring.
  - Until both parents are zero and the child incorporates all relevant operations, repeat the preceding steps.
  - It should be noted that due to the "delete append" scheme, the PPX scheme is not working in a standardized crossover manner.
- Ordered Crossover

  Whenever the problem is ordered based mostly, an ordered two-point crossover is employed. Two random crossover locations are picked from two-parent chromosomes and partitioned into three parts: left portion, center part, and right part. This ordered two-point crossover acts as follows: The left and right parts of child 1 are inherited from their parent's left and right parts, while the child's middle part is determined by the genes in parent 1's middle part, in the order that the values appear in parent 2. In determining child 2, an analogous process is applied (Figure 2.17).
- Partially Matched Crossover

  PMX can be used in a variety of situations, including the traveling salesman problem (TSP). TSP chromosomes are represented by integer sequences, with each integer representing a different city and the order representing the time spent visiting each city. Here, the only point of concern is permutation encoding, which is the labels under this representation. It can be understood as a permutation crossover that ensures that all roles are positioned exactly once in each offspring, i.e., all children receive a full complement of genes, followed by their parents' reciprocal filling of alleles. PMX operates as follows:

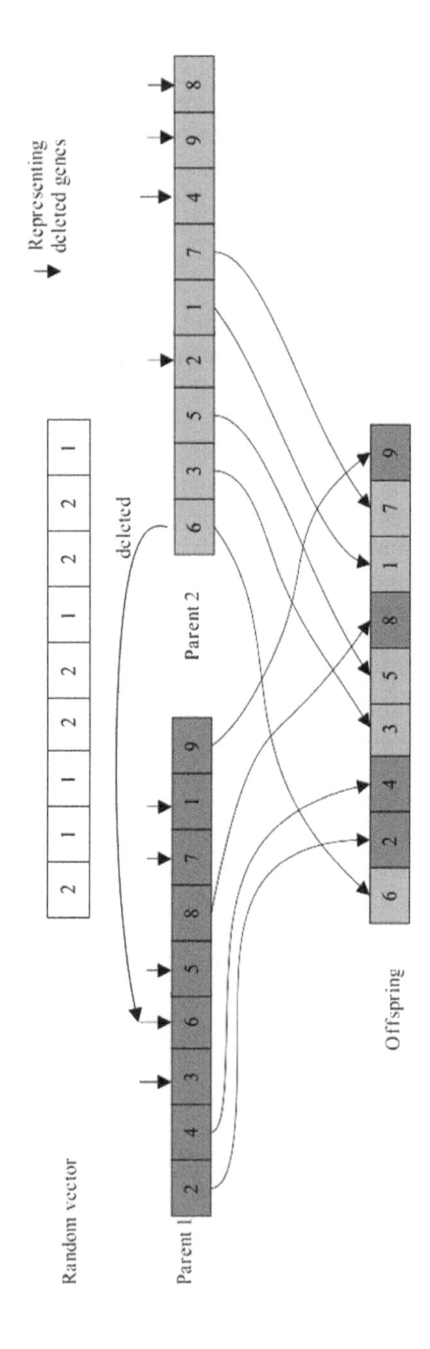

FIGURE 2.16 Precedence preservative crossover.

Parent 1: 4 3| 1 2|5 6 Child 1: 4 3|2 1| 5 6

Parent 2: 3 2| 1 7| 6 5 Child 2: 3 2| 7 1| 6 5

FIGURE 2.17 Ordered crossover.

Name 9 8 4 . 5 6 7 . 1 3 2 10 **Allele** 1 0 1 . 0 0 1 . 1 1 0 0
Name 8 7 1 . 2 3 10 . 9 5 4 6 **Allele** 1 1 1. 0 1 1 . 1 1 0 1

Given Strings

Name 9 8 4 . 3 10 2 . 1 5 7 6 **Allele** 1 0 1 . 1 0 0 . 1 0 10

Name 8 2 1 . 7 5 6 . 9 3 4 10 **Allele** 1 0 1 . 1 1 1 . 1 1 0 1

PMX Crossover

FIGURE 2.18 PMX crossover.

i. The two chromosomes that have been chosen are aligned.
ii. As per the section specified to suit, randomly pick two locations along a string.
iii. A cross-position exchange operation is performed through position-by-position, where the matching section is employed.
iv. The movement of alleles into their new position is involved in their child.

The working steps of PMX are depicted as follows:

- As an example, two strings are taken into account in Figure 2.18.
- Wherever the selected cross points are represented by dots.
- The position-wise exchange that has to occur in each parent to produce offspring described in a subsequent step.
- Scan the swapping from the identical section of a chromosome on any chromosome.
- The place-exchange numbers in this example are 5 and 3, 6 and 10, and 7 and 2. The following are the offspring that resulted:

c. ***Mutation***

The strings endure mutation by the following crossover. Mutation prohibits the algorithm from being captured in a minimum of the locality. The main role of mutation is to retrieve the vanished genetic materials in conjunction with to randomly disturb genetic information. A bit is mutated when it is dynamically switched from 0 to 1 or vice versa [18].

- Flipping

  Flipping off a bit includes shifting 0 to 1 and 1 to 0 based on a created mutant chromosome. A parent is taken into account, and a mutant

Parent	0 1 0 10 1 10
Mutation Chromosome	0 0 1 0 1 0 1 1
Child	0 1 1 1 1 1 0 1

FIGURE 2.19 Mutation flipping.

Parent	1 1 1 0 0 0 0 1
Child	1 1 0 0 0 1 0 1

FIGURE 2.20 Interchanging.

Parent	1 0 0 0 1 0 10
Child	1 0 0 0 1 0 01

FIGURE 2.21 Reversing.

chromosome is generated at random. A child chromosome is created by substituting a 1 in the mutant chromosome for a matched parent chromosome. Binary encoding generally used the concept of flipping (Figure 2.19).

- Interchanging

  The bits analogous in these positions are interchanged, where the randomly chosen position of a string is involved (Figure 2.20).
- Reversing

  When a reversal mutation for binary chromosomes is applied, it selects the random spot and reverses the bits adjacent to it and produces the child's chromosome (Figure 2.21).
- Mutation Probability

  The probability of mutation determines the number of mutant chromosomal segments. The offspring is produced without any modification immediately after the crossover if there is no mutation process involved. One or more parts of the chromosome will be affected or modified if the mutation process happens. If the chance of mutation to be happening is 100 percent, the entire chromosome will change; if it is 0 percent, no change will occur. The major advantage of mutation is that it prevents the IoE-based GA from the extremes of the locality.

2.6 STOPPING CONDITION FOR THE GENETIC ALGORITHM

The various stopping conditions are given below [19–41]:

1. Maximum Generations:
   IoE-based GAs come to a halt after the specified number of generations have passed.
2. Elapsed time:
   Before the stipulated period has passed, the genetic process will come to a stop when the maximum number of generations has been attained.
3. No change in fitness:
   The genetic process will come to an end when the stated number of generations has passed with no change in the population's greatest fitness.
4. Stall generation:
   The IoE-based GA finishes when the objective function does not improve for a sequence of subsequent generations of length "stall generation."
5. Stall time limit:
   The IoE-based GA will come to a halt if the objective function does not improve during the duration (in seconds) equal to the "stall time limit."
   The culmination of the convergence criterion ultimately puts a halt to the search. Some methods of culmination techniques are given below:
   (a) Best Individual:
       This criterion ends the search if the population's minimal fitness falls below the convergence value. This will lead the search to a better and rapid conclusion, ensuring as a minimum one favorable solution.
   (b) Worst individual:
       When the least fit individuals in the population have fitness less than the convergence requirement, the search is terminated. It ensures a minimum norm for the whole population. In any situation, a strict criterion will never be reached, in which situation the search will stop after the limit has been surpassed.
   (c) Sum of Fitness:
       When the sum of fitness in the entire population does not exceed the population record's convergence value, this strategy is used. This parameter ensures that everyone in the population is fit inside a certain range. On setting the convergence value, consideration must be given to population size.
   (d) Median fitness:
       If at least one-half of the individuals must be better than or equal to the convergence value, this requirement offers a wide range of options.

2.7 CONSTRAINTS IN THE GENETIC ALGORITHM

In unconstrained optimization problems, the IoE-based GA is considered to comprise solely an objective function and no knowledge regarding the variable specification. The following is an example of an unconstrained optimization problem [16–41]:

$$\text{Minimize } f(x) = x^2 \tag{2.2}$$

Here, no information is available regarding the "x" range. By using the random specification of its operators, the IoE-based GA minimizes this function.

In constrained optimization problems, the information regarding variable specification will be given. In these problems, constraints are listed below:

(a) Equality relations
(b) Inequality relations

The set of parameters that will be assessed using the IoE-based GA using the system in question, the objective function (to be minimized or maximized), and restrictions. However, the goal function is evaluated on running the program and the constraints are tested to see if there are any breaches. When no violations occur, the fitness value analogous to the objective function measurement is assigned to the specified parameter. The remedy is unreliable and results in no fitness in the case of constraint violation. As a consequence, some knowledge should be derived from the infeasible solution, regardless of their fitness rating in correlation to the extent of the breach of the restriction. With the help of the penalty method, we can achieve this.

By using a penalty approach, which involves the transformation of a problem of constraint optimization into an unconstrained problem of optimization by allying a penalty or expense with entire combinations of constraints. Assessment of the objective function embedded such type of penalty.

Let us examine a problem under constraint as:

$$\text{Maximize } f(x) \tag{2.3}$$

$$\text{Subject to } g_i(x) \geq 0, \quad i = 1,2,3,\ldots,n$$

where, x = k-vector. Converting this to an unrestricted form as:

$$\text{Maximize } f(x) + P \sum_{i=0}^{n} \varnothing\left[gi(x)\right] \tag{2.4}$$

where, $\varnothing$ = penalty function and P = penalty coefficient.

The unconstrained solution becomes confined when the penalty coefficient approaches infinity.

2.8 SIMILARITY AND COMPARISON BETWEEN IOE AND IoT

2.8.1 Similarity between IoE and IoT [58–61]

1. **Decentralization**: The systems are dispersed, such that each node carries out its task itself and functions as a small management center.
2. **Security issues**: Distributed systems are still extremely sensitive to hacking and cyber attacks; the more devices attached to the system, the greater the risk of a breach.

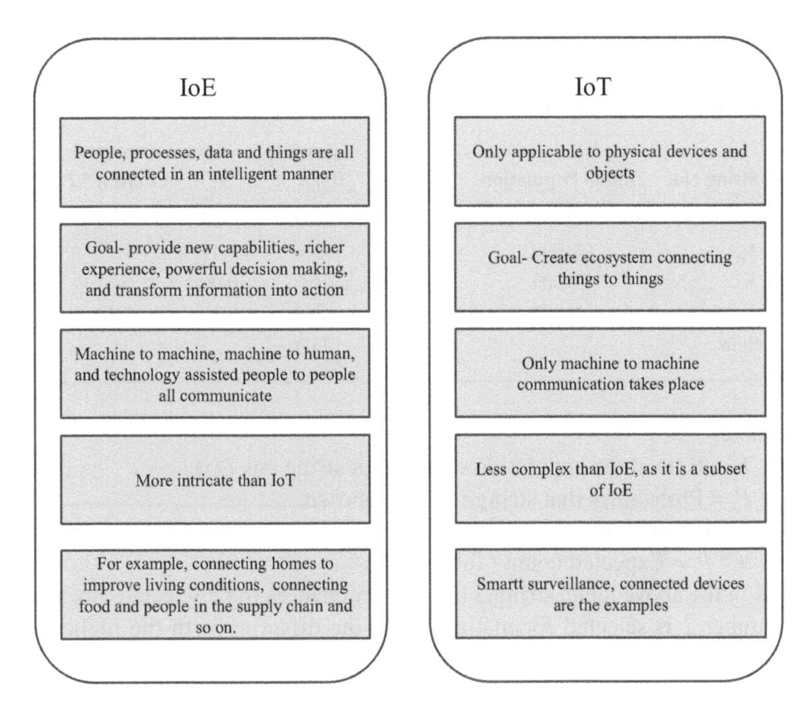

FIGURE 2.22 Distinction between IoE and IoT.

2.8.2 IoE vs IoT: What's the Difference?

The basic difference between IoE and IoT is given below (Figure 2.22):

2.9 PROBLEM-SOLVING USING THE GA

Problem statement – Maximize the function $f(x) = x^2$ with x lying in range [0, 31] i.e., $x = 0, 1, 2, \ldots\ldots\ldots, 31$ [45–51].

(1) Create an initial population at random. Chromosomes and genotypes are the terms for them.
Here, for example, four strings of 5-bit length taken as chromosomes are: 01101 (13), 11000 (24), 01000 (8), and 10011 (19).

(2) Fitness calculation is carried out in two steps:
   (a) Transform chromosomes into an integer known as phenotype.
   01101-->13, 11000-->24, 01000-->8, 10011-->19
   (b) Determine the worth of fitness by using the function given in the problem statement $f(x) = x^2$
   13-->169, 24-->576, 8-->64, 19-->361

(3) Selection operation – Choose two individuals (or parents) depending on their P_i fitness, which is presented as:

$$P_i = F_i \Bigg/ \left(\sum_{j-1}^{n} Fj \right) \tag{2.1}$$

TABLE 2.1
Selection Operation

String No.	Initial Population	*X* value	Fitness F_j $f(x) = x^2$	P_i	Expected Count $n * P_i$
1.	01101	13	169	0.14	0.56
2.	11000	24	576	0.49	**1.97**
3.	01000	8	64	0.06	0.22
4.	10011	19	361	0.31	1.23
Sum			1170		

where,

F_i = Extend the population's fitness for string i as f (x)

P_i = Probability that string i will be chosen.

n = Population size

$n * P_i$ = Expected count (Table 2.1).

As in the above table, string 2 has greater chances to be selected. So, string number 2 is selected for mating to get the offspring with the higher frequency value.

(4) Crossover operation:

It can be of either one point in which there will be one breakpoint and a chosen pair of strings is severed at a random position and then segments are exchanged to generate a new pair of strings or a two-point with two breakpoints.

Here, the one-point crossover is used. We replace the string which has the least value of the expected count with the string which has the highest value of this count.

If we compare the sum of fitness in the below table with the above table, it has increased from 1,170 to 1,754. So, this is done by the genetic algorithm.

(5) Mutation operation

After crossover, it is applied to each child individually. At the randomly chosen position of randomly picked strings, bits are altered from 0 to 1 or from 1 to 0.

We have not done anything with strings 2 and 3 (can do mutation on these two strings) because these strings have the fitness value (see Table 2.2).

So, if we compare the fitness sum after applying mutation operation (Table 2.3), its value is 2,354 from 1,754 (Table 2.2). Since the GA improved the summation of all the individual fatnesses, we can say that we have got better results after applying the GA.

2.10 ADVANTAGES

1. It is effortlessly parallelized, easily modified, and adaptable to different problems.
2. It has massive and extensive solution space searchability.

TABLE 2.2
Crossover Operation

String No.	Mating Pool	Crossover Point	Offspring after Crossover	*X* value	Fitness f(*x*) = x^2
1.	0 1 1 0 | 1 1	4	0 1 1 0 0	12	144
2.	1 1 0 0 | 1 0	4	1 1 0 0 1	25	625
2.	1 1 | 1 0 0 0	2	1 1 0 1 1	27	729
4.	1 0 | 1 0 1 1	2	1 0 0 0 0	16	256
Sum					1754

TABLE 2.3
Mutation Operation

String No.	Offspring after Crossover	Offspring after Mutation	*X* value	Fitness f(*x*) = x^2
1.	_0 1 1 0 0	1 1 1 0 0	26	676
2.	1 1 0 0 1	1 1 0 0 1	25	625
2.	1 1 0 1 1	1 1 0 1 1	27	729
4.	1 0 _0 0 0	1 0 1 0 0	18	324
Sum				2354

3. Rather than using derivatives, it employs payoff (objective function) information.
4. It employs probabilistic rather than deterministic transition principles. It is adequate for a noisy environment.
5. In terms of local minima/maxima, it is very vigorous.
6. It has a straightforward concept.
7. It works exhaustively on mixed discrete/continuous problems.
8. It is stochastic.
9. Rather than a single point, it searches the entire population of points.
10. Multi-objective optimization is bolstered.
11. When compared to the traditional brute force method, it is faster and more efficient.

2.11 IoE-BASED GA APPLICATION

The applications of the IoE-based GA are given in Figure 2.23.

2.12 LIMITATIONS

1. Its implementation is still an art.
2. It is computationally expensive, i.e., time-consuming.

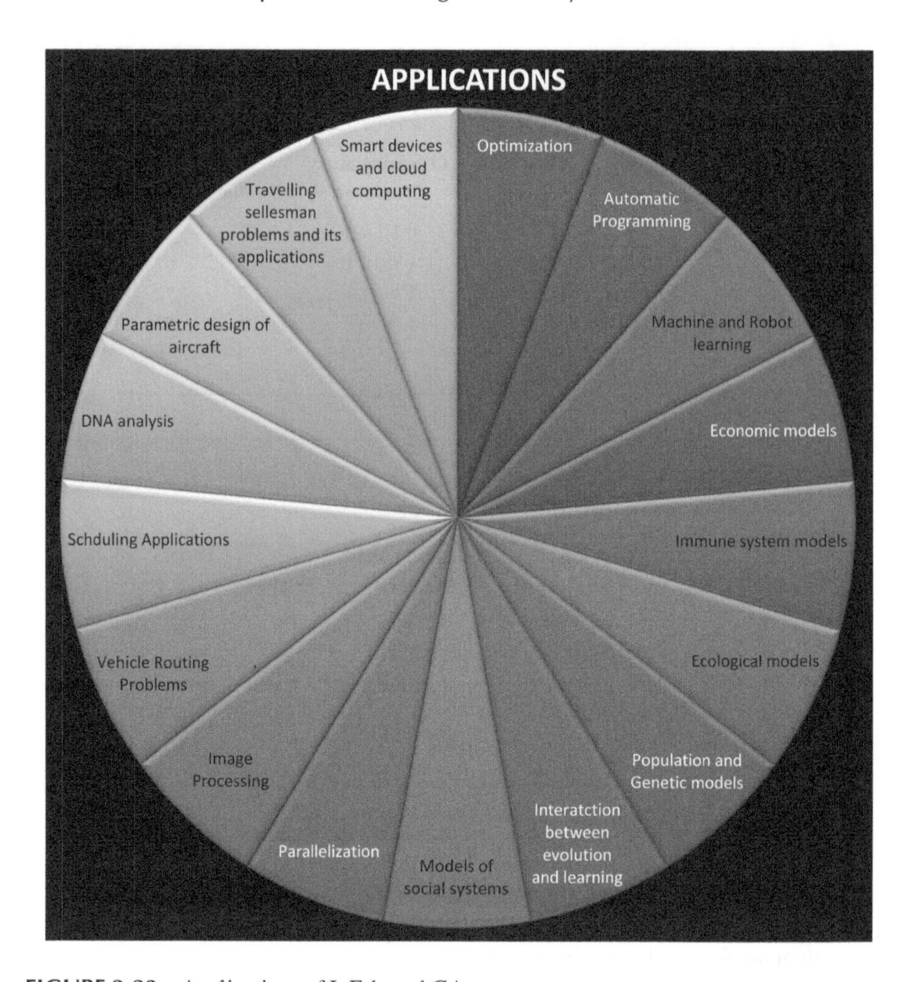

FIGURE 2.23 Applications of IoE-based GA.

3. It is not considered the best solution for simple problems where the derivative information is readily available.
4. If it is not implemented correctly, it may not converge to an optimal solution.
5. The difficult part of the GA is deciding on many factors including population size, crossover rate, mutation rate, selection method, and strength.

2.13 CONCLUSION

The IoE-based genetic algorithm serves as a strong adaptive approach to remedy exploration and optimization issues. It is based on the catechistic searching technique, which solves the problem quickly, i.e., guarantees to give the best solution but the optimal solution may or may not be obtained. They are more robust than conventional algorithms. The GA does not break easily, unlike in older AI systems. Initial population, fitness function calculation, selection, crossover, mutation, and terminating criteria are all part of the general GA. The IoE-based GA relies on four pillars – process, data, people, and things. Encoding, selection, crossover, and mutation are the four primary operators in genetic algorithms.

REFERENCES

1. SN Shivanandam, SN Deepa, *Principles of Soft Computing*, Wiley, 2008.
2. MH Miraz, M Ali, PS Excell, R Picking, "A review on Internet of Things (IoT), Internet of Everything (IoE) and Internet of Nano Things (IoNT)", *Internet Technologies and Applications (ITA)*, pp. 219–224, Sept. 2015.
3. LT Yang, B Di Martino, QC Zhang, *Internet of Everything*, p. 8035421, Hindawi, Jul. 2017.
4. T Škorić, K Katzis, S Jovanović, "Four pillars of IoT in health application", in: IEEE EUROCON 2019 -18th International Conference on Smart Technologies, pp. 1–4, Novi Sad, Serbia, Jul. 2019.
5. X Fan, X Liu. W Hu, C Zhong, J Lu, "Advances in the development of power supplies for the Internet of Everything", *InfoMat*, pp. 130–139, May 2019.
6. DJ Langley, J van Doorn, ICL Ng, S Stieglitz, A Lazovik, A Boonstra, "The Internet of Everything: Smart things and their impact on business models", *Journal of Business Research*, vol. 122, pp. 853–863, Jan. 2021.
7. DJ Feng, WS Wijesoma, "Improving Rao-Blackwellised genetic algorithmic filter SLAM through genetic learning", in: 2008 10th International Conference on Control, Automation, Robotics and Vision, pp. 1200–1205, Hanoi, Vietnam, Dec. 2008.
8. I Moon, J-H Lee, J Seong, "Vehicle routing problem with time windows considering overtime and outsourcing vehicles", *Expert Systems with Applications*, vol. 39, pp. 13202–13213, Dec. 2012.
9. B Singh, RP Payasi, J Sharma, "Effects of DG Operating Power Factor on Its Location and Size by Using GA in Distribution Systems", in: Bansal R. (eds) *Handbook of Distributed Generation*. Springer, 2017.
10. I Nishizaki, M Sakawa, "Computational methods through genetic algorithms for obtaining Stackelberg solutions to two level mixed zero-one programming problems", *Cybernetics and Systems*, vol. 31, pp. 203–221, Oct 2010.
11. J Lin, B Foote, S Pulat, C Chang, JYCheung, "Hybrid genetic algorithm for container packing in three dimensions", in: Proceedings of 9th IEEE Conference on Artificial Intelligence for Applications, pp. 353–359, Orlando, FL, USA, 1993.
12. NS Chaudhari, YS Ong, V Trivedi, "Computational capabilities of soft-computing frameworks: An overview", 2006 9th International Conference on Control, Automation, Robotics and Vision, pp. 1–6, Singapore, Dec. 2006.
13. BL Miller, DE Goldberg, "Genetic algorithms, selection schemes, and the varying effects of noise", *Evolutionary Computation*, vol. 4, no. 2, pp. 113–131, June 1996.
14. N Rathore, I Chana, "Load balancing and job migration algorithm: A survey of recent trends", *Wireless Personal Communication (Q-3)*, vol. 79, no. 3, pp. 2089–2125, IF-2.313, July 2014. Springer Publication-New-York (USA), ISSN print 0929-6212
15. V Sharma, R Kumar, NK Rathore, "Topological broadcasting using parameter sensitivity-based logical proximity graphs in coordinated ground-flying ad hoc networks", *Journal of Wireless Mobile Networks Ubiquitous Computing and Dependable Applications(JoWUA) (Q-2)*, vol. 6, no. 3, pp. 54–72, Sept. 2015. ISSN: 2093-5374 (printed), ISSN: 2093-5382 (online), IF-2.40.
16. N Rathore, I Chana, "Variable threshold based hierarchical load balancing technique in grid", *Engineering with Computers (Q-1)*, vol. 31, no. 3, IF- 3.938, pp. 597–615, June 2015. Springer publication-London (England (UK), ISSN: 0177-0667 (print version).

17. NK Rathore, I Chana, "A cognitive analysis of load balancing technique with job migration in grid environment", in: World Congress on Information and Communication Technology (WICT), Mumbai, pp. 77–82, Dec. 2011. IEEE proceedings paper, ISBN -978-1-4673-0127-5.
18. NK Rathore, I Chana, "A sender initiate based hierarchical load balancing technique for grid using variable threshold value" in International Conference IEEE-ISPC, pp. 1–6, Solan, India, 26–28 Sept. 2013. Paper Presented & Published, ISBN- 978-1-4673-6188-0.
19. N Rathore, I Chana, "Job migration with fault tolerance and QoS scheduling using hash table functionality in social grid computing", *Journal of Intelligent & Fuzzy Systems (Q-3)*, vol. 27, no. 6, pp. 2821–2833, June 2014. IOS Press publication-Netherland, ISSN print 1064-1246, IF- 1.851,.
20. N Rathore, I Chana, "Job migration policies for grid environment", *Wireless Personal Communication (Q-3)*, vol. 89, no. 1, pp. 241–269, July 2016. Springer Publication-New-York (USA), ISSN print 0929-6212, IF- 2.313.
21. NK Rathore, I Chana, "Report on hierarchal load balancing technique in grid environment", *Journal on Information Technology (JIT)*, vol. 2, no. 4, pp. 21–35, Sept.–Nov. 2013. ISSN Print: 2277-5110, ISSN Online: 2277-5250, IF= 2.235.
22. NK Rathore, I Chana, "Checkpointing algorithm in Alchemi.NET", in: Annual Conference of Vijnana Parishad of India and National Symposium Recent Development in Applied Math-ematics & Information Technology, JUET, Guna, MP, Dec. 2009. Abstract Published.
23. N Rathore, "Performance of hybrid load balancing algorithm in distributed web server system", *Wireless Personal Communication (Q-3)*, vol. 101, no. 3, pp. 1233–1246, 2018. Springer Publication-New-York (USA), ISSN print 0929-6212, IF -2.313.
24. N Rathore, "Dynamic threshold based load balancing algorithms", *Wireless Personal Communication (Q-3)*, vol. 91, no. 1, pp. 151–185, Nov 2016. Springer Publication-New-York (USA), ISSN print 0929–6212, ISSN online 1572-834X, IF -2.313.
25. NK Rathore, "Ethical hacking & security against cyber crime", *Journal on Information Technology (JIT)*, vol. 5, no. 1, pp. 7–11, 2016. December 2015–February 2016. ISSN Print: 2277-5110, ISSN Online: 2277-5250, IF= 2.235.
26. NK Rathore, I Chana, "Comparative analysis of checkpointing", *IT Enabled Practices and Emerging Management Paradigm*, pp. 321–327, 2008.
27. NK Rathore, "Efficient agent-based priority scheduling and load balancing using fuzzy logic in grid computing", *Journal on Computer Science (JCOM)*, vol. 3, no. 3, pp. 11–22, Sept.–Nov. 2015. ISSN Print: 2347-2227, ISSN online: 2347-6141, IF= 0.750.
28. NK Rathore, "Map reduce architecture for grid", *Journal on Software Engineering (JSE)*, vol. 10, no. 1, pp. 21–30, July–Sept. 2015. ISSN Print: 0973-5151, ISSN Online: 2230-7168, IF= 3.765.
29. NK Rathore, "Faults in grid", *International Journal of Software and Computer Science Engineering*, ManTech Publication, vol. 1, no. 1, pp. 1–19, 2016.
30. NK Rathore, A Sharma, *Efficient Dynamic Distributed Load Balancing Technique: A Smart Tool & Technology to Balance the Load Among the Network*, LAP LAMBERT Academic Publishing, 19 Oct. 2015. Project ID: 127478, ISBN no-978-3-659-78288-6.
31. NK Rathore, "Efficient hierarchical load balancing technique based on grid", in: 29th MP Young Scientist Congress, vol. 55, Solan, India, 2014.
32. NK Rathore, I Chana, "Fault tolerance algorithm in Alchemi.NET", in: Middleware, National Conference on Education & Research (ConFR10), Third CSI National conference, Jaypee University of Engg. & Tech., Guna, 2010.
33. NK Rathore, "Efficient load balancing algorithm in grid", 30th MP Young Scientist Congress, vol. 56, Bhopal, MP, 2015.

34. R Chouhan, NK Rathore, "Comparision of load balancing technique in grid", in: 17th Annual Conference of Gwalior Academy of Mathematical Science, Jaypee University of Engg. & Tech., Guna, 2012.
35. RI Doewes, AAA Ahmed, A Bhagat, R Nair, PK Donepudi, S Goon, V Jain, NK Rathore, A regression analysis based system for sentiment analysis and a method thereof, Patent Application No: 2021101792, *Australian Official Journal of Patents*, vol. 35, no. 17, 2021.
36. NK Rathore, R Chohan, *An Enhancement of Gridsim Architecture with Load Balancing*. Scholar's Press, 23 Oct. 2016. ISBN: 978-3-639-76989-0, Project id: 4900.
37. NK Rathore, U Rawat, SC Kulhari, "Efficient hybrid load balancing algorithm", *National Academy Science Letters*, vol. 43, no. 2, pp. 177–185, 2020.
38. NK Jain, NK Rathore, A Mishra, "An efficient image forgery detection using biorthogonal wavelet transform and improved relevance vector machine", *Wireless Personal Communications*, vol. 101, no. 4, pp. 1983–2008, 2018.
39. NK Rathore, "Installation of Alchemi.NET in computational grid", *Journal on Computer Science (JCOM)*,vol. 4, no. 2, pp. 1–5, 2016.
40. NK Rathore, P Singh, *An Efficient Load Balancing Algorithm in Distributed Networks*. Lambert Academic Publication House (LBA), 2016.
41. NK Rathore, "Checkpointing: Fault tolerance mechanism", *Journal on Cloud Computing (JCC)*, vol. 4, no. 1, pp. 28–35, 2017.
42. NK Rathore, PK Singh, "A comparative analysis of fuzzy based load balancing algorithm", *i`manager Journal of Computer Science (JCS)*, vol. 5, no. 2, pp. 23–33, 2017.
43. NK Rathore, "GridSim installation and implementation process", *Journal on Cloud Computing (JCC)*, vol. 2, no. 4, pp. 29–40, 2015.
44. NK Rathore, *An Efficient Dynamic & Decentralized Load Balancing Technique for Grid*. Scholars' Press, 2018. Project id: 6621.
45. N Jain, N Rathore, A Mishra, "An efficient image forgery detection using improved relevance vector machine", *Interciencia Journal*, vol. 42, no. 11, pp. 95–120, 2017.
46. NK Rathore, H Singh, "Analysis of grid simulators architecture", *Journal on Mobile Applications and Technologies (JMT)*, vol. 4, no. 2, pp. 32–41, 2017.
47. N Rathore, "A review towards: Load balancing techniques", *i-Manager's Journal on Power Systems Engineering*, vol. 4, no. 4, p. 47, 2016.
48. NK Rathore, I Chana, PIMR ThirdNational IT conference, IT Enabled Practices and Emerging Management Paradigm book and category is Communication Technologies and Security Issues, pp. 32–35, Topic No/Name-46, Prestige Management And Research, Indore, 2008.
49. F Khan, NK Rathore, "Internet of things: A review article", *i-manager's Journal on Cloud Computing*, vol. 5, no. 1, pp. 20–25, 2018.
50. N Jain, NK Rathore, A Mishra, "An efficient image forgery detection using biorthogonal wavelet transform and singular value decomposition", in: 5th International Conference on Advance Research Applied Science, Environment, Agriculture & Entrepreneurship Development (ARABSEED), Bhopal organized & sponsored by Jan Parishad, JMBVSS & International Council of people at Bhopal, pp. 274–281, held on 04-06 December 2017, 2017. ISBN No-978-93-5267-869-3
51. NK Rathore, NK Jain, PK Shukla, US Rawat, R Dubey, "Image forgery detection using singular value decomposition with some attacks", *National Academy Science Letters*, vol. 44, no. 4, pp. 331–338, 2021.
52. D Pandey, U Rawat, NK Rathore, K Pandey, PK Shukla, "Distributed biomedical scheme for controlled recovery of medical encrypted images", *IRBM*, 2020. Innovation andResearch in BioMedical Engineering (Q-3), Elsevier IF=1.09, ISSN no- 1959-0318, Issue-43, pp. 151160, May 2022. https://doi.org/10.1016/j.irbm.2020.07.003

53. NK Rathore, I Channa, "Checkpointing algorithm in Alchemi.NET", *Journal of Information Technology*, vol. 8, no. 1, pp. 32–38, 2010.
54. NK Rathore, "Rupendra tandekar and Alok Gour \Hotel Management" in Scholar's Press, Mauritius, Project id: 12332, ISBN: 978-613-8-95576-4, 2021
55. N Rathore, D Pandey, RI Doewes, A Bhatt, "A novel security technique based on controlled pixel based encryption of image blocks for sharing a secret image", *Wireless Personal Communication (Q-3)*, pp. 191207, 18-June 2021, DOI: 10.1007/s11277-021-08630-w, ISSN print 0929-6212, IF-2.313.
56. NK Rathore, V Jaiswal, V Sharma, S Varma, *A Hybrid Methodology for Flower Images Segmentation & Recognition with extended Deep-Convolution Neural Network.* CNN, 2021.
57. P Laxkar, NK Rathore, "Load balancing algorithm in distributed network", *Solid State Technology*, vol. 63, no. 2s, pp. 6633–6645, 2020. ISSN: 0038–111X, SCOPUS Indexed, IF=0.05.
58. NK Rathore, Load balancing algorithm in distributed network", in: Janparishad 6th International Conference on Science & Environmental, Sustainability for a peaceful Society (SESPS-2018)in association with international cities of Peace (USA), SusTranCon (USA), International Council of people(India), Global Network for Sustainable Development, Center for Global Nonkilling (USA) and JMBVSSat Bhopal (M.P.) India, Conference ID-SESPS 2018:06, ISBN No-978-93-5321-737-2, pp. 02–03, held on 19-21 January 2019
59. NK Rathore, J Rathore, "Efficient checkpoint algorithm for distributed system", *International Journal of Engineering and Computer Science (IJECS)*, vol. 1, no. 2, pp. 59–66, 2019. E-ISSN: 2663-3590, P-ISSN: 2663-3582.
60. NK Rathore, *An Efficient Load Balancing Technique for Grid.* Scholar's Press, Mauritius, 2018. ISBN: 978-3-330-65134-0, Project id: 6621.
61. NK Rathore, F Khan, "Survey of IoT", *International Journal of Computer Science and Soft Computing*, vol. 1, no. 1, pp. 1–13, 2018. ManTech Publication, ISSN no-2319-7242.
62. NK Rathore, presented paper on "Big data analysis", in: 4th International Conference on latest Concept in Science, Technology & Management (ICLCSTM-2017), held in Sinhgad College of Engineering, Vadgao, Pune, 17-August-2017.
63. S Sivanandam, S Deepa, "Terminologies and operators of GA", in: *Introduction to Genetic Algorithms.* Springer, pp. 39–81, 2008.
64. CR Reeves, "Genetic algorithms", in: Gendreau M, Potvin JY (eds) *Handbook of Metaheuristics. International Series in Operations Research & Management Science*, vol 146. Springer, pp. 109–139, 2010.
65. http://www.quora.com/
66. http://mindmajix.com/
67. http://towardsdatascience.com/
68. https://www.youtube.com/watch?v=IZEhdgjZyf0
69. https://www.bbvaopenmind.com/en/technology/digital-world/the-internet-of-everything-ioe/
70. https://www.sam-solutions.com/blog/what-is-internet-of-everything-ioe/
71. J Rathore, VS Chouhan, "Employee job satisfaction: Empirical study of Indian public and private industry", *Turkish Online Journal of Qualitative Inquiry (TOJQI)*, vol. 12, no. 6, pp. 9860 –9875, July 2021.

3 A Framework for Hybrid WBSN-VANET-based Health Monitoring Systems

Pawan Singh, Ram Shringar Raw, and Dac-Nhuong Le

CONTENTS

3.1 Introduction 51
3.2 Literature Review 53
3.3 Proposed Framework 55
3.4 Flowchart and Algorithm for the proposed framework 55
3.5 Conclusion 60
References 60

3.1 INTRODUCTION

Catastrophic disaster event highlights the limitations of current communication technologies and emphasizes the need for a reliable and adaptive remote health monitoring network. The main purpose of such a network is to transfer a patient's health data to a nearby hospital or care provider for emergency treatments. Medical emergencies such as heart attacks, high/low blood pressure, and accidental recovery are all dependent on how quickly the patient receives medical attention. In India, a lack of medical attention at the time of trauma causes nearly 27% of deaths each year. Ambulances provide first assistance to injured people or patients in any situation, including traffic accidents, fires, and unexpected illnesses. Finding a seriously ill or injured individual, dispatching an ambulance with advanced life-saving equipment, and delivering medical treatment on the spot are all challenging tasks. With the advancement of pre-hospital emergency treatment technology and the evolution of first-aid situations in recent years, first-aid equipment has been continually updated. Ambulances are equipped with various emergency equipment, including a multi-function display screen, an electrocardiogram machine, cardiac first-aid instrument, respiratory machine, defibrillator, etc. [1].

However, connectivity between the ambulance and the hospital through 3G/4G or Wi-Fi networks is insufficient to allow all of these devices to communicate vast amounts of data. Medical equipment creates a lot of data, and a significant network

DOI: 10.1201/9781003323426-3

bandwidth is necessary to communicate with the hospital. When the ambulance travels quickly, the network connectivity becomes unpredictable, making it extremely difficult to maintain adequate system performance. Using an existing communication network to send a large volume of medical picture data will incur significant communication expenses.

In this chapter, a prototype that combines the Wireless Body Sensor Network (WBSN) and the Vehicular Ad-hoc Network (VANET) is proposed to solve the above issues. If standard communication networks are unavailable, WBSNs with VANETs can enable remote health monitoring [2]. Figure 3.1 illustrates the VANET health monitoring architecture.

It is advised that a VANET be set up on a city road between the ambulance and the hospital. Using VANETs in place of traditional networks for communication will increase the network bandwidth and accelerate network throughput. The communication mechanism is not the same as a typical cellular network with a base station, but rather multi-hop communication between many moving vehicle nodes on the road. Both VANETs and WBSNs are now the focus of interest for academicians and scientists [3–5]. However, VANETs and WBSNs have significantly distinct properties.

The WBSN deploys highly miniaturized bio-medical sensors on or around the patient's body without hampering the daily life activities and observes vital signs like an electrocardiogram, body temperature, blood pressure, SpO_2 level, heartbeat, etc. These sensors are generally static, use limited resources and energy, and have strong sensing capabilities. These sensors can detect chronic diseases such as heart attacks, asthma, BP, oxygen levels, and diabetes. They can sense environmental parameters such as location, temperature, humidity, light, and have the ability to alert patients in the event of an emergency [6, 7].

The VANET is a class of mobile ad-hoc network that arranges its communication framework itself with no reliance on some other fixed infrastructure. The VANET

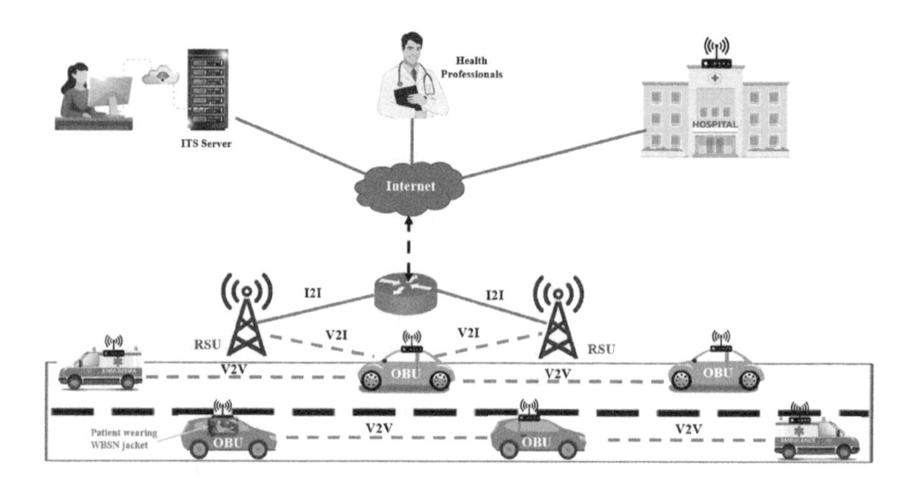

FIGURE 3.1 VANET Health Monitoring Architecture.

consists of running vehicles and fixed infrastructure installed at the roadside. The VANET utilized dedicated short-range communication (DSRC) for high-speed and secure communication between the vehicles and roadside units (RSUs) [8, 9]. RSUs work as a gateway between nodes and servers, providing more coverage to vehicles in their communication range. VANETs have highly dynamic topologies and do not suffer from energy constraints. VANETs have mainly two types of communication that are Vehicle-to-Vehicle Communication (V2V) and Vehicle to Infrastructure Communication (V2I) as depicted in Figure 3.1.

Many of VANET's applications have been developed so far for making the long journey comfortable and more convenient. VANET applications can be classified broadly into two categories: safety application and comfort application. Safety application includes traffic signal violation, intersection collision, turn assistance, blind spot warning, pedestrian crossing, lane change warning, forward collision warning, post-crash alert, emergency service vehicle and curve speed warning, etc. Comfort applications include information about restaurants and free parking slot availability, automatic toll collection, watching a real-time video, route diversion in case of traffic jams, and downloading maps for traveling [10].

In this Hybrid WBSN-VANET Network, miniaturized bio-medical sensors are deployed on the human body in the form of a wearable jacket. These bio-medical sensors constantly communicate their sensor data to the vehicles driving on the road, delivering them accurately and up-to-date Physical Health Information (PHI). Vehicles do V2V multi-hop communication to disseminate this information to intended users over comparatively long distances.

The remainder of this chapter is composed as follows: In Section 3.2, the literature work is reviewed. Section 3.3 discusses the system model used for the Hybrid WBSN-VANET architecture. Section 3.4 presents the health monitoring routing algorithm, flowgraph, and their descriptions. Section 3.5 presents the application scenarios and related challenges, and Section 3.6 presents the conclusions and also suggests future work.

3.2 LITERATURE REVIEW

VANETs can be utilized as an exceptionally helpful technology in transmitting PHI data or some sort of caution messages to the closest clinic, family members, and traffic controlling authorities in the absence of cellular networks or wireless local area networks. VANETs can send intermittent and communicated messages, high need crisis messages, and educational and non-wellbeing application messages to improve traveler security and traffic efficiency (Biswas et al., 2006). VANETs can play a major role in realizing the dream of the Intelligence Transport System (ITS). The ITS is a system in which there will be automatic traffic control, the number of traffic deaths will also be significantly reduced, and this system is not possible without the implementation of VANETs. VANETs have attracted the attention of researchers, academicians, and automobile manufacturers. VANETs can significantly reduce deaths in road accidents. VANETs may be a very beneficial technology in declining road accidents and traffic fatalities. Many VANET projects like NoW (Network on

Wheels) (Festag et al., 2008), FleetNet (Franz et al., 2005), CarTALK (Reichardt et al., 2003), and CarNet (Morris et al., 2000) have been created mainly in Europe with the aim of safer vehicles and roads.

Many contributions in the area of wireless sensor networks (WSNs) and VANETs have been proposed. A new form of network called the Hybrid Sensor and Vehicular Network (HSVN) has been proposed by Carolina et al. in which WSNs and VANETs work together to improve road safety [11]. WSNs and VANETs collaborate to provide a vehicle-to-vehicle communication infrastructure that helps drivers lower road accidents, casualties, and injuries. Sun et al. proposed RescueMe [12], which is a location-based VANET. It helps in safe and trustworthy rescue planning and resource allocation of resources used in rescue operations after disaster.

A healthcare application that uses an RFID-enabled authentication scheme is proposed in [13, 14], which provides medical facilities to traveling patients. It uses RFID technology with a Petri net-based authentication model for the proposed model. A cloud-based health monitoring system is presented in [15]. The Cloud database is used as the central database to upload and download the patient's health information using a mobile phone or a web browser. A health professional may download this uploaded information for monitoring and guidance purposes. To obtain personal PHI from patients, RCare [16] is presented as a delay-tolerant, durable, and long-term healthcare system. To reduce healthcare expenses, RCare provides network connection to rural regions employing regular transport vehicles such as automobiles and buses as relay nodes.

An emergency routing protocol named VehiHealth [17] is proposed to forward the patient's health information to a nearby hospital in a short time. VehiHealth considers the neighboring intersection to forward the data with minimum delay. It selects the next intersection based on the shortest path, vehicle stability, number of link breakage, and delay between neighboring intersections. A VANET-based diagnosis and response system, proposed in [18], used VANET technology to set up a virtual communication network throughout a large rural area with very little infrastructural cost. They proposed a protocol for vehicles equipped with OBUs to communicate with each other using the IEEE 802.11p protocol.

Smart vehicular ad-hoc network (SVANET) [5] architecture is proposed by Prasan et al. that uses WSNs to detect events and vehicles to efficiently broadcast safety and non-safety messages over multiple service channels and a single control channel with varied priorities. The SVANET is data transfer protocol for highway vehicles, which utilizes V2V communication for connected vehicles and nearby WSNs in the absence of connected vehicles. The SVANET protocol is designed to improve driving safety, avoid accidents, and maximize channel utilization by dynamically adjusting control and service channel time intervals.

Kumar et al. proposed a framework for a health monitoring planning for on-board ships using Flying Ad-Hoc Networks (FANETs) and WBAN technology, which is used to provide immediate response to on-board patients in emergency situations where communication is extremely difficult [19]. W-GeoR is proposed in [20] for VANET health monitoring applications, with an emphasis on next-hop node selection for quicker vital sign distribution in urban traffic environments. For the optimal next-hop node selection method, W-GeoR employed traffic-aware parameters such as traffic mobility, inter-vehicle distances, speed variations, communication connection expiry time, channel quality, and proximity features.

3.3 PROPOSED FRAMEWORK

This prototype includes a hazardous health condition warning, which alerts drivers to potentially hazardous road conditions such as heart attack, low SpO_2 level, etc. Driver's vital signs are sensed by different bio-medical sensors like the ECG and heartbeat sensor, blood pressure, EEG sensors, SpO_2 sensor, and temperature sensor. These sensors sense vital signs regularly and communicate this information to the Body Control Unit (BCU). The BCU is a microcontroller or Mote, which monitors the vital sign at regular intervals. The standard value of health parameters is compared by the health monitoring module with the sensed value, and in case of abnormality, a warning signal is sent to the vehicle's OBU. This module provides the wireless connectivity interface to the OBU by utilizing Bluetooth or ZigBee. The data aggregation module averages the extracted PHI and compares with normal PHI values. If average PHI values are not found similar to normal PHI values, only a warning message is sent to the OBU to transmit it to nearby hospital/ ambulance or doctors. These warning messages are composed of the patient's health data in the eXtensible Markup Language (XML) format and the vehicle's IP address and GPS information using the IEEE 802.15.6 standard. To minimize the battery consumption, sensors are put in sleeping mode unless there is an emergency [21]. The vehicle uses carry and forward approaches until some vehicle does not enter its communication range. Whenever another vehicle enters inside its communication range, PHI messages are transmitted to it. Vehicles use appropriate VANET routing protocols to forward PHI messages to nearby ambulances, hospitals, and RSUs. The medical professionals make decisions in response to the received PHI. The nearest ambulance is dispatched to the patient's location with life-saving equipment, and an intelligent health monitoring system completes the testing on patient health [22]. Figure 3.2 represents the entire system model.

3.4 FLOWCHART AND ALGORITHM FOR THE PROPOSED FRAMEWORK

Algorithm 1: *Emergency Event Detection and Route Establishment*

```
Input: PHI values sensed by different bio-sensors
Output: Broadcast PHI messages
Notation:
  • SIP-address: Source Vehicle IPv4 address
  • DIP-address: Destination Vehicle IPv4 address
  • NIP-address: Neighbor Vehicle IPv4 address
  • Next-Hop: Next forwarding vehicle
  • N_Table: Neighbor Table
```

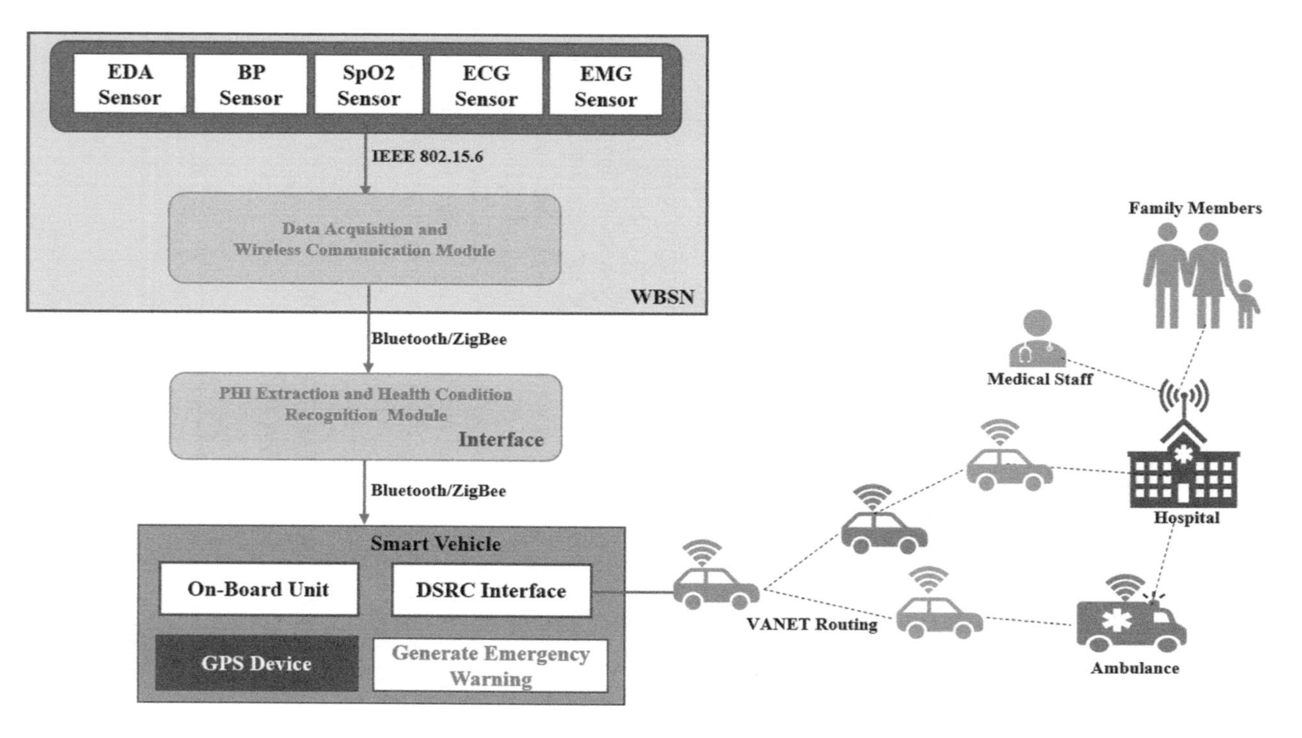

FIGURE 3.2 System Model used for the Hybrid WBSN-VANET architecture.

- Th-Time: Threshold Time
- ID-Time: Sensors Idle Time
- N-Dir: Neighbor Vehicle Direction
- D-Dir: Destination Vehicle Direction

```
Begin Algorithm 1
  1.Computes PHI values of all bio-sensors;
  2.Listen to Mote;
  3.if (Mote is idle)
  4.{
  5. if (ID-Time>=Th-Time)
  6. {
  7.  if (PHI>Standard_PHI)
  8. {
  9.  Mote communicates with OBU;
  10.  OBU calls Route_Discovery_Process (SIP-address,
  DIP-address);
  11. }
  12.}
  13.else
  14.   Bio-Sensors go to sleeping mode;
  15.}
  16.Route_Discovery Process (SIP-address, DIP-address)
  17.{
  18.  OBU receives PHI message and update its N_Table;
  19.   if (NIP-address==DIP-address)
  20.    {
  21.     Stop broadcasting PHI messages;
  22.    }
  23.   else
  24.   {
  25.    while (NIP-address !=DIP-address)
  26.    {
  27.      Calculate Neighbor_Position and Neighbour_Direction
  28.      if (N-Dir towards D-Dir)
  29.      {
  30.        Select NIP-address as Next-Hop by using Algorithm 2;
  31.        SIP-address delivers PHI to NIP-address;
  32.        Next-Hop update its N_Table;
  33.        Broadcast PHI messages;
  34.      }
  35.    }
  36.  }
  37.}
End Algorithm 1
```

The method for generating emergency signals in the event of a PHI abnormality detected by any sensor node is reparented by Algorithm 1. Bio-sensors continuously sense the vital signs in a patient's body and transmit them to Mote. Mote analyzes

PHI and compares it with standard PHI parameters. To preserve the Mote battery, it is put in sleep mode for a specific threshold time. Mote communicates with OBU wirelessly in abnormal PHI readings only. The OBU is now attempting to establish multi-hop ad-hoc communication with moving vehicles to transmit PHI messages to nearby ambulances, hospitals, or RSUs using Algorithm 2. The OBU sends an acknowledgment (ACK) message for each delivered message containing the message sequence number to confirm the message delivery. Mote dispatches another PHI message and waits for the ACK message. Once receiving the ACK message, the PHI record will be marked as delivered (Figure 3.3).

Algorithm 2: *Next-hop Vehicle Selection and Packet Forwarding Procedure*

Input: Position and Velocity information of Sender and Destination Vehicles
Output: The Next-hop IP
Notation:

- SV: Source Vehicle
- DV: Destination Vehicle
- CFV: Current Forwarding Vehicle
- $Set_{CFV:}$ Set of neighbor vehicles within the communication range of the current forwarding vehicle
- WF_{max}: Maximum Weight Factor
- NV: Neighbor Vehicle
- NHV: Next-hop Vehicle

Begin Algorithm 2

```
1. The source vehicle SV generates data to send;
2. Let CFV = SV;
3. CFV broadcast "Hello packet" to its neighbor vehicles
4. CFV node updates the NV Table
5. if (D ε Set_CFV )
6. {
7. A direct link is available.
8.  CFV transmits the PHI data packet to vehicle D;
9.  }
10. else
11. {
12.  for i=1 to sizeof (Set_CFV)
13.  {
14.    Calculate WF_i of all NV using CalWeight( );
15.    if (WFi > WF_max)
16.   {
17.      WF_max = WF_i;
18.      Set NHV = NV_i;
19.    }
20.  }
21. }
22. Set CFV= NHV;
```

End Algorithm 2

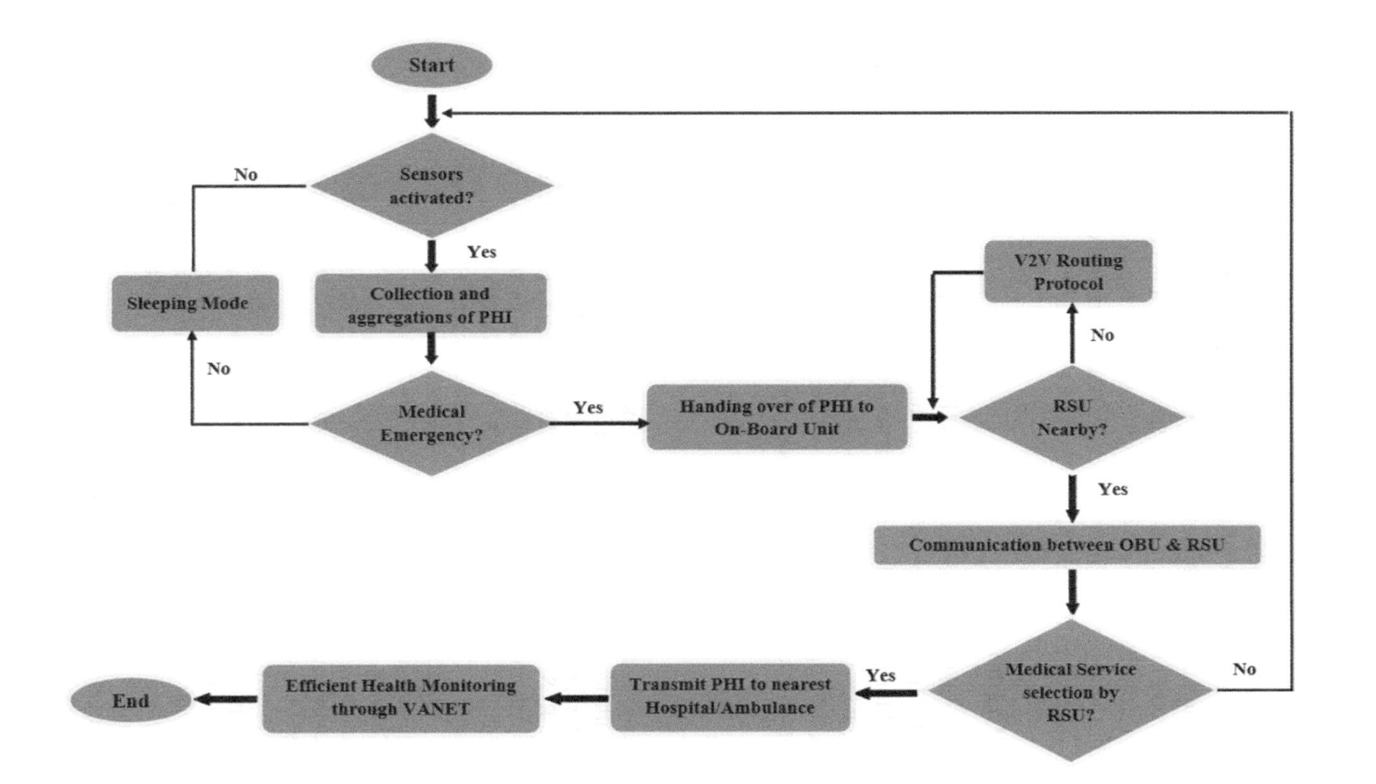

FIGURE 3.3 Flowchart for the Hybrid WBSN-VANET framework.

Algorithm 2 represents the pseudocode of the weight gradient-based VANET routing protocol for next-vehicle selection and packet forwarding procedures. It makes use of GPS services, which offer vehicle position coordinates and velocity. The source node starts this protocol before transmitting a packet; it broadcasts beacon packets to cars within the communication range and changes its neighbor table accordingly. It invokes the *CalWeight( )* procedure, which calculates the relative weights of all nearby cars. The relative importance of neighbor's weights is assigned based on different factors such as vehicle speed, direction, distance, communication link quality, etc. Relative weights of these factors may be optimized by using fuzzy logic and the Analytic Hierarchy Process (AHP). The maximum weight gradient neighbor is selected as the next neighbor to forward the PHI.

3.5 CONCLUSION

We presented some key concepts on a health monitoring system based on a Hybrid WBSN-VANET in this article. We believe that the Hybrid WBSN-VANET could be a valuable tool for warning drivers about severe health conditions like heart attacks and low oxygen levels. Hybrid wireless body sensor networks with VANETs are more reliable than the approaches that address the same application domain but require more complex infrastructure. Simultaneously, availability and accuracy should be adequate.

REFERENCES

1. S. El-Masri and B. Saddik, "An emergency system to improve ambulance dispatching, ambulance diversion and clinical handover communication: A proposed model," *J. Med. Syst.*, vol. 36, no. 6, pp. 3917–3923, Dec. 2012, doi: 10.1007/s10916-012-9863-x.
2. P. Singh, R. S. Raw, and S. A. Khan, "Development of novel framework for patient health monitoring system using VANET: An Indian perspective," *Int. J. Inf. Technol.*, vol. 13, pp. 383–390, 2020, doi: 10.1007/s41870-020-00551-4.
3. H. Noshadi, E. Giordano, H. Hagopian, and W. Universit, "Remote medical monitoring through vehicular ad hoc network," in International Symposium on Wireless Vehicular Communications, Calgary, AB, Canada, (WiVeC 2008), pp. 1–5, 2008, doi: 10.1109/VETECF.2008.456.
4. S. Umamaheswari and R. M. Priya, "An efficient healthcare monitoring system in vehicular ad hoc networks," *Int. J. Comput. Appl.*, vol. 78, no. 7, pp. 45–49, 2013, doi: 10.5120/13505-1254.
5. P. K. Sahoo, M.-J. Chiang, and S.-L. Wu, "SVANET: A smart vehicular ad hoc network for efficient data transmission with wireless sensors," *Sensors*, vol. 14, no. 12, pp. 22230–22260, 2014, doi: 10.3390/s141222230.
6. P. Singh, "Internet of things based health monitoring system: Opportunities and challenges," *Int. J. Adv. Res. Comput. Sci.*, vol. 9, no. 1, Feb. 2018, doi: 10.26483/ijarcs.v9i1.5308.
7. A. Hussain, R. Wenbi, A. L. Da Silva, M. Nadher, and M. Mudhish, "Health and emergency-care platform for the elderly and disabled people in the Smart City," *J. Syst. Softw.*, vol. 110, pp. 253–263, 2015, doi: 10.1016/j.jss.2015.08.041.
8. S. Al-Sultan, M. M. Al-Doori, A. H. Al-Bayatti, and H. Zedan, "A comprehensive survey on vehicular ad hoc network," *J. Netw. Comput. Appl.*, vol. 37, no. 1, pp. 380–392, 2014, doi: 10.1016/j.jnca.2013.02.036.

9. R. S. Raw, M. Kumar, and N. Singh, "Security challenges, issues and their solutions for VANET," *Int. J. Netw. Secur. Its Appl.*, vol. 5, no. 5, pp. 95–105, Sep. 2013, doi: 10.5121/ijnsa.2013.5508.
10. M. R. Ghori, K. Z. Zamli, N. Quosthoni, M. Hisyam, and M. Montaser, "Vehicular ad-hoc network (VANET): Review," in 2018 IEEE International Conference on Innovative Research and Development, Bangkok, Thailand, pp. 1–6, 2018.
11. C. T. Barba, K. Ornelas Aguirre, and M. Aguilar Igartua, "Performance evaluation of a hybrid sensor and vehicular network to improve road safety," in PE-WASUN'10: Proceedings of the 14th ACM Symposium on Performance Evaluation of Wireless Ad Hoc, Sensor, & Ubiquitous Networks, Co-located with MSWiM'10, Bodrum, Turkey, pp. 71–78, 2010, doi: 10.1145/1868589.1868604.
12. J. Sun, X. Zhu, C. Zhang, and Y. Fang, "RescueMe: Location-based secure and dependable VANETs for disaster rescue," *IEEE Journal on Selected Areas in Communications*, vol. 29, no. 3, pp. 659–669, 2011, doi: 10.1109/JSAC.2011.110314.
13. N. Kumar, K. Kaur, S. C. Misra, and R. Iqbal, "An intelligent RFID-enabled authentication scheme for healthcare applications in vehicular mobile cloud," *Peer-to-Peer Netw. Appl.*, vol. 9, no. 5, pp. 824–840, 2016, doi: 10.1007/s12083-015-0332-4.
14. K. Ahed, M. Benamar, A. A. Lahcen, and R. El Ouazzani, "Forwarding strategies in vehicular named data networks: A survey," *J. King Saud Univ.: Comput. Inf. Sci.*, vol. 34, Issue 5, pp. 1819–1835, 2022, doi: 10.1016/j.jksuci.2020.06.014.
15. A. B. Adeyemo, W. O. Adesanya, and O. Ariyo, "Framework for a cloud based health monitoring system," in Proceedings of the 2nd International Conference on Computing Research and Innovations, Ibadan, Nigeria, Sep. 2016.
16. M. Barua, X. Liang, R. Lu, and X. Shen, "RCare: Extending secure health care to rural area using VANETs," *Mob. Networks Appl.*, vol. 19, no. 3, pp. 318–330, 2014, doi: 10.1007/s11036-013-0446-y.
17. S. K. Bhoi and P. M. Khilar, "VehiHealth: An emergency routing protocol for vehicular ad hoc network to support healthcare system," *J. Med. Syst.*, vol. 40, no. 3, pp. 1–12, 2016, doi: 10.1007/s10916-015-0420-2.
18. S. DasGupta, S. Choudhury, and R. Chaki, "VADiRSYRem: VANET-based diagnosis and response system for remote locality," *SN Comput. Sci.*, vol. 2, no. 1, p. 41, 2021, doi: 10.1007/s42979-020-00430-6.
19. S. Kumar, A. Bansal, and R. S. Raw, "Health monitoring planning for on-board ships through flying ad hoc network," *Advances in Intelligent Systems and Computing*, vol. 1089, pp. 391–402, 2020, doi: 10.1007/978-981-15-1483-8_33.
20. P. Singh, R. S. Raw, S. A. Khan, M. A. Mohammed, A. A. Aly, and D.-N. Le, "W-GeoR: Weighted geographical routing for VANET's health monitoring applications in urban traffic networks," *IEEE Access*, vol. 10, pp. 38850–38869, 2022, doi: 10.1109/ACCESS.2021.3092426.
21. H. Noshadi, E. Giordano, H. Hagopian, and W. Universit, "Remote medical monitoring through vehicular ad hoc network," in International Symposium on Wireless Vehicular Communications, Calgary, Canada, (WiVeC 2008), pp. 1–5, 2008.
22. A. Aliyu et al., "Cloud computing in VANETs: Architecture, taxonomy, and challenges," *IETE Tech. Rev. (Institution Electron. Telecommun. Eng. India)*, vol. 35, no. 5, pp. 523–547, Sep. 2018, doi: 10.1080/02564602.2017.1342572.

4 Managing IoT – Cloud-based Security

Needs and Importance

Sarita Shukla, Vanshita Gupta,
Abhishek Kumar Pandey, Rajat Sharma, Yogesh Pal,
Bineet Kumar Gupta, and Alka Agrawal

CONTENTS

4.1 Introduction 63
4.2 Background 64
  4.2.1 Cloud computing 64
    4.2.1.1 Characteristics of Cloud computing 64
    4.2.1.2 Cloud computing deployment models 65
    4.2.1.3 Cloud Computing Service Models 67
  4.2.2 IoT (Internet of things) 69
    4.2.2.1 Evolution 69
    4.2.2.2 Architecture of Internet of Things (IoT) 69
    4.2.2.3 Components of Internet of Things 69
4.3 Security challenges in IoT-Cloud 72
  4.3.1 Challenges of Cloud-Based IoT Integration 72
4.4 Literature survey related to IoT-Cloud Security 75
4.5 Conclusion 75
References 75

4.1 INTRODUCTION

The Internet of things is the digital era's fastest-growing and most widely used information technology (IT) paradigm. The deployment of Internet of things technology in many aspects of life is growing the amount of IoT clients [1]. IoT technologies are currently being used extensively in the health sector [2, 3], smart cities [4], smart homes [5, 6], and so on. However, IoT technology alone may not be able to fulfill the number of clients and their computing needs [7–11]. So, the clients' needs cannot be totally satisfied. So, cloud computing is essential for IoT computations. A network of communication between connected objects and devices via wireless and wired connection over the internet is called "Internet of things" [12–15].

Both the IoT and cloud computing are newly emerging services with distinct properties. The Internet of things (IoT) approach is built on smart devices that

DOI: 10.1201/9781003323426-4

connect with one another over a worldwide network and an active infrastructure. It permits omnipresent computing circumstances. The Internet of things is usually characterized by extensively distributed devices with inadequate processing and storage capacity. Performance, consistency, privacy, and security are all problems that these devices have [16–20]. The Internet of things is made up of three levels, five components, and a variety of applications including smart homes, smart healthcare, smart energy and grid, and so on. Cloud computing refers to a vast network with virtually limitless storage and processing capacity. Furthermore, it provides a flexible and stable environment that allows for dynamic data integration from different data sources [21]. The majority of IoT difficulties may be laid down to cloud computing. In fact, the integrated IoT and cloud computing are changeable for the existing and future environment of the internet services, which is a very comparative and challenging process. Cloud computing is a model of on-demand computing in which we can access any service, resources and application, platform, software from any location. There are five primary characteristics, three service models, and four deployment models in this cloud model that promote accessibility and availability [22–23].

The chapter is arranged as follows: Section 4.2 introduces the backdrop, Section 4.3 discusses the security challenges of IoT-clouds, Section 4.4 discusses the IoT-cloud literature review, and Section 4.5 gives the conclusion.

4.2 BACKGROUND

4.2.1 Cloud computing

Cloud computing offers a variety of services through the internet connection. Tools and applications such as data storage, servers, databases, networking, and software are examples of these services or resources [24–26]. Cloud computing is on-demand computing and clients may access any services like pay per use manner from anywhere. Many clients maintain their business-related data on the cloud. The following are some cloud service providers:

- Google Cloud
- Amazon Web Services (AWS)
- Microsoft Azure
- IBM Cloud

4.2.1.1 Characteristics of Cloud computing

This section discusses several cloud computing models and characteristics. Cloud computing has five distinguishing characteristics [26–29]. They are depicted in the (Figure 4.1).

The following are the fundamental aspects of cloud computing:

- ***Broad network:*** Capabilities are available across the network and may be accessed using standard procedures that encourage the use of diverse types of outdated or cloud-based software programs, as well as client platforms (e.g., mobile phones, computers, and tablets).

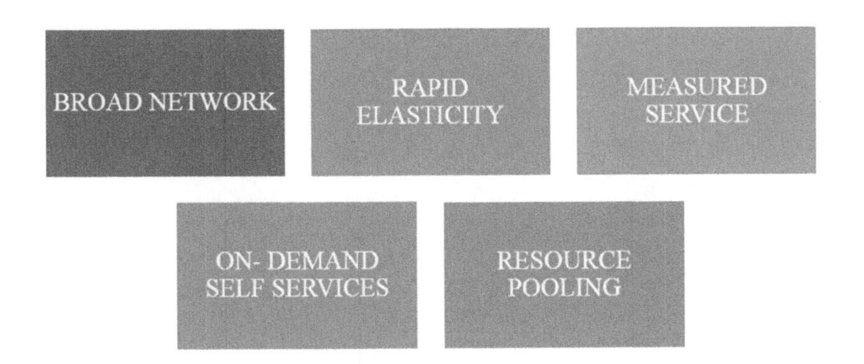

FIGURE 4.1 Characteristics of Cloud Computing.

- ***Rapid elasticity:*** Cloud computing allows you to scale up or down your resources based on your needs [30–33]. For example, throughout the course of a specific activity if you require a significant number of server resources, you can free them after the task is completed.
- ***Measured service:*** A metering capacity at a level of notion suited to the sort is used by cloud frameworks as frequently as practical of administration to control and improve asset use (e.g., capacity, preparing, data transmission, and dynamic client accounts). The use of resources is often inspected, regulated, and communicated, ensuring straightforwardness for both the provider and the client of the administrations [34–36].
- ***On-demand self-service:*** Without having to deal with each service provider individually, a cloud service user may provide computing resources such as server virtualization as needed. Because the cloud provider provides on-demand service, the resources are not part of the customer's IT infrastructure permanently.
- ***Resource pooling:*** Utilizing a multi-tenant approach, the supplier's assets are pooled to help various CSCs, with dissimilar physical and virtual assets being appointed and reallocated in terms of customer interest.

In the current scenario, cloud computing's relevance and role are determined by these different characteristics.

4.2.1.2 Cloud computing deployment models

There are various types of deployment models in cloud computing:

- Public cloud
- Private cloud
- Community cloud
- Hybrid cloud (Figure 4.2)

Public Cloud: The cloud service provider provides the cloud infrastructure as well as it is responsible for the data and services stored within the cloud. A corporation, academic organizations, government agency might own,

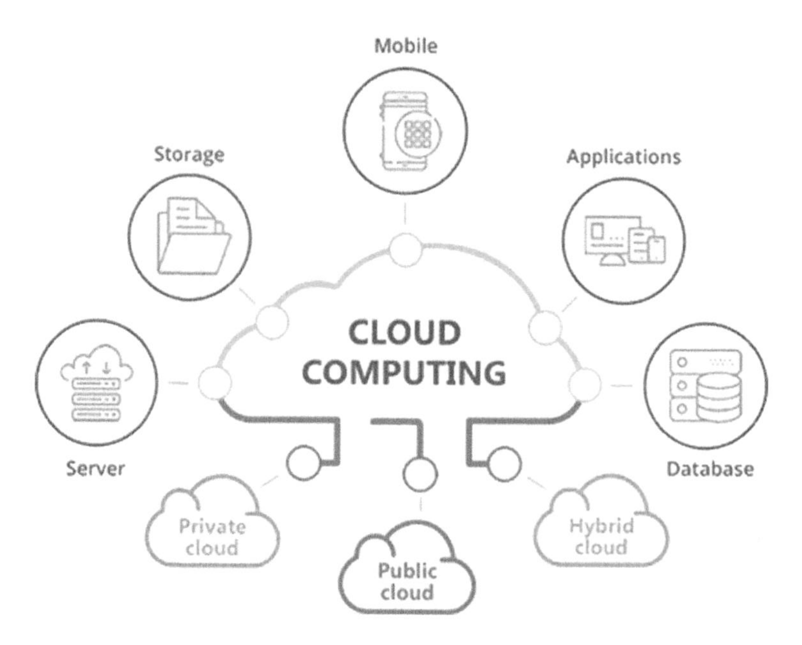

FIGURE 4.2 Cloud Computing Deployment Models.

manage, and operate a public cloud. The cost is the major advantage of a public cloud. When the clients and organization subscribers need the services, they purchase the services and resources by cloud service and they may have to make changes as needed.

Private Cloud: The organization's internal IT infrastructure is where a private cloud is installed. The organization might decide to manage the cloud inside or enlist an outsider to do as such. Cloud workers and capacity devices can likewise be on-premise, off-premise, or both [37–40]. All examples of on-demand services such as database, email, and storage are offered through the private cloud. The security of a private cloud is an important factor to consider. A private cloud architecture gives you more control over where your data is stored and other security concerns. Easy resource sharing and quick deployment to organizational entities are further advantages.

Community Cloud: A community cloud combines the benefits of both private and public clouds. Community cloud, like private cloud, has restricted access. The cloud resources, like those in the public cloud, are shared across a number of different businesses [41–44]. The organizations that use the community cloud have comparable needs and, in most cases, a need to share data. One business that makes use of the community cloud concept is the healthcare industry. The government typically employs a community cloud to meet privacy and other needs. In a controlled manner, data may be shared among community members.

Hybrid Cloud: The combination of two or more clouds, i.e., private, communal, or public is a hybrid cloud that is separate but connected by homogenous or unique technology that allows data and application mobility [45–49]. Critical data is typically housed in a private cloud area using a hybrid cloud system, while less sensitive data might benefit from the public cloud. Smaller organizations may be especially interested in a hybrid public/private cloud solution. Several low-risk apps are commonly offloaded to save money without the company agreeing to shift more essential data and software to the cloud.

4.2.1.3 Cloud Computing Service Models

Every service needs a suitable functioning model. Cloud computing also necessitates the adoption of a management strategy. In cloud computing, there are numerous models and sub-models for its service application [50–53]. In this part of the chapter, the writers cover four of the most important ones in order to provide readers with a technical and instructive discussion. Below is a thorough description of the various models.

Software as a Service: SaaS is a software as a service model that provides services to clients in the form of cloud-based applications and is accessible from anywhere. The user can utilize the cloud provider's apps, which are hosted on the provider's cloud infrastructure. The apps may be accessed from a variety of client devices using a simple interface, such as a web browser. Software installation, maintenance, upgrades, and repairs are all made easier using SaaS. Among the services that are offered are Google Gmail, Microsoft 365, Salesforce, and Cisco WebEx.

Platform as a Service: The PaaS (platform as a service) cloud provides clients with a platform on which to run their apps. Customers can deploy customer-created or customer-acquired apps on cloud infrastructure using PaaS. A PaaS cloud includes software building blocks as well as development tools, programming language tools, runtime environments, and other resources to help with the deployment of new applications.

Infrastructure as a Service: IaaS is an infrastructure as a service that allows customers to use the key cloud infrastructure's resources. The underlying cloud infrastructure resources are not managed or controlled by the user but you do have control over operating systems, installed apps, and maybe some networking devices (e.g., Host, firewalls). Virtual machines (VMs), virtualized hardware, and operating systems are all available through IaaS. Customers may use infrastructure as a service (IaaS) to deploy and run any software, including operating systems and applications, since it offers processing, storage, networks, and other essential computer functions. Table 4.1 summarizes the various cloud computing service models.

Advantages of Cloud Computing

The flexibility to utilize software from any device, whether through a native app or a browser, is one of the many advantages of cloud-built software for businesses across industries [54–57]. As a consequence, customers will be able to seamlessly transfer their data and preferences to other devices.

TABLE 4.1
Cloud Service Models

Cloud Service models	Description
Software as a Service (SaaS)	The SaaS model enables clients to use software applications as a service. For example, the Google App.
Platform as a service (PaaS)	PaaS offers the application runtime environment, as well as development and deployment tools. Consider Google App Engine.
Infrastructure as a service (IaaS)	Physical computers, virtual machines, virtual storage, and other main resources are accessible through IaaS. For instance, Amazon's EC2 and S3 services.

Cloud computing encompasses far more than simply sharing data across several devices. Users may check their email on any computer thanks to cloud computing services, and they can even save files using Dropbox and Google Drive. Customers may also back up their music, data, and photographs using cloud computing services, ensuring that they are quickly accessible in the case of a hard drive failure [58–61].

It also jeopardizes the enormous cost-cutting capabilities of large corporations. Prior to the cloud becoming a viable option, businesses had to acquire, build, and maintain expensive information management systems and infrastructure. Companies may forgo pricey server centers and IT staff in exchange for fast internet connections, allowing employees to collaborate online with the cloud to perform tasks.

Individuals can secure storage space on their PCs or laptops using the cloud framework. It also allows customers to upgrade software more quickly since software businesses can deal with their products online rather than through more traditional, physical means such as disks or flash drives. Customers of Adobe's Creative Suite, for example, can use an internet-based subscription to access apps from the Creative Suite. Customers may now quickly obtain new versions and resolutions for their programs.

Disadvantages of Cloud Computing

There are clearly hazards associated with cloud computing's speed, proficiency, and innovations. When it comes to sensitive medical data and financial information, security has always been a major worry with the cloud. Despite the fact that rules require cloud computing businesses to strengthen their security and compliance practices, it remains a current problem. The encryption protects sensitive data, but if the encryption key is lost, the data is lost forever. Normal catastrophes, internal problems, and power outages may all affect cloud computing servers. Cloud computing's geographical scope cuts both ways: Customers in New York might be affected by a California blackout, while a company in Texas could lose data if its Maine-based supplier goes down. There is a learning curve for both employees and management, as with any technology. However, when a large number of people view and manipulate data through a single gateway, unintentional errors can spread across the system.

4.2.2 IoT (Internet of things)

IoT is a network of physical internet-connected objects. These objects have incorporated technology that allows them to communicate with internal and external states.

The IoT is a system of interconnected computing devices that uses AI, networking, and other technologies to create platforms for internet-based services and the capacity to move data and information over the network without the need of human-to-computer contact.

The section gives a quick rundown of the Internet of things.

4.2.2.1 Evolution

The internet has gone through four stages of development, ending in the Internet of things (IoT):

- ***Information technology:*** PCs, servers, routers, firewalls, and other IT equipment that are subscribed as IT devices by enterprise IT professionals and are generally connected through wired connections are referred to as information technology (IT).
- ***Operational technology (OT):*** Drive OT representatives acquired health machinery, supervisory control and data acquisition as SCADA process control, and generally employing wired communication, are examples of embedded IT created by non-IT firms.
- ***Personal technology:*** Consumers purchase smartphones, tablets, and eBook readers as IT devices that use wireless connectivity exclusively and frequently.
- ***Sensor/actuator technology***: Single-purpose devices that only employ wireless communication, generally in a single form, are purchased by consumers, IT experts, and OT professionals as part of larger systems. The fourth generation of the Internet of things (IoT) is characterized by the utilization of billions of embedded devices (Figure 4.3).

4.2.2.2 Architecture of Internet of Things (IoT)

A three-layer design, as depicted in Figure 4.4, is the most fundamental architecture. Perception, network, and application layers are the three levels of the system [21].

Table 4.2 [8] summarizes the several levels of the Internet of things:

4.2.2.3 Components of Internet of Things

Figure 4.5 shows the main components of an IoT device:

- **Sensor:** The characteristics that are physical, chemical, or biological detected by a sensor device and provides an electrical signal proportionate to the detected characteristic, either as an analog voltage level or as a digital signal. The sensor output is often sent into a microcontroller or other management component in both situations.

FIGURE 4.3 IoT (Internet of things).

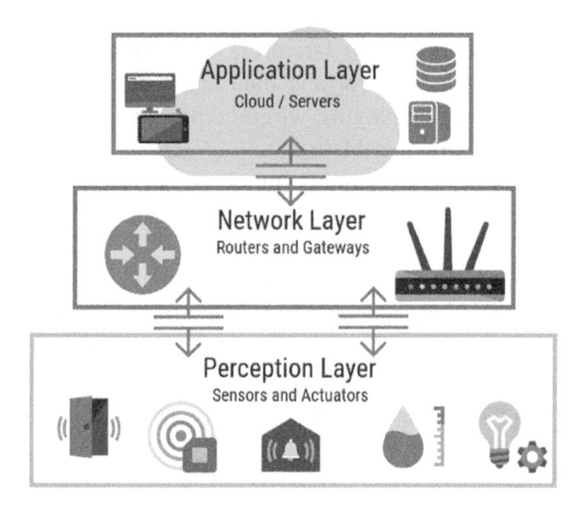

FIGURE 4.4 Architecture of IoT.

- **Actuator:** An electrical signal is received by an actuator from a controller and interacts with its environment as a reaction to change the restriction of a physical, chemical, or biological entity.
- **Microcontroller:** A smart device's "smart" is provided by deeply integrated microprocessors.
- **Transceiver:** A transceiver is an electrical device that sends and receives data. Most IoT devices feature a wireless transceiver that can connect with other IoT devices through wireless protocol such as Wi-Fi or another means.
- **Radio-frequency Identification (RFID):** Radio-frequency Identification, which consumes radio waves to identify items, is quickly gaining traction as an IoT enabler. Tags and readers are the two primary components of an

TABLE 4.2
Layers of the Internet of Things [21]

Layers	Description
Perception Layer	The physical layer with sensors for identifying and obtaining information about the environment is known as the perception layer. It detects some physical characteristics or recognizes other intelligent devices in the environment.
Network Layer	Connecting to other smart objects, network devices, and servers is the responsibility of the network layer. This layer's characteristics are also utilized to communicate and process sensor information.
Application Layer	The application layer is responsible for providing the user with application-specific services. It discusses how the Internet of things may be used in a variety of applications, such as smart homes, smart cities, and smart health.

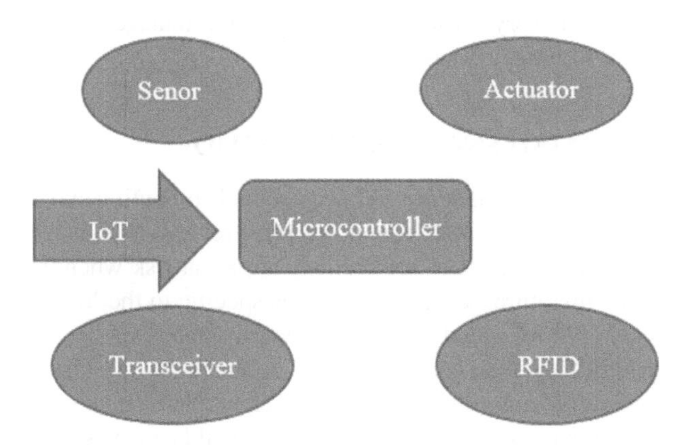

FIGURE 4.5 IoT Components.

RFID system. RFID tags are small programmable devices that are used to track entities, animals, and humans. They come in a wide range of forms, sizes, functions, and prices. Readers of RFID read and, in certain circumstances, inside reading range, RFID tags may be overwritten with new data. Readers are often connected to a computer system that stores information and analyses data for later use.

Advantages of Internet of Things

The following are some of the advantages of the Internet of things: [19]

- **Efficient resource utilization:** We can enhance resource efficiency and monitor natural resources if we understand the functioning and how each device operates.
- **Minimize human effort:** IoT devices, for example, interact and communicate with one another and perform a variety of activities for us, reducing human effort.

- **Save time:** It protects our time since it reduces our effort. Time is the most valuable resource that the IoT platform can help you save.
- **Enhance Data Collection**
- **Improve security:** Now, if we have a system that connects all of these elements, we can make the system even more safe and efficient.

Disadvantages of Internet of Things

The following are some of the disadvantages of IoT:

- **Security:** The Internet of things (IoT) systems are interconnected and interact via networks. Despite any security precautions, the system compromises a little amount of control and may be used to launch a variety of network assaults.
- **Privacy:** Even if the client does not actively participate, the IoT system delivers extensive personal data in great detail.
- **Complexity:** It is very difficult to design, build, manage, and enable huge technologies with the IoT system.

4.3 SECURITY CHALLENGES IN IOT-CLOUD

The authors identify several challenges related to cloud computing and IoT integration based on the preceding study, debate, and literature review. The integrity, confidentiality, and availability of data and applications are all at risk when using IoT-cloud applications. In the literature, security challenges specific to the IoT-cloud paradigm are rarely explored [19]. IoT-cloud application security issues, on the other hand, may arise at the level of IoT devices, communication, and networking. The author [11] discusses the security problems connected with IoT-cloud platforms for the smart home in detail. Security concerns with IoT-cloud-based healthcare systems may also be found in [12, 13]. In their publications, some researchers offered a solution to security issues. Table 4.3 summarizes the literature on IoT-cloud-based security issues.

4.3.1 Challenges of Cloud-Based IoT Integration

The successful integration of the cloud-based IoT paradigm may be constrained by a number of challenges. These challenges are depicted in Figure 4.6.

- ***Security and privacy:*** Data transmission from the real world to the cloud is made easier with cloud-based IoT. Indeed, how to give proper permission rules and regulations is a significant issue that has yet to be answered when it comes to preserving consumer privacy, particularly when data integrity is needed [31]. Furthermore, when critical IoT applications migrate to the cloud, issues arise owing to a lack of trust in the service provider, a lack of understanding of service level agreements (SLAs), and data location [32][34]. Multi-tenancy might potentially result in sensitive data leaks. Furthermore, public-key cryptography cannot be employed at all layers IoT entities have imposed processing power limitations [31]. Session hijacking

TABLE 4.3
Literature on IoT-Cloud Security

Source	Title	Area
12	Intrusion detection in cloud internet of things Environment	Network security
13	A software defined network-based security assessment framework for cloud IoT	Network security
14	Secure and parallel expressive search over encrypted data with access control in multi Cloud IoT	Data security
15	A design of secure communication protocol using RLWE-based homomorphic encryption in IoT convergence cloud environment	Network security
16	Enhancing cloud-based IoT security with a dependable cloud service: a security and reputation-based approach	Access control
21	A lightweight user authentication scheme for Cloud-IoT based healthcare services	Access control
18	Security in lightweight network function virtualization for federated cloud and IoT	Network security

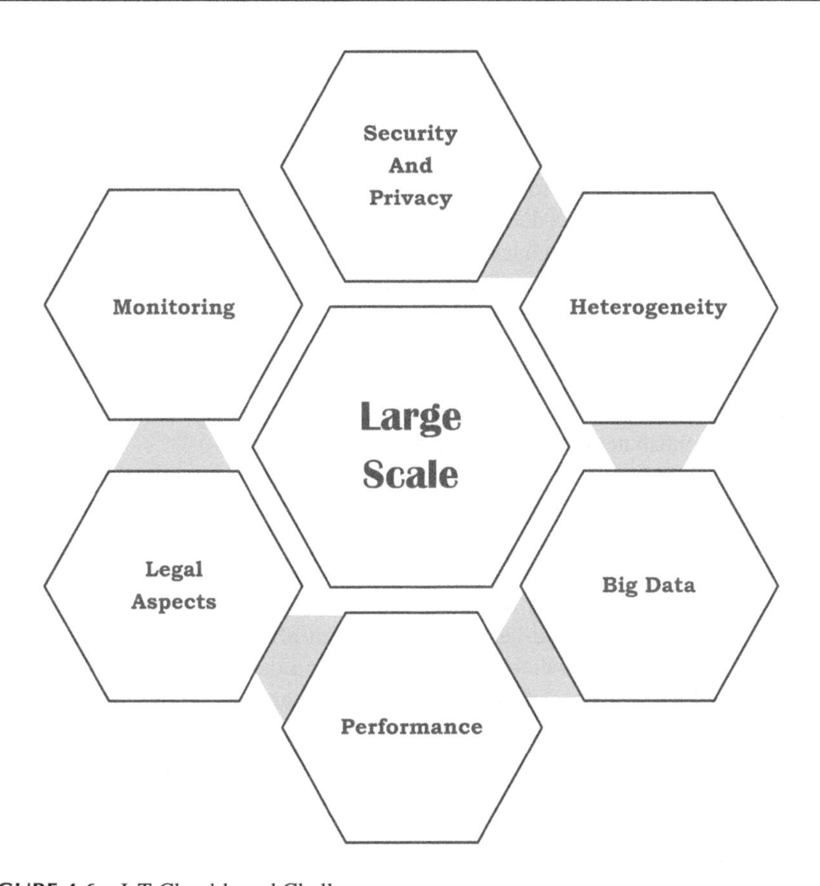

FIGURE 4.6 IoT-Cloud-based Challenges.

and virtual machine escape are two serious vulnerabilities that are difficult to fix [31] [36].

- ***Heterogeneity:*** One of the most significant challenges facing the vast array of devices, platforms, and OS services that exist and may be utilized for new applications is the lack of standardization and is the basis for a cloud-based IoT strategy. Cloud systems have difficulties with heterogeneity; cloud services, for example, frequently have proprietary APIs that enable resource integration based on certain vendors [31]. Furthermore, when end-users use multi-cloud approaches, the heterogeneity problem is worsened, and services become dependent on many providers in order to improve application performance and flexibility [30].
- ***Big data:*** Big data is a precarious challenge that is inextricably linked to the IoT paradigm that is cloud-based despite several contributions, big data remains an important open issue. Many cloud-based IoT applications need real-time execution of complex tasks [31], [35]. With many experts projecting that 50 billion IoT devices will use big data by 2020, it is important to focus on how the vast volumes of data will be transported, accessed, stored, and processed. Given recent developments in technology, the Internet of things (IoT) will undoubtedly be one of the most significant big data sources, and the cloud may assist with both long-term storage and complex analysis [38].
- ***Performance:*** High bandwidth is required to transport the massive amounts of IoT device-transmitted data to the cloud. Finally, because broadband growth is exceeding storage and computing development, finding optimal network performance for data transfer to cloud settings is a big problem [31]. Services and data delivery should be done with caution in a variety of scenarios and with a high level of responsiveness [38]. This is due to the fact that impulsive variables can affect timeliness, and real-time applications rely heavily on performance efficiency [31].
- ***Legal aspects:*** In recent research on specific applications, legal aspects have become increasingly relevant. Service providers, for example, must comply with a number of international rules. Customers, on the other hand, should contribute to the data collection effort [32].
- ***Monitoring:*** In cloud computing, monitoring is the most essential activity in terms of performance, resource management, security, SLAs, and troubleshooting, all aspects of capacity planning. Finally, the cloud-based IoT strategy inherits the monitoring responsibilities of the cloud, but there are still certain challenges exacerbated by the IoT's velocity, volume, and variety [38], [31].
- ***Large scale:*** The cloud-based IoT paradigm allows developers to create new apps with the objective of integrating and determining data from all around the IoT devices, transforming the world around them. This includes engaging with billions of devices located in various locations [35]. The consequent enormous scale systems introduce a slew of new challenges. It is becoming increasingly difficult, for example, to meet computational and storage capacity needs. Because IoT devices must deal with connectivity difficulties and delay dynamics, the monitoring process is likely to have made the deployment of IoT devices more challenging.

4.4 LITERATURE SURVEY RELATED TO IOT-CLOUD SECURITY

Arasteh H. et al. [1] explain IoT technologies for smart cities as well as the major components and aspects of a smart city. In their study, Alababy et al. recommend using security settings on network devices to secure and shield the network from attacks. The international conference on intelligent computing and communication technologies [3] published the Internet of things (IoT) augmentation in healthcare: application analytics, which gives an overview of the benefits and drawbacks of IoT in healthcare, as well as recent attempts, future scopes, and challenges. The results acquired in this study by Da Silveira F. et al. [4] contribute to closing the knowledge gap on industry 4.0 in the health sector. Five evaluators examine the proposed framework using four types of performance tests [5]. Hany F. Atlam et al. [7] provide an overview of cloud integration with the IoT, noting both the benefits and limitations of doing so. The architecture of the final product will also be discussed.

4.5 CONCLUSION

The Internet of things (IoT) is a cutting-edge technology that connects things to things and humans to things over the internet to offer a variety of applications such as smart health, smart homes, smart energy, smart cities, and smart environments. The background concepts of IoT-cloud are examined in this paper, and IoT-cloud security solutions suggested by other researchers in their existing publications are discussed. As a consequence, we have given an up-to-date theoretical and well-explained study on the security problems and solutions of IoT applications in the IoT era, such as the healthcare industry, etc. More research on security difficulties methods that are much more capable of protecting a large volume of data and information in IoT application sectors should be conducted in the future.

REFERENCES

1. S. A. Alabady, et al., "A novel security model for cooperative virtual networks in the IoT era", *International Journal of Parallel Programming*, p. 1–16, 2018.
2. T. Adhikary, et al., "The internet of things (IoT) augmentation in healthcare: An application analytics", International Conference on Intelligent Computing and Communication Technologies, Beijing. 2019, Springer.
3. D. Silveira., et al., "Analysis of industry 4.0 technologies applied to the health sector: Systematic literature review", *Occupational and Environmental Safety and Health*. 2019, Springer. pp. 701–709.
4. H. Arasteh, et al., "Iot-based smart cities: A survey", IEEE 16th International Conference on Environment and Electrical Engineering (EEEIC), New Delhi, 2016.
5. D.-M. Park, et al., "Smart home framework for common household appliances in IoT", *Network Journal of Information Processing Systems*, 15(2), pp. 56–65, 2019.
6. S. Mahmud, et al.,"A smart home automation and metering system using internet of things (IoT)", International Conference on Robotics, Electrical and Signal Processing Techniques (ICREST), Lucknow, 2019.
7. H. F. Atlam, et al., "Integration of cloud computing with internet of things: Challenges and open issues", International Conference on Internet of Things (iThings), Paris, 2017.

8. Dr. M. L. Sharma, et al., "Internet of Things Application, Challenges and Future Scope", *International Research Journal of Engineering and Technology (IRJET)*, Vol. 3, pp. 156–169, 2018.
9. A. K. Mani, et al., "A review: IoT and cloud computing for future internet", *International Research Journal of Engineering and Technology (IRJET)*, 2019.
10. M. Chopra, et al., "A comparative study of cloud computing through IoT", *International Journal of Engineering Development and Research (IJEDR)*, 2019.
11. W. Zhou, et al., "Discovering and understanding the security hazards in the interactions between IoT devices, mobile apps, and clouds on smart home platforms", 28th {USENIX} Security Symposium ({USENIX} Security 19), Tokyo, 2019.
12. A. Ahmed, et al., "Malicious insiders attack in IoT based multi-cloud e-healthcare environment: A systematic literature review", *Multimedia Tools and Applications*, 77(17), pp. 21947–21965, 2018.
13. A. N. Moussa, et al., "A consumer-oriented cloud forensic process model", IEEE 10th Control and System Graduate Research Colloquium (ICSGRC), Noida, 2019.
14. Z. Han, et al., "A software defined network-based security assessment framework for cloudIoT", *IEEE Internet of Things Journal*, 5(3), pp. 1424–1434, 2018.
15. Guechi, et al., "Secure and parallel expressive search over encrypted data with access control in multi-cloudIoT", 3rd Cloudification of the Internet of Things (CIoT), Frankfurt, 2018.
16. Y. Mo, "A data security storage method for IoT under Hadoop cloud computing Platform", *International Journal of Wireless Information Networks*, pp. 1–6, 2019.
17. G. Sharma et al., "Advanced lightweight multi-factor remote user authentication scheme for cloud-IoT applications", *Journal of Ambient Intelligence and Humanized Computing*, pp. 1–24, 2019.
18. Choi, et al., "Ontology-based security context reasoning for power IoT-cloud security service", *IEEE Access*,7, pp. 110510–110517, 2019.
19. N. Almolhis, et al., *The Security Issues in IoT-Cloud: A Review*, IEEE, 2020.
20. V. Sharma, et al., "A review paper on 'IOT' & it's smart applications", *International Journal of Science, Engineering and Technology Research (IJSETR)*, 5(2), p. 472, February 2016.
21. M. Mahmud, et al., "A brain-inspired trust management model to assure security in a cloud based IoT framework for neuroscience applications", *Cognitive Computation*, 10(5), pp. 864–873, 2018.
22. A. Skaržauskienė, et al., "The future potential of internet of things", *Social Technologies*, 2(1), pp. 102–113, 2012.
23. D. A. Vyas, et al., "IoT: Trends, challenges and future scope", *IJCSC*, 7, pp. 26–35, March 2016.
24. S. M. Babu, et al., "A study on cloud based internet of things: Cloud-IoT", Global Conference on Communication Technologies, no. GCCT, Berlin, 2015, pp. 60–65.
25. A. Alenezi, et al., "The impact of cloud forensic readiness on security", 7th International Conference on Cloud Computing and Services Science, Beijing, 2017, pp. 1–8.
26. J. Zhou, et al., "Security and privacy for cloud-based IoT: Challenges, countermeasures, and future directions", *IEEE Communications Magazine*, 55, pp. 26–33, January 2017.
27. K. S. Dar, et al., "Enhancing dependability of cloud-based IoT services through virtualization", 1st International Conference on Internet IEEE, Hamburg, 2016.
28. J. Zhu, et al., "Research on supply chain simulation system based on internet of things", *Advances in Internet of Things*, 5, pp. 1–6, 2015.
29. Y. Yang, et al., "A survey on security and privacy issues in internet-of-things", *IEEE Internet of Things Journal*, vol. 4, no. 5, pp. 1250–1258, Oct. 2017, doi: 10.1109/JIOT.2017.2694844.

30. S. Jun, "Technology analysis for Internet of Things Using Big Data Learning", *International Journal of Research in Engineering and Technology*, Vol. 263, No. 4, p. 042–070. IOP Publishing. eISSN: 2319-1163 | pISSN: 2321-7308.
31. G. Suciu, et al., "Smart cities built on resilient cloud computing and secure internet of things", 19th International Conference on Control Systems and Computer Science (CSCS), Chicago, 2013, pp. 513–518.
32. M. Díaz, et al., "State-of-the-art, challenges, and open issues in the integration of Internet of things and cloud computing", *Journal of Network and Computer Applications*, 2016, pp. 99–117.
33. C. Doukas et al., "*Bringing IoT and Cloud Computing Towards Pervasive Healthcare*", Proceedings of the 6th International Conference on Innovative Mobile and Internet Services in ubiquitous computing IMIS, Beijing, 2012, pp. 922–926.
34. H. F. Atlam, et al., "An overview of risk estimation techniques in risk-based access control for the internet of things", 2nd International Conference on Internet of Things, Big Data and Security, New Delhi, 2017, pp. 1–8.
35. A. Botta, et al., "On the integration of cloud computing and internet of things", International Conference on Future Internet of Things and Cloud, Barcelona, 2014, pp. 23–30.
36. R. Buyya, et al., "Cloud computing and emerging IT platforms: Vision, hype and reality for delivering computing as the 5th utility", *Future Generation Computer Systems*, vol. 2, pp. 599–616, 2009.
37. M. Armbrust, et al., "A view of cloud computing", *Communications of the ACM*, vol. 53, no. 4, pp. 50–58, 2010.
38. P. Velmurugadass, et al., "Enhancing blockchain security in cloud computing with IoT environment using ECIES and cryptography hash algorithm", *Materials Today: Proceedings*, vol. 37, pp. 2653–2659, 2021.
39. A. Wilczyński et al., "Modelling and simulation of security aware task scheduling in cloud computing based on blockchain technology", *Simulation Modelling Practice and Theory*, vol. 99, p. 102038, Feb. 2020.
40. J. Cha, et al., "Blockchain-empowered cloud architecture based on secret sharing for smart city", *Journal of Information Security and Applications*, vol. 57, p. 102686, Mar. 2021.
41. H. Huang, et al., "Blockchain-based eHealth system for auditable EHRs manipulation in cloud environments", *Journal of Parallel and Distributed Computing*, vol. 148, pp. 46–57, Feb. 2021.
42. Y. Ren, et al., "Multiple cloud storage mechanism based on blockchain in smart homes", *Future Generation Computer Systems*, vol. 115, pp. 304–313, Feb. 2021.
43. J. Li, et al., "Blockchain-based public auditing for big data in cloud storage", *Information Processing and Management*, vol. 57, no. 6, p. 102382, Nov. 2020.
44. N. Eltayieb, et al., "A blockchain-based attribute-based signcryption scheme to secure data sharing in the cloud", *Journal of Systems Architecture*, vol. 102, p. 101653, Jan. 2020
45. M. Zhaofeng, et al., "Blockchain-enabled decentralized trust management and secure usage control of IoT big data", *IEEE Internet of Things Journal*, vol. 7, no. 5, pp. 4000–4015, May 2020.
46. S. Algarni, et al., "Blockchain-based secured access control in an IoT system", *Applied Sciences*, vol. 11, no. 4, p. 1772, Feb. 2021
47. R. Kumar, et al., "Revisiting software security: durability perspective", *International Journal of Hybrid Information Technology*, vol. 8, no. 2, pp. 311–322, 2015.
48. R. Kumar, et al., "Durability challenges in software engineering", *Crosstalk-The Journal of Defense Software Engineering*, pp. 29–31, 2016.

49. K. Sahu, et al., "Risk management perspective in SDLC", *International Journal of Advanced Research in Computer Science and Software Engineering*, vol. 4, no. 3, pp. 1–15, 2014.
50. K. Sahu, et al., "Hesitant fuzzy sets based symmetrical model of decision-making for estimating the durability of Web application", *Symmetry*, vol. 12, no. 11, p. 1770, 2020
51. R. Kumar, et al., "Analytical network process for software security: a design perspective", *CSI Transactions on ICT*, vol. 4, no. 2, pp. 255–258, 2016.
52. K. Sahu, et al., "Evaluating the impact of prediction techniques: Software reliability perspective", *Computers Materials and Continua*, vol. 67, no. 2, pp. 1471–1488, 2021.
53. R. Kumar, et al., "Durable security in software development: Needs and importance", *CSI Communication*, vol. 39, no. 7, pp. 34–36, 2015.
54. M. T. J. Ansari, et al., "P-STORE: Extension of STORE methodology to elicit privacy requirements", *Arabian Journal for Science and Engineering*, vol. 46, no. 9, pp.8287–8310, 2021.
55. R. Kumar, et al., "Software security testing: A pertinent framework", *Journal of Global Research in Computer Science*, vol. 5, no. 3, pp. 23–27, 2014.
56. A. Attaallah, et al., "Analyzing the big data security through a unified decision-making approach", *Intelligent Automation and Soft Computing*, vol. 32, no. 2, pp. 1071–1088, 2022.
57. A. H. Almulihi, et al., "Analyzing the implications of healthcare data breaches through computational technique", *Intelligent Automation and Soft Computing*, pp. 1763–1779, 2022.
58. A. K. Pandey, et al., "Analyzing the implications of COVID-19 pandemic through an intelligent-computing technique", *Computer Systems Science and Engineering*, pp. 959–974, 2022.
59. R. Kumar, et al., "An integrated approach of fuzzy logic, AHP and TOPSIS for estimating usable-security of web applications", *IEEE Access*, vol. 8, pp. 50944–50957, 2020.
60. R. Kumar, et al., "Measuring security durability of software through fuzzy-based decision-making process", *International Journal of Computational Intelligence Systems*, vol. 12, no. 2, p. 627, 2019.
61. R. Kumar, et al., "A knowledge-based integrated system of hesitant fuzzy set, ahp and topsis for evaluating security-durability of web applications", *IEEE Access*, vol. 8, pp. 48870–48885, 2020.

5 Predictive Maintenance in Industry 4.0

Manoj Devare

CONTENTS

5.1 Introduction 79
5.2 Related Work 80
5.3 Materials and Methods 81
  5.3.1 Data Set 81
  5.3.2 Data Insights 84
  5.3.3 Outlier Detection 85
5.4 ML Model Evaluation Metrics 87
5.5 Results with the Multinomial Naïve Bayes Algorithm 89
5.6 Markov Chain 91
5.7 Weibull Process for Predictive Maintenance 95
5.8 Discussion and Conclusion 96
References 96

5.1 INTRODUCTION

The high availability of equipment is a requirement to save time and money in the manufacturing industry. The reduction in unexpected failure due to prediction of the failure states is the core expectation in predictive maintenance (PdM). Automation industry expects intelligent solutions based on an amalgamation of machine learning (ML), a range of sensors, data storage, and intelligent data processing techniques. The techniques for data processing, exploration, and ML results presented on the sensor data set collected from Internet of things (IoT)-enabled industrial equipment.

Due to the high use of the handheld devices, internet connectivity, and sensors configured at the asset located in the industry premises, the communication happens using the various industry standard protocols. Some of these protocols are low power consumption protocols, and some are designed to work in particular environments. The communication regarding the health of the equipment was reported to the subscriber using the application layer protocols working in coordination with the cloud-based micro-services.

PdM has high demand in industries like oil and gas, construction, and vessels industry. The continuous running equipment during the active labor shifts achieves the daily targets of the production. If the machinery is not working, the labor goes into an idle state; this wastes the resources and leads to financial, time, and reputation loss. The delay in the delivery of parts may lead to further cancellation of contracts

DOI: 10.1201/9781003323426-5

with the clients. The cause of equipment failure can be related to the mechanical, electrical, and equipment conditions.

The success rate of PdM depends on the availability of the correct data, data preparation techniques, and expertise in the domain. PdM is a facilitation to the manufacturers to schedule the maintenance activity as per the need. A comprehensive PdM uses the most cost-effective vibration monitoring, thermography imaging, ultrasonic lubricators, and tribology. The operating condition of critical systems based on sensor data, maintenance schedules, the equipment operating conditions, the damage of the machine parts, spindle speed, hours running, temperature, and vibration are important factors in PdM. The change in the vibration patterns indicates the change in the condition of the machine. Abnormal condition is proportional to the amount of variation detected.

The outcome of predictive analytics results in an increase in availability, asset performance, asset utilization, and machine life. The rise in the Overall Equipment Effectiveness (OEE) such as availability, effectiveness, and quality. The Total Effective Equipment Performance (TEEP) takes into account both equipment and schedule losses. It takes Overall Production Effectiveness (OPE) parameters into consideration.

5.2 RELATED WORK

PdM tries to identify the failure of equipment in the provided time window. Anomaly detection, finding the probability of the common failures also comes in PdM. PdM is known as system failure diagnosis, prognostics, and health management. The Remaining Useful Lifetime (RUL) is a technique to find the hidden state of the components of the system. The approach of failure prediction is on feature extraction and status classification. The other approach is on time series modeling and anomaly detection using statistical techniques.

The regression equation is developed (Holladay et al., 2006) for voltage regulators using the dependent variables as the number of failures and the different types of failures. The indistinguishable and distinguishable production units and their problem with parallel liquid filling in the bottles are detected using the Hidden Markov Model (HMM) to detect machine failure (Tai et al., 2006). Salfner (Salfner 2005) mentions that identifying the pattern of errors leads to predicting the failure of the machines. The previous error events are helpful in calculating the failure probability during the prediction interval (Salfner, 2005). Hidden Semi-Markov Models (HSMMs) and Akaike information criterion-based RUL of the machine are calculated by Cartella, F. et al., (Cartella et al., 2015). The problem addressed here by Yuan (Yuan, 2015) is similar to the problem addressed here for the machine under observation in this case.

As per the Markov property, the system's future state can be identified on the current state. The past state may not be useful for finding the future state. The failure rate increases as the time elapsed increases. As time increases, the reliability rate decreases (Kalaiarasi et al., 2017). Deka et al. used the Markov chain for building low-power robust systems (Deka et al., 2014). Hofmann developed the mixed membership function and the Hidden Markov Model (HMM) with the help of past data and live readings to estimate the degradation of an asset and detect the failure rates (Hofmann and Tashman, 2020). Kalaiarasi et al. discuss the system reliability, mean time to failure (MTTF), and failure probability using the Markov Model under human error conditions (Table 5.1).

5.3 MATERIALS AND METHODS

The problem statement is to predict the states and failure of the machine. The rim welding machine creates vehicle rims from iron plates. The machine under observation is the first machine in the assembly line. The succeeding machines are dependent on the welded rim from the machine. If the machine fails then the entire assembly line stops. Hence, predicting the reliability based on the state of the machine is of high importance.

5.3.1 DATA SET

The IoT-based sensors are embedded on the rim welding machine in the manufacturing industry. The data used is of one-month duration with 56981 entries. The working machine readings are recorded using nine temperature and three voltage sensors attached to the various parts of the machine. There is a label column called production count. The machine states are prepared using the production count, which is converted into the running and idle states. After understanding the shop-floor activities, the states of the machine are labeled.

As per the requirement of the ML models, the data sets are preprocessed and prepared. The subset of the features is used for the particular ML model. The ML-based probabilistic models Naïve Bayes and Markov Chain are selected as a methodology for predicting the states of the machine.

The problem of PdM needs a sufficient amount of data set. The data set can be created through historical documents, interviewing the assembly line engineer, purchase history, and maintenance history of the machine. In this research work, the domain knowledge is collected through interviewing the assembly line engineer. When the data is unlabeled, the choice is an unsupervised ML model or an attempt to create the labeled data set. When the actual failure events are not available, it must appoint a person to record failure events. Another option is to create the labeled data set manually by finding the idle points of the machine or equipment. The idle points are the superset of failure events. Although manual labeling is a time-consuming task, it is helpful to understand the nature of the data series. In this work, the sensors provided the data thousands of times in a second. However, the data collection system is recording a few of the readings out of the thousands.

$$M_i \subseteq R \subseteq S \tag{5.1}$$

where S = the set of actual sensor values produced by the sensors, S is unavailable.

R = the set of values recorded in the database.

M_i = is the data set prepared for the particular ML model, where i varies from 1 to 4.

TABLE 5.1
Predictive Maintenance Relevant Literature Review

Paper	Data Used	Equipment	Method
Yuan, C. (Yuan, 2015)	Unlabeled data sets of gas turbine failure, honeybees, and truck sensors	Turbine, truck	Segmental HMM
Aisong Qin (Qin et al., 2017)	Vibration data	Rotating machinery	Wiener process-based method, RUL, genetic programming algorithm
Changhua HU (Hu et al., 2018)	Imperfect maintenance activities data	Gyroscopes in inertial navigation systems	Maximum likelihood estimation, RUL, Bayesian method
Jean Nakamura (Nakamura, 2007)	temporal vibration data, 10,000 machines and over 142,000 tests are part of the data set REDI-PRO	Design of a generalized framework to detect Time to Failure (TTF)	Case-based reasoning, Time to Failure, KNN, logistic regression, multi-layer perception
Ameeth et al., (Kanawaday, n.d.).	Sensor data	Slitting machine	ARIMA time series, Naïve Bayes, support vector machine (SVM), CART, and deep neural network
Peng et al, (Peng, 2021).	run-to-failure simulation	Turbofan engine data set	Neural Network
Mishra et al., (Mishra and Manjhi, 2018).	380 features and 360 error types	ATM	Youden J. cutoff probability
Gugulothu et al., (Delhi et al., 2017).	NASA Ames Prognostics Data Repository has 24 sensor data	Turbofan engine data set	RUL, recurrent neural network, and multivariate time series
Ramos et al.,(Ramos, n.d.).	An extensive set of time series, each describing a key sensor of a refiner machine	Wood product refiner machine	Neural network and ARIMA, ACF, PACF
Gupta, R et al., (Gupta and Bhardwaj, 2014)	Transition probabilities	Two-unit cold standby system	Discrete parametric Markov model
Ahmadi et al., (Ahmadi et al., 2020).	Repairing history data	Tunneling equipment	Markov chain, fault tree analysis, MTTF, mean time to repair (MTTR), and mean time between failure (MTBF)
Kuzin et al., (Kuzin and Boroviˇ, 2016).	Sensor data of outlier, spike, high noise, stuck-at	Sensor part data	ARIMA time series modeling, autoencoder, feedforward neural network

The labeling is done after understanding the shop-floor processes and understanding the constituent parts of the overall machine assembly. The preprocessing and initial statistical analysis of the data is performed, such as finding the null values, missing values, outliers, and away points. The box-plot, five-point summary is used in the data set.

Failure and running are mainly two classes for any PdM problem. However, the data set labeled with these two labels is unequal in proportion. The running state of the machine has more record sets than failure states. The failure state of the machine is a rare case. The labeling of the records is text or numerical values, such as the 0 for stopped and 1 for the running. The stopped states are prepared from the production count available with time series. Initially, the data is converted as stopped, running, and change point labels.

$$M_f \subseteq M_s \subseteq M_a \tag{5.2}$$

$$M_r \subseteq M_a \tag{5.3}$$

where,

M_f, M_s, M_r, and M_a are sets of machine's failed, stopped, running, and all states, respectively.

The change point is point in time when either before or after that point, something happened due to which machine has changed the State. The reasons can be a raw material issue, idle labor, tea time, shift change, or machine failure. Figure 5.1 shows the shop-floor schedule useful to understand the idle states of the machine. The probability of the machine being in the stopped state during lunch, dinner, handover, cleaning and tea time is higher, due to idle working. However, the probability of failure during this timing is less. Hence, the running state of the machine is analyzed to find the events of failures.

The transitions of the machine states with reference to PdM are shown in Figure 5.1. There are two main assumptions in this research work. The first assumption is that the machine failed state can be a subset of the stopped states. The second assumption is, the reason behind the machine failure can be detected by investigating its running state.

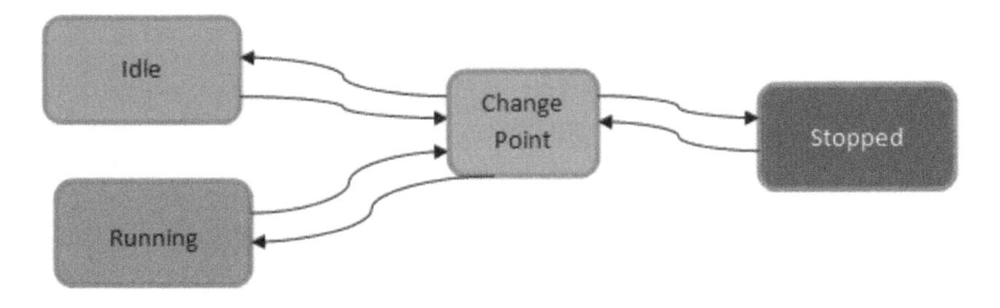

FIGURE 5.1 Machine State Transitions in PdM.

5.3.2 Data Insights

After observing the data for two consecutive days, the running, idle, and change points are plotted visually in Figure 5.2. The good points, bad points, and their transitions are distinguished herewith in Figure 5.2. The shift activities are carried out in one day and connected to the next day too. The sensor data is collected continuously. All algorithms are implemented using the Sklearn library (Pedregosa, 2011).

```
plt.subplots(figsize = (18, 18))
machine_data.corr()
figure_object = sns. heatmap (machine_data. corr(),annot = True,
vmin = 0,vmax = 1,center = 0,linewidth = "0.5", square = True)
down_edge, upper_edge = figure_object.get_ylim()
   figure_object.set_ylim(down_edge + 0.5, upper_edge - 0.5)
```

The correlation between each pair of the sensor values is plotted in Figure 5.4. The correlation between each pair of sensor values is observed and visible across the diagonal. The correlation coefficient values spread from –1 to 1. The zero value of the correlation coefficient (ρ) means that there is no relationship between the variables. The diagonal value of ρ is 1. The ρ values near to 1 show a high correlation. It is observed that the ρ of the three voltage sensor values with other sensors is less than zero. Hence, there is a negative relationship. There are ten temperature sensors connected to the different parts of the machine. There is a high ρ value between each pair of temperature sensors. The value of ρ between the current sensor and the

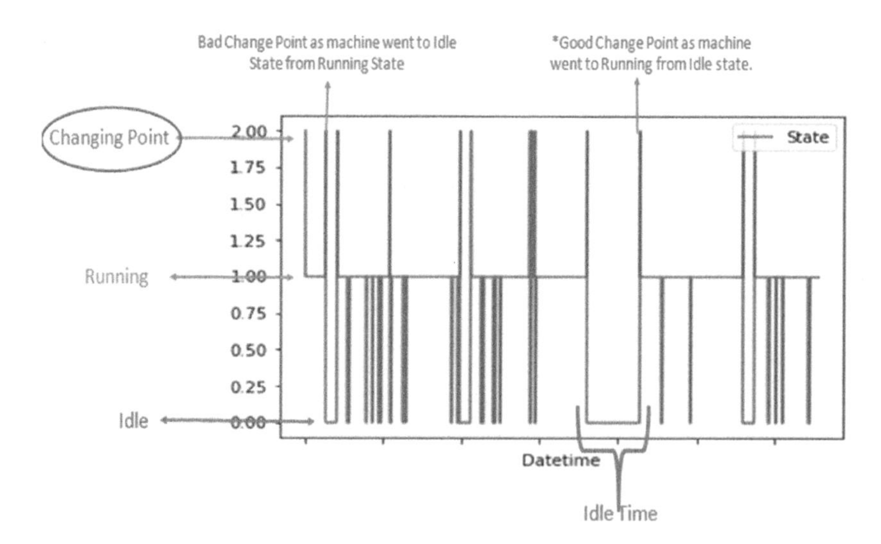

FIGURE 5.2 Dividing the Data into Good and Bad Change Points for Two Consecutive Days.

temperature sensor is less than 0.5, except for the right jaw temperature. The ρ value for the right jaw temperature with current is 0.62.

$$\rho = \frac{\sum_{i=1}^{n}\left(x_i - \bar{x}\right)\left(y_i - \bar{y}\right)}{\sqrt{\sum_{i=1}^{n}\left(x_i - \bar{x}\right)^2\left(y_i - \bar{y}\right)^2}} \quad (5.4)$$

where,

R = correlation coefficient
x_i = value of x sensor in the sample
$\bar{x}$ = mean of the individual sensor feature
y_i = value of the second sensor feature
$\bar{y}$ = mean of the second sensor feature.

Figure 5.3 shows the state count for the idle, running, and change points. The entire data set is considered for counting the states of the machine. There are 426 state points called the change points and 8956 idle states (Figure 5.4).

5.3.3 Outlier Detection

The outliers and anomaly detection can provide insights into the data set. The individual features spread across the permissible values are investigated using statistical techniques such as outlier Inter-Quartile Range (IQR) and Z-score. The box and whisker plot and the IQR are calculated. The box and whisker plot for the temperature and voltage sensor is plotted in Figure 5.5 and Figure 5.6, respectively

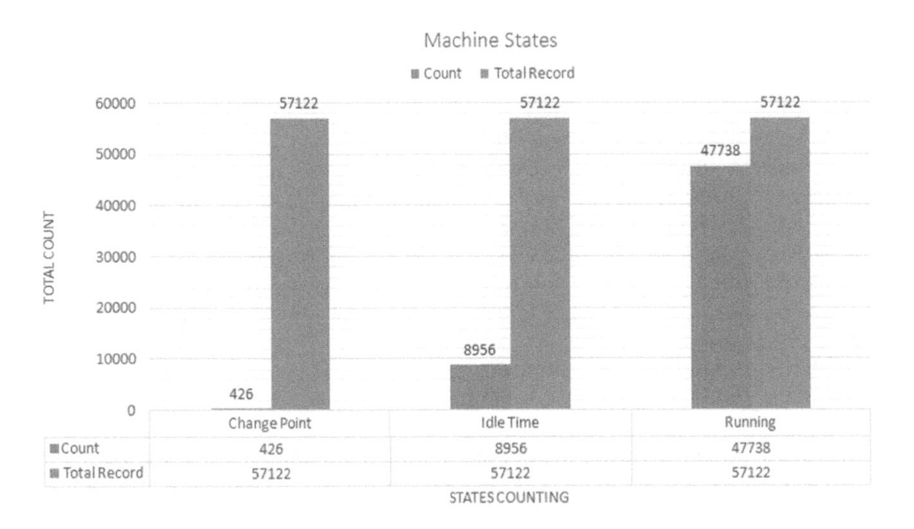

FIGURE 5.3 Counting the States in One-Month Duration.

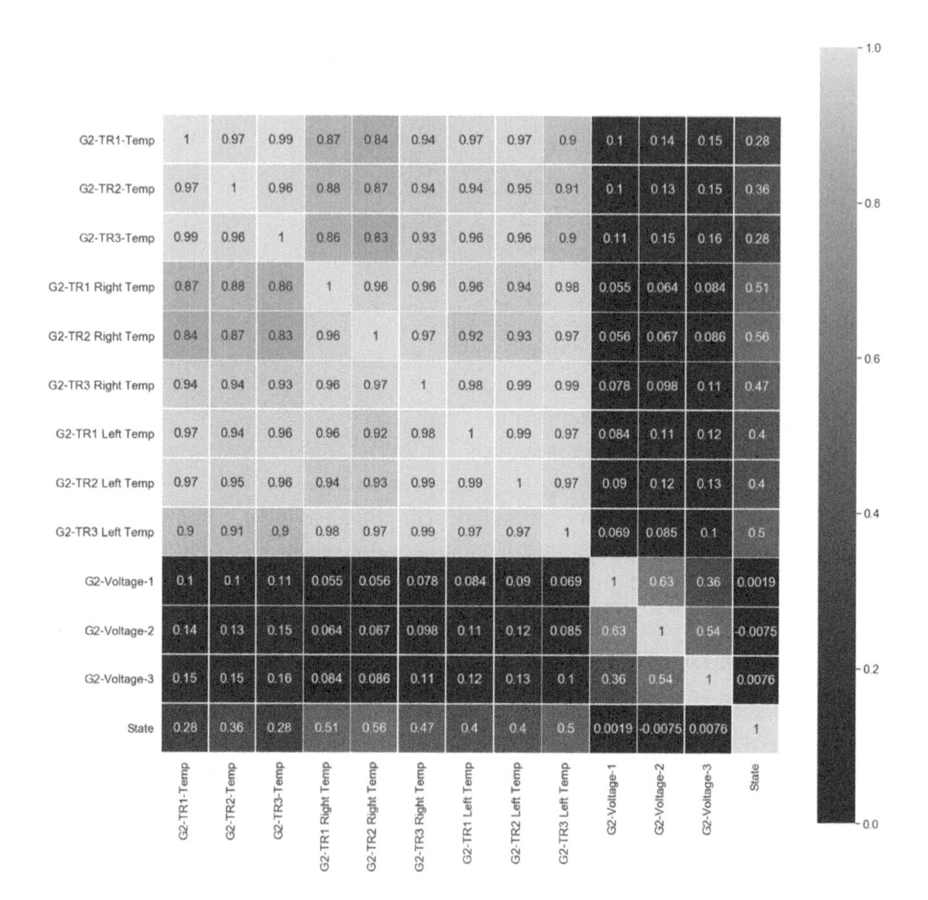

FIGURE 5.4 Correlation Matrix and the Value of ρ among Each Pair of 16 Features.

```
Quartile1 = machine_data.quantile(0.25)
Quartile3 = machine_data.quantile(0.75)
IQ_Range = Quartile3 - Quartile1
print(IQ_Range)
print(machine_data < (Quartile1 - 1.5 * IQ_Range))|
(machine_data > (Quartile3 + 1.5 * IQ_Range))
machine_data_outlier = machine_data[((machine_data <
(Quartile1 - 1.5 * IQ_Range))|(machine_data > (Quartile3 +
1.5 * IQ_Range))).any(axis = 1)]
machine_data_outlier.shape
machine_data.shape

machine_data_without_outlier = machine_data[~ ((machine_
data < (Quartile1 - 1.5 * IQ_Range))|(machine_data > (Qu
artile3 + 1.5 * IQ_Range))).any(axis = 1)]
machine_data_without_outlier.shape
machine_data.shape
machine_data_outlier.to_csv("outlier_IQR.csv")
```

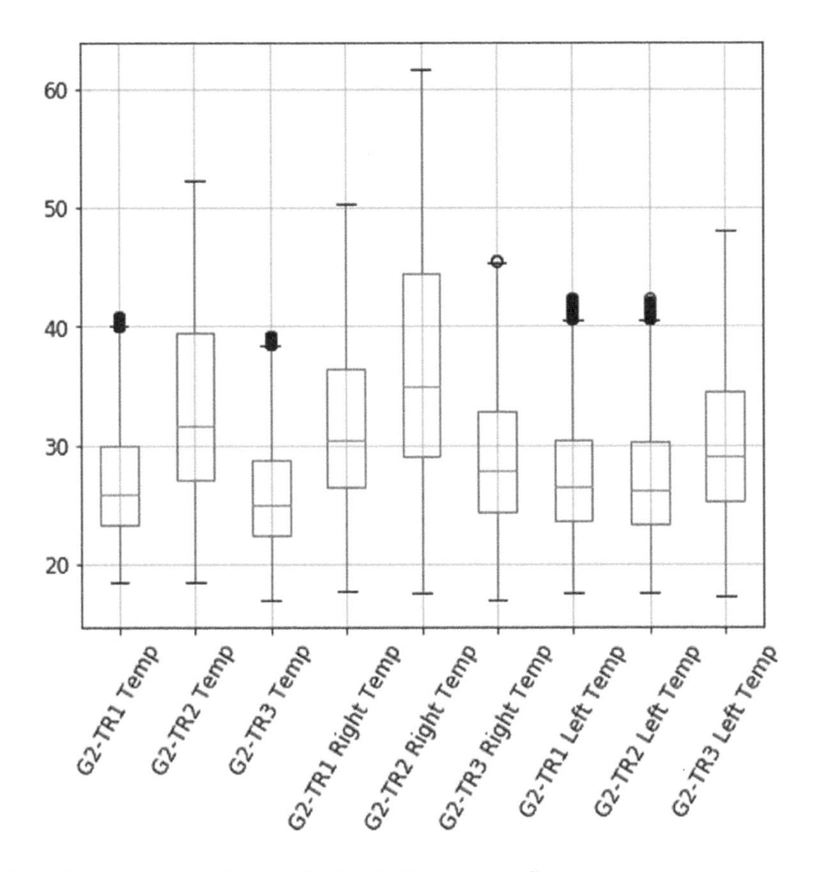

FIGURE 5.5 Box and Whisker Plot for the Temperature Sensor.

Software environment and library installations can be useful to execute the program and understand the sensor data. The Python code can be executed on Jupyter Notebook on Windows 10 OS. The common installations and programming libraries Sklearn, Pandas, NumPy, Matplotlib, and PyDotPlus are useful to build, evaluate, and visualize the results.

5.4 ML MODEL EVALUATION METRICS

The ML models are evaluated using accuracy, precision, recall, and F1 score. The choice of implementation can be decided using an area under curve (AUC) and receiver operating characteristics (ROC) curve. The supervised ML classification algorithm performance can be analyzed using the confusion matrix. In this case, the binary classification of the states of the machine is taken into consideration (Kubat, 2017)(Bramer, 2016).

$$\text{Accuracy} = \frac{TP + TN}{TP + TN + FP + FN} \tag{5.5}$$

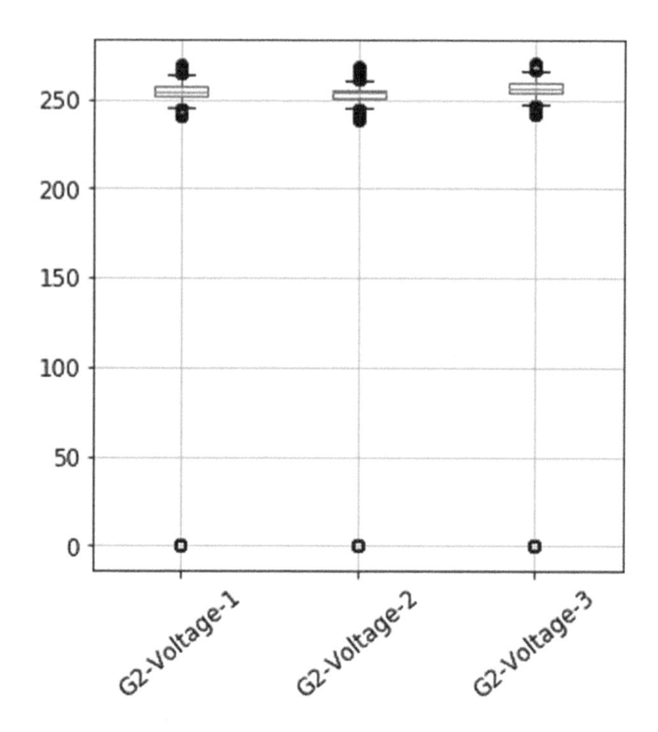

FIGURE 5.6 Box and Whisker Plot for the Voltage Sensor.

where TP = true positives, TN = true negatives, FP = false positives, and FN = false negatives. Here, the TP is the correctly classified running state, and TN is the correctly classified stopped state. FP is the state classified as 1, but the true label is 0. FN is the predicted state 0, but the true label is 1. The denominator of the accuracy equation is the count of all states of the data set.

- Precision is how precise the model is out of the predicted positive, how many of them are actually positive. Precision is a good measure to determine when the costs of false positive are high. The false positive is the state of the machine identified by the ML model as running, but actually, the state is stopped. While preparing the final probability of failure of the machine, this probability should be considered.

$$\text{Precision} = \frac{TP}{TP + FP} \tag{5.6}$$

Recall or sensitivity calculates how many of the actual positives our model captures through labeling it as true positive. Recall shall be the model metric to select the best model when there is a high cost associated with false negative. Here, the false negative is the state of the machine identified by the ML model as stopped, but actually, the state is running.

$$\text{Recall} = \frac{TP}{TP + FN} \tag{5.7}$$

TABLE 5.2
AUC Score and Meaning

AUC Score	Meaning
1	A classifier is able to perfectly distinguish between two states of the machine.
$0.5 < AUC < 1$	A high chance that the classifier distinguishes the two states of the machine.
0.5	A classifier is not able to distinguish between the states of the machine.

F1 score is useful to find a balance between precision and recall. F1 score is also useful in case of uneven class distribution. Here, the stopped state count is 16.27% of the total states of the machine, and running states are 83.72%. Hence, there is uneven class distribution. The receiver has to use the F1 score effectively while selecting the ML model for the implementation in the production environment.

$$F1 = 2 * \frac{\text{Precision} * \text{Recall}}{\text{Precision} + \text{Recall}} \tag{5.8}$$

ROC is a probability curve that plots the true positive rate (TPR) against the false positive rate (FPR). The TPR is sensitivity, and FPR is 1 – specificity (Table 5.2).

5.5 RESULTS WITH THE MULTINOMIAL NAÏVE BAYES ALGORITHM

Naïve Bayes algorithm's important assumption is that all features are equally important and independent (Dinov, 2018). The independence of the features is in terms of their distributions. Although the practical independence of the features is not true, the Naïve Bayes algorithm works well for complex applications.

$$P(A|B) = \frac{P(B|A) \cdot P(A)}{P(B)} \tag{5.9}$$

where,

$P(A \mid B)$ is posterior probability, the probability of class A when predictor B is already provided.

$P(B \mid A)$ is likelihood, occurrence of predictor B given class A probability.

$P(A)$ is prior probability of class A.

$P(B)$ is prior probability of class B

The multinomial Naïve Bayes is a specific case where each feature follows the multinomial distribution. Here, the nine temperature sensors and three voltage sensors are attached to the machine. The chiller temperature is not a part of the features for NB. There are 12 features and one label for NB. The multinomial distribution is discrete distribution and not continuous distribution.

```
from sklearn.naive_bayes import MultinomialNB
import pandas as panda
import matplotlib.pyplot as GraphPlot

NB_Model=MultinomialNB()
TVDataFrame=panda.read_csv("TempAndVoltageFF.csv")
OutputLabel=TVDataFrame.State
InputData=TVDataFrame.drop('State', axis=1)
InputData
InputTrain, InputTest, OutputTrain, OutputTest=train_test_
split(InputData, OutputLabel, test_size=0.2)
NB_Model.fit(InputTrain, OutputTrain)
ModelPrediction=NB_Model.predict(InputTest)
confusion_matrix(OutputTest, ModelPrediction)
plot_confusion_matrix(NB_Model, InputTest, OutputTest,
cmap=GraphPlot.cm.Blues)
GraphPlot.show()
print ('NB Model Accuracy:', accuracy_score(OutputTest,
ModelPrediction))
print ('NB Model F1 score:', f1_score(OutputTest,
ModelPrediction))
print ('NB Model Recall:', recall_score(OutputTest,
ModelPrediction))
print (' NB Model Precision:', precision_score(OutputTest,
ModelPrediction))
print ('\n NB Model classification report:\n',
classification_report(OutputTest, ModelPrediction))
print ('\n Confusion Matrix:\n', confusion_
matrix(OutputTest, ModelPrediction))
PredictionProbability=NB_Model.predict_proba(InputTest)
FPR, TPR, thresh=roc_curve(OutputTest,
PredictionProbability[:,1], pos_label=1)
AUC_Score=roc_auc_score(OutputTest,
PredictionProbability[:,1])
GraphPlot.subplots(figsize=(6,6))
GraphPlot.title('Multinomial Naive Bayes ROC curve')
GraphPlot.xlabel('False Positive Rate')
GraphPlot.ylabel('True Positive Rate')
GraphPlot.legend(fontsize=12);
GraphPlot.plot(FPR, TPR, linestyle='-',color='red',
label='MultinomialNB')
GraphPlot.plot(np.linspace(0, 1, 100), np.linspace(0, 1,
100), label='baseline', linestyle='--')
AUC_Score
```

The Gaussian NB is useful in the case of the features available in decimal form. GNB assumes features to follow a normal distribution. Bernoulli NB is helpful in the case of features with binary or Boolean values like true or false. The multinomial

NB is helpful in case of the features with discrete values. Here, multinomial NB is used for the machine state classifications. The confusion matrix is shown for multinomial NB in Figure 5.7 and Table 5.3.

The AUC-ROC curve shown in Figure 5.8 shows 91% score and accuracy with minimum 80% scores for multinomial Naïve Bayes.

5.6 MARKOV CHAIN

The discrete Markov chain memoryless model is helpful in finding the next state of the machine. There are two important factors that contribute to the decision of the probability that equipment moves to the next state. One factor is the current state and the times spent in the current state. The Markov state also explicitly mentions that the sequence of the states preceding the current state does not contribute to the decision of the next state.

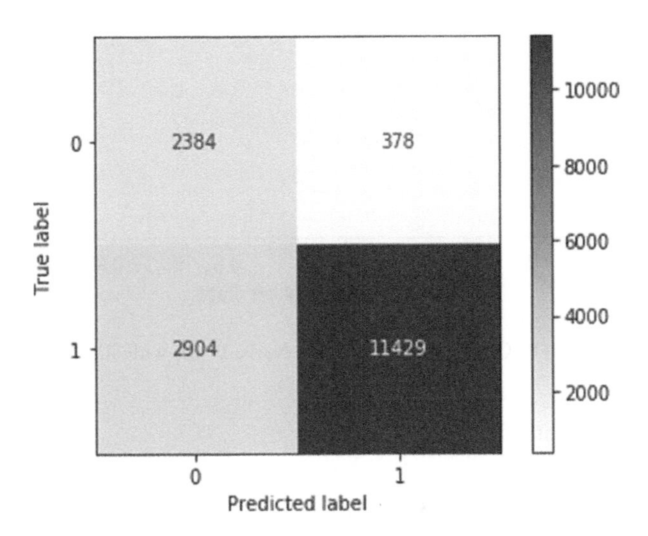

FIGURE 5.7 Confusion Matrix for Multinomial Naïve Bayes with 70-30 Train-Test Data Divide.

TABLE 5.3
Multinomial Naïve Bayes Model Evaluation and Data Divide

	60-40	70-30	80-20
Accuracy	0.8059	0.8039	0.8108
Precision	0.9661	0.8080	0.9675
Recall	0.7971	0.7973	0.8009
F1 Score	0.8735	0.8744	0.8764
ROC-AUC-Score	0.9099	0.9117	0.9130

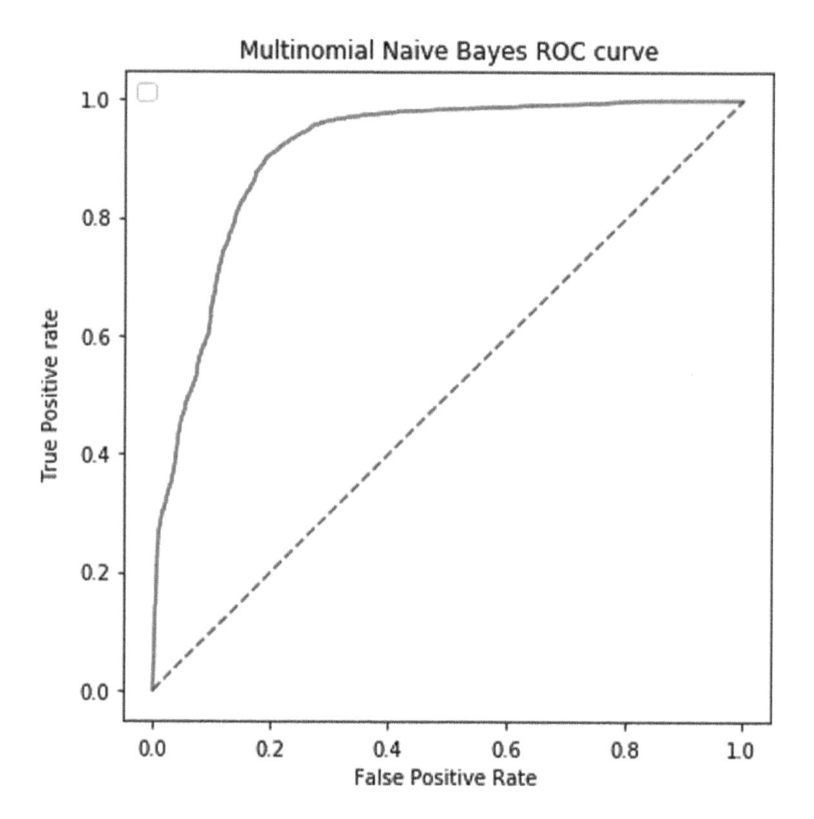

FIGURE 5.8 AUC-ROC Curve for Multinomial Naïve Bayes with 70-30 Data Divide.

Note that $P_{ij} \geq 0$ and for all i,

$$\sum_{k=1}^{r} P_{iK} = \sum_{K=1}^{r} P_i\left(X_{m+1} = k \mid X_m = i\right) = 1. \tag{5.10}$$

The production count column is converted into the state label. As the running production is represented as 1 and the stopped production count is 0, the probability of moving from state I to J is calculated over the entire one-month data set (Table 5.4 to 5.6).

```
=AND(M2=0,M3=0)
=AND(M2=0,M3=1)
=AND(M2=1,M3=0)
=AND(M2=1,M3=1)
```

The total number of stopped states and running states are counted.

```
=COUNTIF(M2:M56982,"=0")
=COUNTIF(M2:M56982,"=1")
```

TABLE 5.4
Welding Machine State Counts

State Count	Total
Running state count	47705
Idle state count	9276
Total entries	56981

TABLE 5.5
Welding Machine State Transition Counts

State	To Idle	To Running
From idle	9042	234
From running	234	47471

TABLE 5.6
Probabilities of States Transition

State	To Idle (Probability)	To Running (Probability)	Summation
From idle	=9042/9276 =0.9747	=234/9276 =0.0252	≈1
From running	=234/47705 =0.0049	=47471/47705 =0.9950	≈1

```
=COUNTIF(S2:S56982,"TRUE")
=COUNTIF(T2:T56982,"TRUE")
=COUNTIF(U2:U56982,"TRUE")
=COUNTIF(V2:V56982,"TRUE")
```

The state transition diagram of the welding machine for the Markov chain is shown in Figure 5.9. The critical point of failure is the switching from the running to idle state with a probability of less than 0.49%.

Numerical 1: Starting from the stopped state at time T_n, what will be the probability is that the machine will be in the running state at time T_{n+2}.

Solution: Consider the case at time T_n, the machine will move to the stopped state at time T_{n+1} with probability 0.9747 and then it will move to the running state with probability 0.0252 at time $T_{n+2.}$

Hence, total probability =0.9747 * 0.0252=0.02456.

Alternatively, consider the case at time T_n; the machine will move to the running state at time T_{n+1} with probability 0.0252 and then it will continue to the running state with probability 0.9950 at time T_{n+2}.

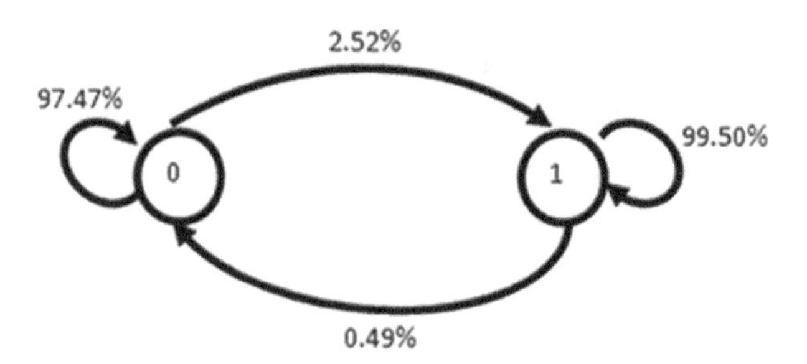

FIGURE 5.9 State Transition diagram of a Welding Machine for the Markov Chain.

TABLE 5.7
For Markov Chain Numerical 2

T_n	T_{n+1}	T_{n+2}	T_{n+3}	Total Probabilities
Running	Running	Running	Stopped	
	0.9950	0.9950	0.0049	=0.0048
Running	Running	Stopped	Stopped	
	0.9950	0.0049	0.9747	=0.0047
Running	Stopped	Running	Stopped	
	0.0049	0.0252	0.0049	=0.000
Running	Stopped	Stopped	Stopped	
	0.0049	0.9747	0.9747	=0.00465
Σ				**0.01415**

Hence, total probability =0.0252 * 0.9950=0.025.

The probability: ((0.9747 * 0.0252)+(0.0252 * 0.9950))=0.0496. The machine has a 4.9% probability to move to the Running State at T_{n+2}.

Numerical 2: Starting from the running state at time T_n, calculate the stopping probability of the machine at T_{n+3}.

The solution is summarized in Table 5.7.

There is a probability of 1.41% that it will be in the stopped state at time T_{n+3}, provided that the machine was in the running state at T_n.

Numerical 3: Starting from the running state at time T_n, calculate the running probability at T_{n+3}.

Solution: There is a probability of 98.63% that it will be in the running state at time T_{n+3}, provided that the machine was in the running state at T_n. In this manner, the time to failure for starting of the machine at the running or stopped state and ending at the stopped state can be calculated (Table 5.8).

Here, the probability is of the machine going into the stopped state, which internally leads to the calculation of the probability of the machine to go into the failure state. The probability of the machine going into the failure state is a fraction of the

TABLE 5.8
For Markov Chain Numerical 3

T_n	T_n+1	T_n+2	T_n+3	Total Probabilities
Running	Running	Running	Running	
	0.9950	0.9950	0.9950	=0.9850
Running	Running	Stopped	Running	
	0.9950	0.0049	0.0252	=0.000122
Running	Stopped	Running	Running	
	0.0252	0.0252	0.9950	=0.00063
Running	Stopped	Stopped	Running	
	0.0252	0.9747	0.0252	=0.00061
Σ				**=0.9863**

probability of the machine going to the stopped state. The time to failure and probability to failure is correlated.

5.7 WEIBULL PROCESS FOR PREDICTIVE MAINTENANCE

Weibull distribution was invented by Mr. Waloddi Weibull in 1937. Life data analysis can be performed through the Weibull process. During the lifetime of a product, the data can be collected. Weibull analysis is a technique to find reliability characteristics of an equipment in a number of hours or a number of cycles.

Providing the input as in-service time in days and future run-time in days, the Weibull reliability can provide the output as unconditional reliability, current reliability, conditional reliability, and the conditional probability of failure. There are two important parameters that are required to be calculated in the Weibull process, i.e., beta β and eta η.

The Time to Failure (TTF) is required to be arranged in the ascending order.

All values of the Time to Failure arranged in the ascending order can be indexed from 1 to n.

The median rank can be calculated as

```
=(IndexedValue - 0.3)/(MAX(IndexedValues) + 0.4).
Calculate the natural logarithm of TTF.
Calculate linear median rank using
log(-log (1 - MedianRank))
Calculate the intercept using
the sum of the linear median rank divided by the sum of the
log of TTF.
Calculate the slope using the sum of the linear median rank
divided by the sum of the log of TTF.
The slope value is the β value.
The η value can be calculated as = 1/(exp(intercept/slope))
The η value is in terms of the number of days.
```

```
The unconditional reliability can be calculated as
=EXP(-POWER(((FutureRuntime + InServiceTime)/η), β)))
Current reliability can be calculated as
   = EXP(-POWER(((inServiceTime)/η), β)))
Conditional reliability can be calculated as
=unconditional reliability/current reliability
The conditional probability of failure can be calculated as
=1 - (unconditional reliability/current reliability).
```

5.8 DISCUSSION AND CONCLUSION

Prediction of the state of the manufacturing machines is useful in the prediction of the failure of the machine. The learning parameters of the ML models applied on the IoT-based sensor data set are useful to find the Time to Failure and the probability associated with it. The AUC-ROC curve assures the prediction of the welding machine's state with 91% accuracy for multinomial Naïve Bayes technique. The model evaluation criteria recall, precision, F1-score will be useful to the receiver. The probability-based Markov chain provides the future time window to predict the state of a welding machine. Three numerical solutions demonstrated the working of the Markov chain to predict the future state.

REFERENCES

Ahmadi, S., Hajihassani, M., Moosazadeh, S., & Moomivand, H. (2020). An overview of the reliability analysis methods of tunneling equipment. *The Open Construction and Building Technology Journal*, 14(1), 218–229. https://doi.org/10.2174/1874836802014010218

Bramer, M. (2013). *Introduction to Data Mining. In: Principles of Data Mining. Undergraduate Topics in Computer Science.* Springer, London. https://doi.org/10.1007/978-1-4471-4884-5_1

Cartella, F., Lemeire, J., Dimiccoli, L., & Sahli, H. (2015). Hidden semi-markov models for predictive maintenance. *Mathematical Problems in Engineering*, 1–23. 2015 https://doi.org/10.1155/2015/278120

Deka, B., Birklykke, A. A., Duwe, H., Mansinghka, V. K., & Kumar, R. (2014). Markov chain algorithms: A template for building future robust low-power systems. *Philosophical Transactions of the Royal Society A: Mathematical, Physical and Engineering Sciences*, 372, (2018), 1–16. https://doi.org/10.1098/rsta.2013.0277

Gugulothu, N., Gugulothu, N., Tv, V., Malhotra, P., Vig, L., Agarwal, P., & Shro, G. (2017). *Predicting Remaining Useful Life using Time Series Embeddings based on Recurrent Neural Networks* International Journal of Prognostics and Health Management∗. https://doi.org/10.1145/nnnnnnn.nnnnnnn

Dinov, I. D. (2018). *Data Science and Predictive Analytics: Biomedical and Health Applications using R.*, Springer, Cham. https://doi.org/10.1007/978-3-319-72347-1

Gupta, R., & Bhardwaj, P. (2014). Analysis of a discrete parametric Markov-chain model of A two unit cold standby system with repair machine failure. *International Journal of Scientific & Engineering Research*, 5(2), 924–927.

Hofmann, P., & Tashman, Z. (2020). Hidden markov models and their application for predicting failure events. In Computational Science – ICCS 2020. ICCS 2020, Amsterdam, The Netherlands. Lecture Notes in Computer Science, vol 12139. Springer, Cham. https://doi.org/10.1007/978-3-030-50420-5_35

Holladay, D. W., Dallman, B. D., & Grigg, C. H. (2006). Reliability centered maintenance study on voltage regulators. In Proceedings of the IEEE International Conference on Transmission and Distribution Construction and Live Line Maintenance, Albuquerque, NM, USA, ESMO, 2–6. https://doi.org/10.1109/TDCLLM.2006.340728

Hu, C., Pei, H., Wang, Z., Si, X., & Zhang, Z. (2018). A new remaining useful life estimation method for Equipment subjected to intervention of imperfect maintenance activities. *Chinese Journal of Aeronautics*, 31(3), 514–528. https://doi.org/10.1016/j.cja.2018.01.009

Peng, C., Chen, Y., Chen, Q., Tang, Z., Li, L., & Gui, W. (2021). A Remaining Useful Life Prognosis of Turbofan Engine Using Temporal and Spatial Feature Fusion. *Sensors* (Basel, Switzerland), 21(2), 418. https://doi.org/10.3390/s21020418

Kalaiarasi, S., Merceline Anita, A., & Geethanjalii, R. (2017). Analysis of system reliability using markov technique. *Global Journal of Pure and Applied Mathematics*, 13(9), 5265–5273. http://www.ripublication.com

Kanawaday, A. (2017). *Machine Learning for Predictive Maintenance of Industrial Machines using IoT Sensor Data*. IEEE. Figures 2, 4–7.

Kubat, M. (2017). An introduction to machine learning. In *An Introduction to Machine Learning* (Vol. 2). Cham, Switzerland: Springer International Publishing. https://doi.org/10.1007/978-3-319-63913-0

Kuzin, T., & Borovi, T. (2016). Early failure detection for predictive maintenance of sensor parts. ITAT 2016 Proceedings, CEUR Workshop Proceedings, 1649, 123–130.

Mishra, K., & Manjhi, S. K. (2018). Failure prediction model for predictive maintenance. In 2018 IEEE International Conference on Cloud Computing in Emerging Markets (CCEM), Bangalore, India, 72–75. https://doi.org/10.1109/CCEM.2018.00019

Nakamura, J. (2007). *Predicting Time to Failure of Industrial Machines with Temporal Data Mining*. Masters of Science, University of Washington.

Pedregosa. (2011). *JMLR*. https://scikit-learn.org/stable/about.html#citing-scikit-learn

Qin, A., Zhang, Q., Hu, Q., Sun, G., He, J., & Lin, S. (2017). Remaining useful life prediction for rotating machinery based on optimal degradation indicator. *Shock and Vibration*, 2017, 1-12. https://doi.org/10.1155/2017/6754968

Ramos, P., Oliveira, J. M. S., & Silva, P. (2014). Predictive maintenance of production equipment based on neural network autoregression and ARIMA. In 21st International EurOMA Conference-Operations Management in an Innovation Economy. https://core.ac.uk/download/pdf/143396566.pdf

Salfner, F. (2005). Predicting failures with hidden Markov models. In Proceedings of 5th European Dependable Computing Conference (EDCC-5), 41–46. http://www.rok.informatik.hu-berlin.de/Members/Members/salfner/publications/salfner05predicting.pdf

Tai, A. H., Ching, W. K., & Chan, L. Y. (2006). Hidden Markov model for the detection of machine failure. In 36th International Conference on Computers and Industrial Engineering, ICC and IE 2006, Taipei, Taiwan, 2009–2020.

Yuan, C. (2015). Unsupervised machine condition monitoring using segmental hidden Markov models. In IJCAI International Joint Conference on Artificial Intelligence, Buenos Aires, Argentina, 2015-Janua(IJCAI), 4009–4016.

6 Fast and Efficient Lightweight Block Ciphers Involving 2d-Key Vectors for Resource-Poor Settings

Shirisha Kakarla, Geeta Kakarla, D. Narsinga Rao, and M. Raghavender Sharma

CONTENTS

6.1 Introduction ... 100
6.2 Fast and Lightweight Block Cipher Model Development for the Resource-poor Healthcare System ... 102
6.3 Mathematical Models and Design Details ... 102
6.3.1 Pseudo-codes for the Cryptic Procedures ... 105
6.4 Exemplification and the Outcomes ... 107
6.4.1 Avalanche Effect Analysis ... 110
6.5 Simulation and Performance Analysis ... 112
6.5.1 Comparative Analysis of Popular Ciphers ... 112
6.6 Cryptanalysis ... 115
6.7 Conclusive Remarks and Future Directions ... 117
Acknowledgments ... 118
Funding ... 118
Contributions ... 118
Corresponding Author ... 118
Ethics Declarations ... 118
References ... 118

DOI: 10.1201/9781003323426-6

6.1 INTRODUCTION

With the rise in the digitization of patients' health records [1, 2], the secrecy and privacy of the patient's information becomes the most overbearing barrier in the healthcare information exchange [3]. With the sweeping changes brought about by the adoption and proliferation of digital computers and digital record-keeping, the healthcare sector is no exception. For the healthcare workers' pervasive use of digital devices in recording, analyzing, and sharing of the patient's medical condition among the stakeholders of the healthcare ecosystem, in providing the course of treatment effectively, security has become the concern.

The developed and developing nations are adopting technologies and security practices to better conserve confidentiality and to maintain integrity of health information systems [4–6]. Studies suggest that a number of periodic assessments take place in the healthcare organizations in developed countries [7] to review the security loopholes. From time to time, the computer and cyber personnel (re)design the models to address the vulnerabilities and the security gaps traced. The optimized security approach is adopted from a string of security solutions proposed to enable vulnerability reduction and offer deterrence against cyberattacks and integrity violations [8]. In many of the security solutions, robust and advanced encryption techniques are the essential components with enhanced size of the parametric inputs. Generally, in any cryptographic system, key(s) and chunks of the input data to be encrypted are fundamental. The increase in key size and input data size in the encryption brings in the higher degree of security in data. The computing resources in terms of enhanced processing power and storage media are correspondingly desired to implement the security protocols with bigger key sizes.

The availability of computing resources for the security system in the healthcare sector of developed nations is prioritized [9], very often, thereby meeting the demand. On the other hand, in resource-poor nations, there is a dearth of high-powered computing machinery for deploying software security solutions. The budgetary allocations toward healthcare are mainly for the medical infrastructure for providing healthcare services to the patients. Although a number of hospitals and medical care centers in underdeveloped regions are vouching for software applications for registering the patient's details, recording the prognosis and diagnostic details, and updating the course of treatment followed, digitizing the data is the primary task. Exchanging of private and sensitive health-related information also takes place among the immediate stakeholders connected with the patient. However, installing software security solutions to the digitized data becomes secondary in the line of direct services. As a matter of fact, the confidentiality and privacy of the electronic health records is equally important in the case of resource-poor regions, for the internet is not location specific. Usage of the robust security solutions is not viable [10] due to the want of the high-end computing infrastructure in implementing and maintaining them. Therefore, for securing the digital assets in resource-poor regions, lightweight cryptosystems are desired with less processing and storage needs.

In this chapter, a novel lightweight procedure is presented for securing the data stored in the storage units, especially in the resource-poor settings of the healthcare domain. The lightweight security procedure is integrated [11] by encrypting the patients' and others' data staged in units with the help of the key bunch matrix, interchangeably addressed as the 2d-key vector. Apart from this, an additional supplementary key block is also used to further enhance the strength of the cipher. The size of the key bunch matrix as well as the supplementary 2-dimensional key vector in this prototype is chosen as 2048 binary bits, for illustration. The operations being used in encryption and decryption are defined in such a way that the computational power required of the processing units is low, thereby performing very well even with the low configured resources. The time required for carrying out the enciphering and deciphering of the data in the low configured devices of the resource-poor settings is relatively less as compared to the conventional encryption and decryption methods for data sets. Furthermore, the robustness of this cipher is found to be comparable with the other popularly used block ciphers of this sort. The encryption model of the cipher is presented below:

$$C = \left(\left[P \bullet Key\_Enc\right] \bmod 256\right) \oplus Sup\_Key,$$

where Key_Enc is the encryption 2d-key vector and the Sup_Key is the additional supplementary key vector, and $\forall x \in \left[Key\_Enc\right], 0 \le x \le 255, \forall x = 2k+1 | 0 \le k \le 127$. $[P]$ and $[C]$ are the plain data set characters and the corresponding cryptic characters, arranged in the form of the blocks, respectively. The Sup_Key contains elements z, such that $\forall z \in \left[Sup\_Key\right], 0 \le z \le 255$. Each of the above blocks chosen are of size $n \times n$. The dot product of the corresponding elements of the two operands is represented by the binary operator "•". The impact of introducing the supplementary key is also discussed in Section 6.4. The model of the decryption procedure is

$$P = \left[(C \oplus Sup\_Key) \bullet Key\_Dec\right] \bmod 256,$$

where the decryption vector is Key_Dec, such that $\forall y \in \left[Key\_Dec\right], 0 \le y \le 255$, $\forall y = 2k+1 | 0 \le k \le 127$. The governing principle: $\left[Key\_Enc \bullet Key\_Dec\right] \equiv 1 \bmod 256$ relates the two key bunch matrices of encryption and decryption. In this cryptosystem, as the corresponding elements of the block matrices undergo "•" operation, the constrained key size of the erstwhile block ciphers can be overcome for enhancing the security.

This chapter is sectioned in the following manner. The framework development for the fast and lightweight block cipher for the resource-poor healthcare system is discussed in the Section 6.2. In Section 6.3, the preliminary mathematical models and the pseudo-codes for the cryptic procedures are detailed. Using these models, the exemplification is demonstrated with outcomes in Section 6.4 along with the avalanche effect analysis, aka *related key attack*. In the subsequent Section 6.5, the simulation parameters and the comparative metrics of the ciphers are mentioned. In Section 6.6, the theoretical analyses of the widely implemented attacks are summarized. Finally in Section 6.7, concluding remarks are drawn with future directions.

6.2 FAST AND LIGHTWEIGHT BLOCK CIPHER MODEL DEVELOPMENT FOR THE RESOURCE-POOR HEALTHCARE SYSTEM

The focus is to develop a block cipher involving a 2d-key matrix which encrypts the voluminous data with a high number of dimensions. The cryptosystem under discussion can also be implemented for securing the information either at rest in the storage units or in motion in the transmission media. A key vector is essentially a chosen two-dimensional vector containing a set of keys [12]. For each of the keys, a multiplicative inverse is determined for decrypting the ciphered data. Besides, an additional key is also introduced in the cryptosystem for enhancing the randomness in the encipherment of the data set contents. In the encrypting procedure, each and every character input is manipulated with a key from the key vector as well as an element from the additional key vector. The operations performed are iteratively used to bring in adequate randomness in the corresponding ciphered information. During the decryption phase, the multiplicative inverse of the key element is identified and applied with the reverse operations to obtain the original input. The objective is to present the procedure and the cryptanalysis to institute the cipher's strength and its resilience in countering the popular attacks.

The architectural model as depicted in Figure 6.1 consists of the components meant to secure the patients' information while in transmission as well as can also be fairly used to encrypt the whole of the information in the storage media. The patient's information is enciphered upon enrollment, with the block cipher technique discussed here and transmitted to the healthcare system storage units. The received information of the patients from the different sources of electronic media are decrypted and re-encrypted with different sets of keys. These encrypted contents may be searched using the customized query in an encrypted form and the information retrieved is further transmitted to the remote location of the query initiation over public channels, which is deciphered at the receiver.

The discussion here is of presenting a strong block cipher which can encrypt and decrypt loads of data even with the low availability of resources in the resource-poor settings of the healthcare system or the legacy installations where data is generated and needs to be protected from misuse. The apparent strength of the cipher is to withstand the popular active and the passive attacks.

6.3 MATHEMATICAL MODELS AND DESIGN DETAILS

To begin with the building of the cryptosystem, the contents of the records, transmitted from the patients' side as well as the loads of the staged data in the healthcare system, are to be presented in a commonly exchangeable columnar file format, this being the widely transferable arrangement of bulk data across the networks. Here, the data contents are considered to have the format of the comma-separated values (CSV), with the comma as the delimiter between the values of a record. The healthcare data set *HC_DS* containing the medical records of the patients is considered for illustration and is represented as

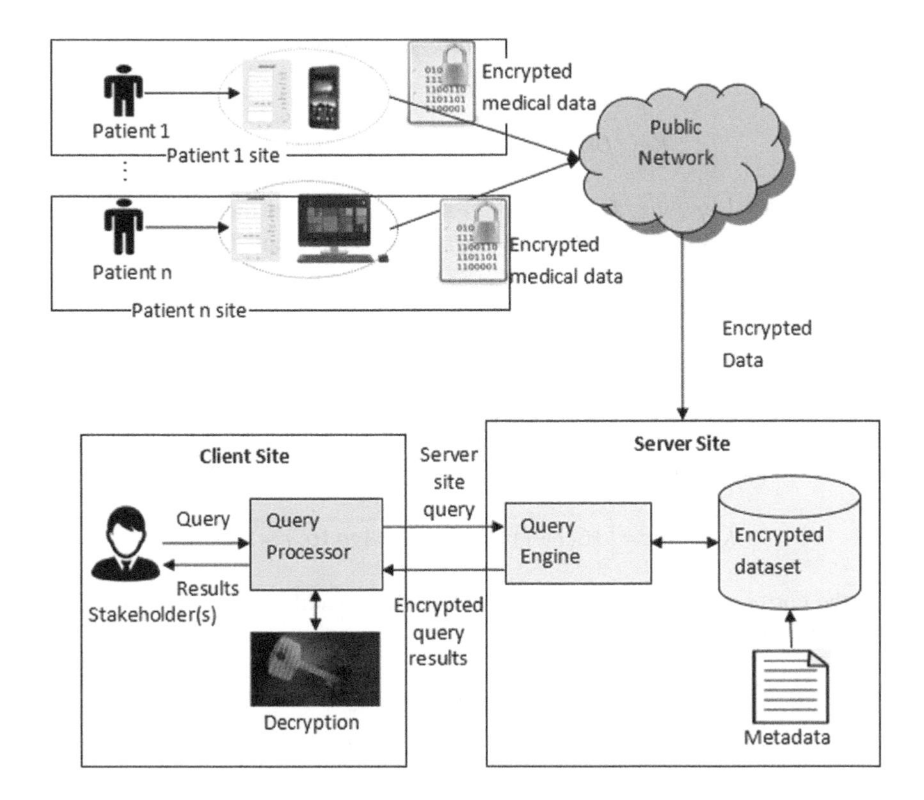

FIGURE 6.1 Architectural components and the interactions among the stakeholders (client), server, and patient sites.

$$HC\_DS = \left[rec_{uvw} \right], u = 1 \; to \; r, v = 1 \; to \; d, w = 1 \; to \; c, \tag{6.1}$$

representing the total number of records with r and the total number of dimensions with d. The number of characters is symbolized by c. rec_{uvw} represents the wth character in the value vth dimension of the uth record in the HC_DS.

For computations, each character of the record of the above mentioned HC_DS is converted into its equivalent $EBCDIC$ code of eight binary digits. The binary digits so obtained of the HC_DS are arranged row wise in the square matrices P with each of size n, as shown in the equation 6.2.

$$P = \left[p_{ij} \right], i = 1 \; to \; n, j = 1 \; to \; n, p_{ij} = EBCDIC\left(HC\_DS\left(rec_{uvw} \right) \right), \tag{6.2}$$

where [0, 255] is the domain of each p_{ij}.

Let Key_Enc be the key vector used for enciphering the block elements and can be represented as

$$Key\_Enc = \left[ke_{ij} \right], i = 1 \; to \; n, j = 1 \; to \; n. \tag{6.3}$$

Similarly, the decryption key vector, to reverse the ciphered elements, Key_Dec can written as

$$Key\_Dec = \left[kd_{ij}\right], i = 1\,to\,n, j = 1\,to\,n. \tag{6.4}$$

The individual keys, viz. ke_{ij} and $kd_{ij} \mid 1 \le i, j \le n$, used in the cryptosystem are bound by the relation of multiplicative inverse and is represented as

$$\left(ke_{ij} \times kd_{ij}\right) \equiv 1 \bmod 256 \mid i = 1\,to\,n, j = 1\,to\,n. \tag{6.5}$$

The value chosen for each ke_{ij} is an odd integer and lies in the range of [1, 255] and the corresponding value computed for kd_{ij} lies in the same interval of [1, 255] and remains an odd and unique integer. For Key_Enc, if all the odd values of range [1, 255] are arranged in a matrix of 16 × 8, mathematically represented as

$$Key\_Enc = \left[16(i-1)+2j-1\right], i = 1\,to\,16, j = 1\,to\,8, \tag{6.6}$$

then the corresponding $kd_{ij} \mid i = 1\,to\,16, j = 1\,to\,8$ is readily computed by (6.5) and arranged in the 16 × 8 matrix form as

$$Key\_Dec = \begin{bmatrix} 1 & 171 & 205 & 183 & 57 & 163 & 197 & 239 \\ 241 & 27 & 61 & 167 & 41 & 19 & 53 & 223 \\ 225 & 139 & 173 & 151 & 25 & 131 & 165 & 207 \\ 209 & 251 & 29 & 135 & 9 & 243 & 21 & 191 \\ 193 & 107 & 141 & 119 & 249 & 99 & 133 & 175 \\ 177 & 219 & 253 & 103 & 233 & 211 & 245 & 159 \\ 161 & 75 & 109 & 87 & 217 & 67 & 101 & 143 \\ 145 & 187 & 221 & 71 & 201 & 179 & 213 & 127 \\ 129 & 43 & 77 & 55 & 185 & 35 & 69 & 111 \\ 113 & 155 & 189 & 39 & 169 & 147 & 181 & 95 \\ 97 & 11 & 45 & 23 & 153 & 3 & 37 & 79 \\ 81 & 123 & 157 & 7 & 137 & 115 & 149 & 63 \\ 65 & 235 & 13 & 247 & 121 & 227 & 5 & 47 \\ 49 & 91 & 125 & 231 & 105 & 83 & 117 & 31 \\ 33 & 203 & 237 & 215 & 89 & 195 & 229 & 15 \\ 17 & 59 & 93 & 199 & 73 & 51 & 85 & 255 \end{bmatrix} \tag{6.7}$$

The governing principle followed in the cryptosystem for enciphering the textual contents of the *HC_DS* arranged in blocks *P* to obtain the ciphered elements block *C* is as follows:

$$C = [c_{ij}] = ([p_{ij} \times ke_{ij}] \bmod 256) \oplus Sup\_Key, \tag{6.8}$$

where $i = 1\ to\ n$, $j = 1\ to\ n$ and Sup_Key are the additional key vectors brought in to further introduce the confusion and diffusion in the ciphered elements. The integers of Sup_Key are chosen to contain the value in the range [1, 255]. The resulting integer of the $[p_{ij} \times ke_{ij}] \bmod 256$ is XOR with the corresponding element of the Sup_Key block matrix.

In the reverse process to get back the original contents of the *HC_DS* in the *EBCDIC* form, the decryption procedure is adopted as

$$P = [p_{ij}] = \left[kd_{ij} \times (C \oplus Sup\_Key)_{ij}\right] \bmod 256,\ i = 1\ to\ n,\ j = 1\ to\ n. \tag{6.9}$$

6.3.1 Pseudo-codes for the Cryptic Procedures

The representative pseudo-codes for the encryption and the decryption processes used in the cryptosystem are as follows:

Pseudo-code for Encryption

```
Read and translate all elements of data sets HC_DS to
HC_DS_CSV
HC_DS_CSV = EBCDIC(HC_DS_CSV)
Select size n and itr
Read Key_Enc of square size n
Read Sup_Key of square size n
Enclose the characters into matrices; each of square
sized n, say P
//If the last block is incomplete, then add 0s to make
it a complete block
For each block P in total_blocks do begin
For k = 1 to itr do begin
For i = 1 to n do begin
For j = 1 to n do begin
p_ij = (ke_ij × p_ij) mod 256
loop j end
loop i end
Assemble P by combining [p_ij]
P = P ⊕ Sup_Key
P = mixBits(P)
Loop k end
```

```
C = P
Write C
end of P in total_blocks for loop
```

In the above pseudo-code, *itr* is the number of rounds in the iterative process to perform the thorough confusion and diffusion of the bits in each block of *total_blocks* formed out of the *HC_DS_CSV*.

Pseudo-code for Decryption

```
Read C,Key_Enc,Sup_Key,n,itr
Key_Dec = obtainMultiInverse(Key_Enc)
For each block C in total_blocks do begin
For k = 1 to itr do begin
C = inverseMixBits(C)
C = C ⊕ Sup_Key
//Sup_Key = [sk_ij] | i = 1 to n, j = 1 to n
For i = 1 to n do begin
For j = 1 to n do begin
Calculate c_ij = (d_ij × c_ij) mod 256
Loop j end
Loop i end
Assemble C by combining [c_ij]
Loop k end
P = C
Write P
end of C in total_blocks for loop
```

Each of the squared blocks created out of the *HC_DS_CSV* is subjected to the *mixBits*( ) procedure after manipulating with the key. This procedure would introduce diffusion of the binary bits within the block, thereby inducing more randomness and thus the strength to the cryptosystem. In the *mixBits*( ) procedure, the plaintext block, say $P = [p_{ij}], i = 1\ to\ n, j = 1\ to\ n$ has the data set contents in any round of the iteration. Considering $n = 2m$, the block of *P* can be expressed in the following form as (6.10).

$$P = \begin{bmatrix} p_{11} & p_{12} & ..p_{1m} & p_{1(m+1)} & \cdot & \cdot & p_{1(n-1)} & p_{1n} \\ p_{21} & p_{22} & ..p_{2m} & p_{2(m+1)} & \cdot & \cdot & p_{2(n-1)} & p_{2n} \\ \cdot & \cdot & \cdot & \cdot & \cdot & \cdot & \cdot & \cdot \\ \cdot & \cdot & \cdot & \cdot & \cdot & \cdot & \cdot & \cdot \\ \cdot & \cdot & \cdot & \cdot & \cdot & \cdot & \cdot & \cdot \\ p_{n1} & p_{n2} & ..p_{nm} & p_{n(m+1)} & \cdot & \cdot & p_{n(n-1)} & p_{nn} \end{bmatrix} \quad (6.10)$$

Upon converting each of the above elements of block *P* to its equivalent *EBCDIC* form and further to its corresponding binary form, a vector of n rows and $8n$ columns is obtained as represented in (6.11).

$$\begin{bmatrix} b_{111}b_{112}..b_{118} & \cdots & b_{1m1}b_{1m2}..b_{1m8} & b_{1(m+1)1}b_{1(m+1)2}..b_{1(m+1)8} & \cdots & b_{1n1}b_{1n2}..b_{1n8} \\ b_{211}b_{212}..b_{218} & \cdots & b_{2m1}b_{2m2}..p_{2m8} & b_{2(m+1)1}b_{2(m+1)2}..b_{2(m+1)8} & \cdots & b_{2n1}b_{2n2}..b_{2n8} \\ . & \cdots & . & . & \cdots & . \\ . & \cdots & . & . & \cdots & . \\ b_{n11}b_{n12}..b_{n18} & \cdots & b_{nm1}b_{nm2}..b_{nm8} & b_{n(m+1)1}b_{n(m+1)2}..b_{n(m+1)8} & \cdots & b_{nn1}b_{nn2}..b_{nn8} \end{bmatrix} \tag{6.11}$$

The decimal integers are generated by following the unique reading operation of the binary bits, which is summarized here. The first eight bits of the first column are read and converted to its equivalent decimal integer. The subsequent eight bits of the same column if $n > 8$ are read similarly to form the second integer, and this continues until the end of the first column of the (6.11) vector. For creating the next integer, the first eight binary digits of the $(m+1)^{th}$ column of the vector are considered and the same procedure is followed until the end of this column. Further bunch of integers are obtained by following the mentioned procedure with the second column, $(m+2)^{th}$ column, third column, $(m+3)^{th}$ column, and so on. The decimal integers so obtained are arranged row wise in a matrix of square size n. In case, the matrix is containing the number of rows less than eight then for obtaining an integer, the digits of the binary form of the first column in the first half and the first column of the second half are considered. This step ensures the thorough mix-up of the binary bits, thus inducing the diffusion in the generated ciphertext elements in each of the iterations. The reverse procedure is performed in the *inverseMixBits()* function used in the pseudo-code of the decryption procedure.

6.4 EXEMPLIFICATION AND THE OUTCOMES

For the illustrative practical implementation, the healthcare data set [13], designated as HC_DS_CSV, is considered of approximately 57,000 characters and having 29 dimensions, mostly with descriptive and categorical data. The initial two records of the data set are hereby mentioned below for reference:

Provider ID,Hospital Name, Address,City,State,ZIP Code,County Name,Phone Number,Hospital Type,Hospital Ownership,Emergency Services,Meets criteria for meaningful use of EHRs,Hospital overall rating,Hospital overall rating footnote,Mortality national comparison,Mortality national comparison footnote,Safety of care national comparison,Safety of care national comparison footnote,Readmission national comparison,Readmission national comparison footnote,Patient experience national comparison,Patient experience national comparison footnote,Effectiveness of care national comparison,Effectiveness

of care national comparison footnote,Timeliness of care national comparison,Timeliness of care national comparison footnote,Efficient use of medical imaging national comparison,Efficient use of medical imaging national comparison footnote,Location

10005, MARSHALL MEDICAL CENTER SOUTH,2505 U S HIGHWAY 431 NORTH,BOAZ,AL,35957,MARSHALL,2565938310,Acute Care Hospitals,Government – Hospital District or Authority,TRUE,TRUE,3,,Below the national average,,Same as the national average,,Above the national average,,Same as the national average,,Same as the national average,,Above the national average,,Below the national average,,"2505 U S HIGHWAY 431 NORTH BOAZ, AL (6.12)

Out of (6.12), the initial 256 characters, shown in (22.13), are chosen for the plaintext matrix of 16 rows and 16 columns. Upon converting these 256 characters into their corresponding *EBCDIC* values, a row-wise plaintext matrix is constructed. For the encryption procedure, the plaintext matrix is considered as the raw input and presented in the following format of (6.14).

Provider ID,Hospital Name,Address,City,State,ZIP Code,County Name,Phone Number,Hospital Type,Hospital Ownership,Emergency Services,Meets criteria for meaningful use of EHRs,Hospital overall rating,Hospital overall rating footnote,Mortality national compare (6.13)

$$P = \begin{bmatrix} 215 & 153 & 150 & 165 & 137 & 132 & 133 & 153 & 64 & 201 & 196 & 107 & 200 & 150 & 162 & 151 \\ 137 & 163 & 129 & 147 & 64 & 213 & 129 & 148 & 133 & 107 & 193 & 132 & 132 & 153 & 133 & 162 \\ 162 & 107 & 195 & 137 & 163 & 168 & 107 & 226 & 163 & 129 & 163 & 133 & 107 & 233 & 201 & 215 \\ 64 & 195 & 150 & 132 & 133 & 107 & 195 & 150 & 164 & 149 & 163 & 168 & 64 & 213 & 129 & 148 \\ 133 & 107 & 215 & 136 & 150 & 149 & 133 & 64 & 213 & 164 & 148 & 130 & 133 & 153 & 107 & 200 \\ 150 & 162 & 151 & 137 & 163 & 129 & 147 & 64 & 227 & 168 & 151 & 133 & 107 & 200 & 150 & 162 \\ 151 & 137 & 163 & 129 & 147 & 64 & 214 & 166 & 149 & 133 & 153 & 162 & 136 & 137 & 151 & 107 \\ 197 & 148 & 133 & 153 & 135 & 133 & 149 & 131 & 168 & 64 & 226 & 133 & 153 & 165 & 137 & 131 \\ 133 & 162 & 107 & 212 & 133 & 133 & 163 & 162 & 64 & 131 & 153 & 137 & 163 & 133 & 153 & 137 \\ 129 & 64 & 134 & 150 & 153 & 64 & 148 & 133 & 129 & 149 & 137 & 149 & 135 & 134 & 164 & 147 \\ 64 & 164 & 162 & 133 & 64 & 150 & 134 & 64 & 197 & 200 & 217 & 162 & 107 & 200 & 150 & 162 \\ 151 & 137 & 163 & 129 & 147 & 64 & 150 & 165 & 133 & 153 & 129 & 147 & 147 & 64 & 153 & 129 \\ 163 & 137 & 149 & 135 & 107 & 200 & 150 & 162 & 151 & 137 & 163 & 129 & 147 & 64 & 150 & 165 \\ 133 & 153 & 129 & 147 & 147 & 64 & 153 & 129 & 163 & 137 & 149 & 135 & 64 & 134 & 150 & 150 \\ 163 & 147 & 150 & 163 & 133 & 107 & 212 & 150 & 153 & 163 & 129 & 147 & 137 & 163 & 168 & 64 \\ 149 & 129 & 163 & 137 & 150 & 149 & 129 & 147 & 64 & 131 & 150 & 148 & 151 & 129 & 153 & 137 \end{bmatrix} \tag{6.14}$$

The encryption key bunch square matrix of size 16 is chosen to contain the random odd integers [0–255] and represented here as (6.15)

$$Key_Enc = \begin{bmatrix}
217 & 169 & 61 & 255 & 65 & 65 & 111 & 7 & 33 & 73 & 133 & 71 & 223 & 53 & 173 & 43 \\
75 & 163 & 103 & 207 & 205 & 107 & 191 & 113 & 245 & 123 & 143 & 129 & 205 & 227 & 43 & 51 \\
125 & 103 & 117 & 209 & 57 & 221 & 159 & 213 & 181 & 199 & 13 & 113 & 95 & 121 & 21 & 143 \\
9 & 241 & 137 & 41 & 73 & 93 & 167 & 97 & 207 & 79 & 185 & 47 & 35 & 25 & 147 & 23 \\
171 & 65 & 39 & 107 & 123 & 73 & 51 & 7 & 15 & 125 & 77 & 71 & 75 & 235 & 47 & 117 \\
137 & 37 & 149 & 93 & 11 & 197 & 187 & 107 & 37 & 87 & 233 & 195 & 217 & 17 & 9 & 111 \\
61 & 129 & 5 & 189 & 63 & 17 & 35 & 199 & 223 & 45 & 49 & 191 & 209 & 241 & 9 & 97 \\
179 & 123 & 229 & 193 & 75 & 117 & 79 & 197 & 107 & 99 & 219 & 113 & 15 & 123 & 35 & 79 \\
33 & 179 & 219 & 53 & 253 & 131 & 169 & 213 & 45 & 221 & 59 & 67 & 197 & 29 & 171 & 113 \\
89 & 105 & 225 & 159 & 173 & 135 & 213 & 43 & 149 & 75 & 83 & 149 & 7 & 69 & 163 & 71 \\
29 & 91 & 203 & 155 & 7 & 97 & 187 & 85 & 199 & 163 & 89 & 219 & 19 & 223 & 93 & 139 \\
93 & 81 & 83 & 73 & 175 & 57 & 249 & 243 & 105 & 31 & 39 & 9 & 205 & 215 & 73 & 103 \\
131 & 129 & 65 & 163 & 163 & 11 & 195 & 243 & 135 & 171 & 25 & 15 & 113 & 197 & 109 & 215 \\
105 & 179 & 215 & 89 & 209 & 245 & 223 & 193 & 141 & 75 & 121 & 157 & 7 & 197 & 41 & 11 \\
189 & 51 & 245 & 99 & 227 & 193 & 235 & 155 & 73 & 107 & 35 & 223 & 213 & 115 & 29 & 43 \\
191 & 33 & 63 & 13 & 59 & 23 & 161 & 169 & 31 & 133 & 73 & 251 & 39 & 13 & 41 & 131
\end{bmatrix} \quad (6.15)$$

The additional key, named *Sup_Key* as shown in (6.16), is constructed to contain the random integers in the range [0–255] arranged in the square size of 16, for enhancing the degree of diffusion during the encipherment process of the plaintext data.

$$Sup_Key = \begin{bmatrix}
223 & 13 & 0 & 233 & 56 & 23 & 140 & 234 & 161 & 55 & 98 & 31 & 119 & 196 & 44 & 56 \\
180 & 230 & 62 & 107 & 111 & 136 & 8 & 247 & 25 & 36 & 224 & 170 & 138 & 224 & 84 & 0 \\
31 & 243 & 145 & 205 & 29 & 252 & 62 & 109 & 127 & 125 & 62 & 10 & 179 & 103 & 18 & 125 \\
65 & 89 & 47 & 241 & 119 & 217 & 145 & 149 & 10 & 171 & 246 & 68 & 100 & 32 & 151 & 57 \\
182 & 19 & 4 & 165 & 67 & 40 & 193 & 53 & 65 & 178 & 136 & 87 & 157 & 163 & 194 & 105 \\
105 & 208 & 126 & 193 & 53 & 212 & 125 & 237 & 165 & 209 & 39 & 105 & 166 & 114 & 40 & 110 \\
50 & 134 & 56 & 136 & 54 & 189 & 1 & 7 & 111 & 34 & 136 & 129 & 7 & 39 & 97 & 2 \\
200 & 60 & 10 & 210 & 35 & 43 & 48 & 36 & 214 & 184 & 79 & 71 & 38 & 68 & 13 & 145 \\
137 & 143 & 244 & 135 & 114 & 16 & 170 & 89 & 114 & 172 & 158 & 77 & 36 & 19 & 37 & 58 \\
141 & 115 & 117 & 215 & 87 & 171 & 112 & 53 & 33 & 166 & 71 & 99 & 89 & 36 & 5 & 245 \\
138 & 168 & 181 & 198 & 207 & 1 & 122 & 45 & 13 & 104 & 27 & 115 & 16 & 229 & 184 & 104 \\
12 & 107 & 26 & 123 & 27 & 178 & 219 & 243 & 54 & 193 & 226 & 100 & 239 & 205 & 12 & 222 \\
129 & 221 & 188 & 215 & 55 & 157 & 220 & 53 & 23 & 141 & 101 & 136 & 141 & 168 & 221 & 240 \\
136 & 168 & 54 & 218 & 184 & 239 & 157 & 70 & 210 & 26 & 45 & 22 & 188 & 115 & 39 & 80 \\
37 & 90 & 94 & 76 & 155 & 246 & 185 & 49 & 118 & 149 & 81 & 36 & 7 & 92 & 89 & 59 \\
55 & 247 & 88 & 53 & 141 & 94 & 102 & 176 & 211 & 235 & 238 & 182 & 216 & 80 & 101 & 32
\end{bmatrix} \quad (6.16)$$

Upon following the principle of the multiplicative inverse, the corresponding decryption key bunch vector, say Key_Dec, is calculated for the given (6.15) and represented as follows:

$$Key\_Dec = \begin{bmatrix} 105 & 153 & 21 & 255 & 193 & 193 & 143 & 183 & 225 & 249 & 77 & 119 & 31 & 29 & 37 & 131 \\ 99 & 11 & 87 & 47 & 5 & 67 & 63 & 145 & 93 & 179 & 111 & 129 & 5 & 203 & 131 & 251 \\ 213 & 87 & 221 & 49 & 9 & 117 & 95 & 125 & 157 & 247 & 197 & 145 & 159 & 201 & 61 & 111 \\ 57 & 17 & 185 & 25 & 249 & 245 & 23 & 161 & 47 & 175 & 137 & 207 & 139 & 41 & 155 & 167 \\ 3 & 193 & 151 & 67 & 179 & 249 & 251 & 183 & 239 & 213 & 133 & 119 & 99 & 195 & 207 & 221 \\ 185 & 173 & 189 & 245 & 163 & 13 & 115 & 67 & 173 & 103 & 89 & 235 & 105 & 241 & 57 & 143 \\ 21 & 129 & 205 & 149 & 191 & 241 & 139 & 247 & 31 & 165 & 209 & 63 & 49 & 17 & 57 & 161 \\ 123 & 179 & 237 & 65 & 99 & 221 & 175 & 13 & 67 & 75 & 83 & 145 & 239 & 179 & 139 & 175 \\ 225 & 123 & 83 & 29 & 85 & 43 & 153 & 125 & 165 & 117 & 243 & 107 & 13 & 53 & 3 & 145 \\ 233 & 217 & 33 & 95 & 37 & 55 & 125 & 131 & 189 & 99 & 219 & 189 & 183 & 141 & 11 & 119 \\ 53 & 211 & 227 & 147 & 183 & 161 & 115 & 253 & 247 & 11 & 233 & 83 & 27 & 31 & 245 & 35 \\ 245 & 177 & 219 & 249 & 79 & 9 & 73 & 59 & 217 & 223 & 151 & 57 & 5 & 231 & 249 & 87 \\ 43 & 129 & 193 & 11 & 11 & 163 & 235 & 59 & 55 & 3 & 41 & 239 & 145 & 13 & 101 & 231 \\ 217 & 123 & 231 & 233 & 49 & 93 & 31 & 65 & 69 & 99 & 201 & 181 & 183 & 13 & 25 & 163 \\ 149 & 251 & 93 & 75 & 203 & 65 & 195 & 147 & 249 & 67 & 139 & 31 & 125 & 187 & 53 & 131 \\ 63 & 225 & 191 & 197 & 243 & 167 & 97 & 153 & 223 & 77 & 249 & 51 & 151 & 197 & 25 & 43 \end{bmatrix} \tag{6.17}$$

On using (6.15) and (6.16) as input keys and the enciphering procedure as discussed in the preceding section, upon the (6.14), the resulting ciphertext block of squared size 16 is obtained and is shown in (6.18).

$$C = \begin{bmatrix} 91 & 250 & 17 & 39 & 180 & 73 & 80 & 215 & 70 & 58 & 144 & 249 & 36 & 18 & 37 & 114 \\ 135 & 64 & 145 & 81 & 22 & 161 & 130 & 252 & 144 & 56 & 226 & 190 & 164 & 100 & 25 & 111 \\ 90 & 13 & 118 & 221 & 195 & 202 & 225 & 215 & 9 & 148 & 97 & 200 & 110 & 196 & 120 & 225 \\ 225 & 85 & 159 & 69 & 209 & 102 & 142 & 34 & 154 & 114 & 155 & 163 & 148 & 241 & 53 & 76 \\ 5 & 104 & 252 & 73 & 27 & 137 & 51 & 246 & 150 & 12 & 204 & 238 & 83 & 232 & 147 & 45 \\ 188 & 125 & 255 & 178 & 127 & 167 & 89 & 104 & 186 & 127 & 84 & 106 & 73 & 191 & 18 & 175 \\ 121 & 199 & 224 & 52 & 252 & 35 & 121 & 233 & 29 & 132 & 23 & 80 & 202 & 225 & 4 & 51 \\ 46 & 162 & 141 & 133 & 47 & 24 & 15 & 83 & 111 & 185 & 224 & 208 & 119 & 43 & 196 & 117 \\ 89 & 50 & 0 & 181 & 26 & 184 & 68 & 65 & 121 & 214 & 36 & 112 & 15 & 239 & 201 & 255 \\ 36 & 154 & 150 & 216 & 183 & 63 & 220 & 144 & 221 & 191 & 8 & 78 & 178 & 230 & 130 & 117 \\ 213 & 143 & 49 & 166 & 15 & 45 & 122 & 25 & 114 & 66 & 178 & 31 & 200 & 148 & 41 & 18 \\ 246 & 39 & 76 & 40 & 111 & 230 & 46 & 194 & 183 & 144 & 96 & 141 & 3 & 91 & 223 & 70 \\ 201 & 97 & 63 & 151 & 11 & 81 & 122 & 249 & 17 & 36 & 80 & 170 & 55 & 127 & 137 & 157 \\ 174 & 215 & 153 & 178 & 176 & 69 & 44 & 156 & 36 & 239 & 7 & 138 & 3 & 45 & 122 & 211 \\ 223 & 254 & 94 & 102 & 86 & 25 & 213 & 138 & 39 & 82 & 44 & 15 & 112 & 42 & 159 & 190 \\ 190 & 24 & 254 & 72 & 53 & 52 & 237 & 34 & 166 & 168 & 1 & 155 & 20 & 199 & 67 & 77 \end{bmatrix} \tag{6.18}$$

At the receiver side, the decryption procedure is followed along with the communicated Sup_Key as in (6.16) and the calculated (6.17) on the ciphertext block to get back the original P, as shown in (6.14).

6.4.1 Avalanche Effect Analysis

In the block cipher systems, the avalanche effect is one of the sought-after properties to determine the impact of the one-bit flip in either of the inputs supplied to the encryption procedure, discussed in the earlier section. More the number of bits changed in the resulting ciphertext block as compared to the original ciphertext block, apart from the corresponding integers, more robust is the cryptosystem. Observations are made by flipping one bit in either of the Key_Enc block, (6.13), or the Sup_key block to calculate the percentage of change.

Case 1. Change in Key_Enc: Initially, the integer at position 15th row and 12th column of the Key_Enc is changed from 223 to 207, and keeping other inputs and the procedure intact, the ciphertext block is obtained, as mentioned in the (6.19). On examining (6.18) and (6.19), out of the total 2048 bits, 1057 bits have reversed leading to an avalanche effect of 51.61%. And 255 out of 256 total number of elements changed, thus indicating the cipher to be predictably good.

$$C'_{key\_Enc} = \begin{bmatrix} 95 & 24 & 225 & 142 & 181 & 72 & 5 & 114 & 147 & 249 & 34 & 80 & 76 & 155 & 146 & 20 \\ 54 & 203 & 214 & 237 & 242 & 66 & 27 & 188 & 153 & 142 & 245 & 33 & 166 & 34 & 86 & 42 \\ 128 & 120 & 255 & 63 & 177 & 135 & 179 & 114 & 108 & 155 & 151 & 250 & 70 & 145 & 54 & 134 \\ 142 & 108 & 65 & 89 & 9 & 89 & 144 & 68 & 92 & 126 & 168 & 236 & 238 & 34 & 132 & 28 \\ 237 & 211 & 126 & 244 & 2 & 169 & 109 & 53 & 204 & 236 & 163 & 82 & 196 & 111 & 126 & 75 \\ 129 & 13 & 214 & 113 & 23 & 184 & 203 & 216 & 108 & 136 & 227 & 19 & 65 & 113 & 105 & 116 \\ 254 & 80 & 71 & 214 & 210 & 167 & 170 & 96 & 36 & 68 & 126 & 27 & 91 & 47 & 181 & 206 \\ 164 & 9 & 243 & 114 & 208 & 143 & 185 & 53 & 29 & 250 & 41 & 238 & 168 & 23 & 99 & 219 \\ 50 & 124 & 67 & 138 & 222 & 241 & 161 & 252 & 129 & 173 & 2 & 83 & 167 & 13 & 123 & 218 \\ 24 & 31 & 212 & 139 & 205 & 83 & 35 & 139 & 145 & 88 & 222 & 112 & 198 & 214 & 41 & 61 \\ 109 & 177 & 133 & 215 & 216 & 69 & 199 & 44 & 124 & 40 & 73 & 96 & 141 & 47 & 85 & 17 \\ 110 & 11 & 83 & 228 & 189 & 89 & 233 & 178 & 209 & 185 & 87 & 27 & 101 & 226 & 223 & 226 \\ 14 & 153 & 138 & 117 & 240 & 143 & 93 & 250 & 223 & 243 & 224 & 63 & 244 & 200 & 196 & 213 \\ 170 & 150 & 47 & 142 & 250 & 237 & 113 & 103 & 128 & 19 & 212 & 230 & 249 & 8 & 197 & 25 \\ 121 & 151 & 249 & 98 & 134 & 223 & 37 & 255 & 188 & 168 & 98 & 129 & 89 & 50 & 203 & 249 \\ 40 & 156 & 193 & 153 & 5 & 110 & 29 & 72 & 112 & 80 & 9 & 59 & 24 & 13 & 110 & 229 \end{bmatrix} \tag{6.19}$$

Case 2. Change in Plaintext: Instead of changing a bit in Key_Enc, a single bit is reversed randomly in the plaintext block to study the avalanche effect. The character "e" is changed to "n" in the word: Hospital Type in (6.13), i.e., the integer 133, an *EBCDIC* value of "e" is replaced by 149 in the 6th row 12th column. With the modified raw input block, the original Key_Enc, Sup_Key blocks and the encryption procedure, the resultant ciphertext C is determined, shown in the form (6.20). On relating the corresponding binary bits of the (6.18) and (6.20) matrices, the number of binary bits changed is 1016 out of 2048 and the total number of integers changed is 255 out of 256 in *EBCDIC* forms, indicating the good avalanche effect.

$$C'_p = \begin{bmatrix} 102 & 130 & 89 & 212 & 179 & 232 & 237 & 239 & 67 & 168 & 245 & 204 & 76 & 197 & 7 & 23 \\ 199 & 40 & 88 & 135 & 183 & 118 & 207 & 193 & 200 & 232 & 17 & 102 & 89 & 248 & 204 & 239 \\ 52 & 24 & 131 & 161 & 62 & 233 & 99 & 128 & 228 & 52 & 86 & 21 & 192 & 176 & 204 & 44 \\ 44 & 44 & 77 & 43 & 56 & 26 & 228 & 136 & 51 & 60 & 227 & 156 & 153 & 255 & 157 & 160 \\ 13 & 103 & 107 & 46 & 31 & 153 & 187 & 125 & 15 & 101 & 223 & 226 & 132 & 190 & 145 & 154 \\ 135 & 246 & 70 & 222 & 236 & 47 & 104 & 182 & 134 & 206 & 209 & 156 & 179 & 106 & 100 & 143 \\ 209 & 29 & 248 & 103 & 98 & 164 & 129 & 16 & 82 & 67 & 153 & 152 & 15 & 86 & 229 & 175 \\ 1 & 69 & 129 & 192 & 99 & 127 & 64 & 251 & 182 & 108 & 160 & 35 & 254 & 162 & 110 & 127 \\ 147 & 39 & 165 & 236 & 69 & 86 & 113 & 116 & 200 & 174 & 77 & 158 & 187 & 243 & 11 & 140 \\ 179 & 56 & 236 & 12 & 254 & 149 & 38 & 102 & 136 & 69 & 210 & 139 & 24 & 59 & 191 & 23 \\ 41 & 31 & 16 & 93 & 207 & 248 & 165 & 89 & 129 & 65 & 251 & 8 & 205 & 140 & 7 & 98 \\ 130 & 255 & 208 & 70 & 98 & 46 & 9 & 46 & 81 & 245 & 31 & 124 & 48 & 177 & 14 & 130 \\ 223 & 95 & 99 & 62 & 3 & 72 & 44 & 222 & 179 & 164 & 214 & 50 & 21 & 174 & 130 & 254 \\ 31 & 222 & 103 & 87 & 233 & 46 & 34 & 30 & 12 & 148 & 230 & 77 & 201 & 110 & 50 & 191 \\ 142 & 135 & 135 & 24 & 134 & 198 & 208 & 100 & 12 & 232 & 137 & 226 & 81 & 130 & 95 & 255 \\ 107 & 16 & 221 & 213 & 246 & 230 & 3 & 238 & 44 & 82 & 232 & 121 & 30 & 71 & 195 & 77 \end{bmatrix} \tag{6.20}$$

Case 3. Change in Sup_Key: On randomly changing one bit in the *Sup_Key* vector, say element 90, located in the 15th row and 2nd column to 91, the cryptic procedures are performed. The ciphertext so obtained from the original *Key _Enc* matrix, *P* block, and the changed *Sup_Key* is mentioned in (6.21) and when compared with the actual ciphertext of (6.18), in binary forms, is found to differ by 1040 bits making 50.78% of the avalanche effect. In *EBCDIC* formats, the total count of integers changed is 254 out of 256. From the above three cases, it can be inferred that this cipher is stronger.

$$C_{Sup\_Key'} = \begin{bmatrix} 197 & 10 & 95 & 48 & 155 & 46 & 196 & 78 & 239 & 57 & 38 & 34 & 90 & 96 & 176 & 92 \\ 134 & 12 & 219 & 247 & 1 & 168 & 212 & 70 & 13 & 215 & 109 & 245 & 147 & 139 & 165 & 38 \\ 126 & 0 & 125 & 58 & 69 & 72 & 139 & 162 & 123 & 203 & 17 & 13 & 227 & 169 & 28 & 250 \\ 221 & 163 & 117 & 126 & 121 & 37 & 48 & 254 & 48 & 240 & 115 & 104 & 120 & 180 & 223 & 102 \\ 157 & 98 & 163 & 48 & 114 & 141 & 133 & 163 & 87 & 123 & 79 & 186 & 153 & 144 & 213 & 155 \\ 167 & 10 & 85 & 245 & 177 & 207 & 80 & 92 & 225 & 198 & 0 & 88 & 79 & 244 & 81 & 70 \\ 64 & 57 & 134 & 96 & 159 & 4 & 226 & 105 & 60 & 30 & 240 & 164 & 255 & 29 & 237 & 62 \\ 127 & 131 & 38 & 183 & 144 & 24 & 112 & 6 & 40 & 180 & 99 & 164 & 140 & 204 & 182 & 102 \\ 110 & 132 & 135 & 238 & 125 & 175 & 24 & 215 & 108 & 207 & 172 & 15 & 208 & 12 & 193 & 139 \\ 108 & 7 & 93 & 190 & 60 & 0 & 200 & 252 & 118 & 111 & 11 & 244 & 174 & 13 & 7 & 225 \\ 186 & 183 & 52 & 38 & 130 & 146 & 164 & 26 & 38 & 193 & 125 & 77 & 200 & 94 & 83 & 190 \\ 168 & 248 & 195 & 143 & 20 & 9 & 16 & 118 & 62 & 116 & 217 & 10 & 136 & 100 & 124 & 202 \\ 39 & 237 & 32 & 171 & 236 & 36 & 170 & 171 & 201 & 37 & 168 & 28 & 178 & 129 & 122 & 8 \\ 25 & 193 & 143 & 140 & 72 & 90 & 58 & 154 & 47 & 53 & 44 & 101 & 143 & 111 & 2 & 143 \\ 137 & 70 & 98 & 51 & 82 & 161 & 73 & 136 & 204 & 153 & 162 & 76 & 244 & 160 & 19 & 34 \\ 43 & 45 & 170 & 87 & 30 & 215 & 33 & 132 & 194 & 254 & 70 & 104 & 238 & 205 & 20 & 237 \end{bmatrix} \tag{6.21}$$

6.5 SIMULATION AND PERFORMANCE ANALYSIS

For simulating the encryption, decryption procedures and key generations, a processor with the Intel i5-8250U model of 1.80 GHz, RAM of 8.00 GB, and 64-bit Windows 10 OS is used. The size of each block considered for plaintext and the individual keys involved in the illustration is of 256 integers, arranged in 16 × 16 vector, thus enclosing 2048 binary digits. The number of rounds in each of the encryption and decryption procedures is taken as 16 to thoroughly merge the bits and cause confusion and diffusion to a large extent. Besides, the *mix*() function is used in each round to further diffuse the binary bits within a 2d vector. The algorithms are practically executed in Java 1.8.0_0171(64 bit) with the ECB mode. Time taken by the key generation module for outputting the 256 odd and random integers for a key of square size 16 is 0.16×10^{-4} seconds. The programs are executed on a few data sets available in .csv format in the different data-sourced websites for comparing the elapsed times in the encryption–decryption procedures for a 2d vector size of 2048 bits. In Table 6.1 and Figure 6.2, the elapsed time durations are presented for the fast block cipher model developed.

6.5.1 Comparative Analysis of Popular Ciphers

The popular symmetric encryption methods like data encryption standard (DES), advanced encryption standard (AES), DES variant, and Blowfish are also

TABLE 6.1
Time elapsed in the cryptic procedures of fast block cipher (on considering 2d-vector size: 2048-bit, key size: 2048-bit, and 16 rounds of iteration for a block)

Data set size	Time elapsed in seconds for encryption	Time elapsed in seconds for decryption
1.02 MB	1.18	1.23
2.1 MB	2.16	3.01
5.05 MB	5.02	5.87
10.8 MB	9.12	9.28
20.5 MB	16.8	17.12

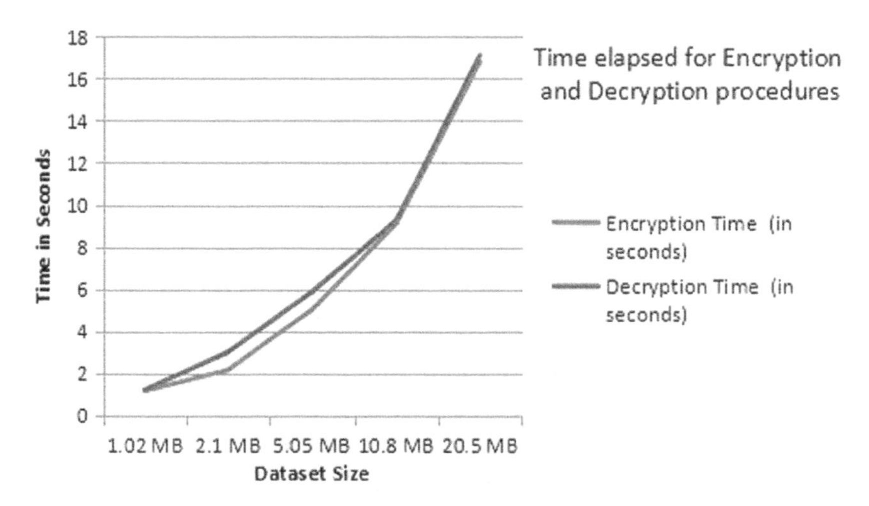

FIGURE 6.2 Line graph for elapsed time for cryptic procedures for the fast block cipher.

programmed using the abovementioned setup to analyze the variations and similarities with respect to the performance and the other parametric features against the proposed and developed fast block cipher model. The observations obtained are presented in Table 6.2 for providing the distinctive features like size of key(s) and block, time expended in key generation and data block encryption, flexibility in key and block size expansion, comprehensively.

The fast data set block cipher model involves the scalar matrix multiplications among the corresponding block elements of the Key_Enc, encryption key, and the raw data set matrices and the resulting block is XORed with an additional key in each round, besides the *mixBits* operation. The block size of the raw data set block, encryption key bunch block, and the additional key is 2048 bits, each. These operations provide a flexibility to go beyond the 2048 bits and choose the block size conveniently with very little computational overhead, comparatively, for implementing the key generation module and the cryptic procedures involved.

TABLE 6.2
Comparative analysis of block cipher encryption procedures

Features	Block size*	Key size*	Operations used in the encryption	Number of iterations for each block	Key generation time#	Time elapsed in encryption (in seconds) for one round and one data block#	Flexibility in expanding block size
DES	64	56	S-box generation, subkey generation	16	2.4	1.58	No
AES	64	128	shift rows, substitute bytes, add round key, mix columns	10	3.71	6.73	No
Triple DES	64	168 (for three different keys)	S-box generation, subkey generation	16	7.2	3.8	Partial
Blowfish	64	64	Key expansion, addition, XOR	16	2.41	0.84	No
Fast Data set Block Cipher (proposed cipher)	2048	4096 (two keys: *Key_Enc, Sup_Key*)	Scalar matrix multiplications, XOR, mixbits	16	2.36×10^{-6}	7.5×10^{-5}	Yes

***In binary bits # In seconds**

The given raw data set, in the form of .csv file format, is divided into the blocks, each of size 16×16. In case, the last block contains the number of elements less than the considered block size, then the sufficient number of additional 0's is appended to make the block complete. As presenting the ciphertext matrices of all the input blocks formed out of the raw data is beyond the scope of this chapter, the first enciphered square matrix is presented in (6.18) and analyzed in this section.

6.6 CRYPTANALYSIS

Essentially, the strength of the cipher model needs to be analyzed for the vulnerabilities and breaches that pose a threat to the confidentiality and integrity of the cryptic messages. Cryptanalysis is an investigative methodology to gauge the robustness of the cryptosystem in the event of leaked cryptic procedures with unknown cryptographic keys(s) and the degree to which the attacker gains access to the enciphered messages. The desirable aspect is to keep intact the confidentiality of the cryptic key(s) and the information shared across the transmission channels. The popular attacks that are commonly exercised by the cyber criminals worldwide and articulated in the cryptographic literary sources [14] are enlisted hereunder:

- Ciphertext-only attack, aka the brute force attack
- Known-plaintext attack
- Chosen-plaintext attack
- Chosen-ciphertext attack
- Related-key attack

In the proposed and developed fast data set block cipher model, the cryptic key size and block size are considered to contain a larger number of binary bits. In this chapter, the objective is to demonstrate robust cryptic procedures to preserve the confidentiality and privacy of the sensitive information of the patients' ailments and healthcare, transmitted in the form of records among the stakeholders over public networks and/or stored in voluminous data sets, with less computational cost under resource-poor settings. As the raw data is principally considered in the .csv format and then encrypted with Key_Enc and Sup_Key, the latter two attacks, chosen-plaintext and chosen-ciphertext, can be leveled to the enciphered contents. The intuitive analysis for these two attacks, although can be presented. However, the cipher's sustainability must be analyzed against the initial two attack types, namely, ciphertext-only and known-plaintext [15].

The theoretical proofs for the former two attacks are presented here. For carrying out the ciphertext-only attack, the attacker gains access to the enciphered texts as well as the cryptic procedures used from which he attempts to retrieve the plaintext by decoding using various combinations and permutations of the keys from the key space. For speculating the original information and ascertaining the correct key(s) used in the cryptosystem, brute force technique is the most popular one which is widely used by the attackers [16]. In the generic discussion of the present cipher model, the size of the encryption key bunch matrix: Key_Enc is n^2

and each of the random odd integers chosen is in the range [1–255], thus making it 128 different ways. Besides, the *Sup_Key* is having the block size of n^2 with each decimal chosen in the range [0–255]. Thus, the entire key space of *Key_Enc* and *Sup_Key* matrices is

$$128^{n^2} \times 256^{n^2} = 2^{7n^2} \times 2^{8n^2} = 2^{15n^2} = \left(2^{10}\right)^{1.5n^2} \approx 10^{4.5n^2} \tag{6.22}$$

Assuming the time needed to execute the cryptic algorithm with a single value of the key space of (6.22) is 10^{-7} seconds, the total time required for execution using all keys of the (6.22) would be unassumingly and approximate to

$$\frac{10^{4.5n^2} \times 10^{-7}}{365 \times 24 \times 60 \times 60} = 3.17 \times 10^{4.5n^2-15} \; years. \tag{6.23}$$

The time required is directly proportional to the size of the block and the possibilities of the number of integers selected. With the n value chosen as 16 for the illustrative purpose, the value for the above (6.23) converges to 3.17×10^{1137} years, which is convincingly very large, thus downsizing the viability if the brute force approach.

In the second attack type, i.e., known-plaintext, the attacker is holding the limited number of pairs of the original data set contents, also termed as crib, and the corresponding cryptic messages. The task is to unravel the entire raw information derivable only after tracing the secret key(s). The robust cipher tends to provide minimalistic details to the attacker for finding the secret key(s) from the cryptic procedures and pairs of the cribs and the encrypted snippets available with him. In the present context, the sequence of sub-procedural tasks operating on the input elements and functions manipulating the binary bits, considered for analyzing a single iterative round in the enciphering procedure are mentioned hereunder:

$$P = \left(ke_{ij} \times p_{ij}\right) \bmod 256, 1 \le i \le n, 1 \le j \le n, \tag{6.24}$$

$$P = P \oplus Sup\_Key \tag{6.25}$$

$$P = mixBits\left(P\right), and \tag{6.26}$$

$$C = P \tag{6.27}$$

In the above case, the pair of $\left[p_{ij}\right]$ matrix of the right side of (6.24) and the corresponding block C on the left side of (6.27) are known. As the encryption key block $\left[ke_{ij}\right]$ of $n \times n$ size is unknown, the P on the left side of (6.24) is difficult to derive. By using the brute force strategy, the number of combinations to be tried for correctly identifying the encryption key vector used is approximately $10^{2.1n^2}$ with the time requirement of $3.17 \times 10^{537.6}$ years. On the other hand, considering C and using the inverse of the *mixBits*() procedure, the block P on the left side of the (6.25) can be determined. However, the elements of the Sup_Key being unknown, the exhaustive search method, if used, would require approximately $10^{2.4n^2}$ different combinations to conclude. The time required would be accordingly $3.17 \times 10^{599.4}$ years, for $n = 16$, which is formidably a long time. Either way, the total time required to break the cipher with one round of the iteration is very large. As the number of rounds is 16, the computing cost to unravel the keys used and the raw information is very high, which makes the cryptic procedure developed resistant to the known-plaintext attack.

In the further set of the two attacks: chosen-plaintext and chosen-ciphertext, the attacker gains access to the cryptic information of the arbitrarily chosen few plaintext records from the healthcare data set and plaintext records for the few randomly chosen enciphered information, respectively. The objective is to correctly speculate the complete raw information of the patients' medical health records from the available pairs of the cryptic and corresponding plaintext information and gain insight into the keys used for the information transfer confidentially over the public channel. There is no scope, by any means, to generate the correct healthcare data set or its enciphered blocks.

For performing the related-key attack, the attacker is in possession of the ciphertexts for two different key sets, although varying by exactly binary bit. On comparing the ciphertexts pairs (6.18) with (6.19) and (6.18) with (6.21), the variation which is recorded as the avalanche effect is certainly very large indicating the strength offered by the cryptic procedures.

Conclusively, the block cipher developed for encrypting the sensitive healthcare data in resource-poor settings is robust to withstand any of the above cited attacks.

6.7 CONCLUSIVE REMARKS AND FUTURE DIRECTIONS

A cryptosystem is developed in this chapter, especially to work efficiently and effectively in resource-poor settings, for preserving the security and privacy of the patients' sensitive information shared among the stakeholders as well as stored in voluminous and high dimensional data sets of the healthcare sector. The cryptic procedures developed perform the tasks using the encryption key block having a bunch of keys, an additional key vector, and multiplicative inverses. The procedures developed are appropriate for the low configured machinery of the legacy systems. To further improve upon the security, the size of the encryption key bunch block, additional key vector, and plaintext block can be further increased with very little change in the computing cost. A detailed investigation is presented in the Cryptanalysis section to establish the robustness of the cryptic procedures against the popular attacks by the manner in which the binary bits of the raw data are impacted in every cycle of the iteration. The binary bits of each character undergo transformations several times in the iterative process with the help of modular multiplication and XORed operations among the input blocks and matrices, thus resulting in the cryptic text being resistant to the cryptanalytic attacks. As the procedures are highly effective even in the low resource settings, the developed cryptosystem can be efficiently adapted to securely transmit the imagery and spatial data sets of healthcare as well as the other formatted information, with trivial change in the computing cost and pre-formatting performed of the raw information in the appropriate forms.

ACKNOWLEDGMENTS

Not applicable

FUNDING

This research was not funded.

CONTRIBUTIONS

The ideas presented in this manuscript are based on discussions of all authors. Shirisha and Geeta implemented the system. Shirisha and Narsinga Rao wrote the first draft of this manuscript. Raghavender Sharma and S. Phani Kumar revised the article. All authors reviewed and improved the manuscript. All authors have read and approved the final manuscript.

CORRESPONDING AUTHOR

Correspondence to Shirisha Kakarla.

ETHICS DECLARATIONS

Competing interests

The authors declare that they have no competing interests.

REFERENCES

1. Adler-Milstein J, DesRoches CM, Kralovec P et al. 2015. Electronic health record adoption in US hospitals: progress continues, but challenges persist. *Health Aff.* 34:2174–80.
2. Adler-Milstein J, Lin SC, Jha AK. 2016. The number of health information exchange efforts is declining, leaving the viability of broad clinical data exchange uncertain. *Health Aff.* 35:1278–85.
3. Atasoy H, Greenwood BN, McCullough JS. 2019. The digitization of patient care: A review of the effects of electronic health records on health care quality and utilization, *Annu Rev Public Health.* 40(1):487–500
4. Braa J, Hanseth O, Heywood A, et al. 2007. Developing health information systems in developing countries: the flexible standards strategy. *MIS Quarter.* 31:381–402.
5. Braa J, Sahay S. 2012. *Integrated Health Information Architecture: Power to the Users: Design, Development and Use.* New Delhi: Matrix Publishers.

6. Gebre-Mariam, M., & Fruijtier, E. 2017. Countering the 'dam effect': The case for architecture and governance in large scale developing country health information systems. *Inf Technol Develop.* 24(2): 1–26.
7. Mihailescu M, Mihailescu D, Schultze U. 2015. The generative mechanisms of healthcare digitalization. In Thirty Sixth International Conference on Information Systems. Fort Worth.
8. Manda TD 2015. Developing capacity for maintenance of HIS in the context of loosely coordinated project support arrangements. In IST-Africa Conference (pp. 1–10). 2015 IST-Africa Conference.
9. Mingers J, Standing C 2017. Why things happen–developing the critical realist view of causal mechanisms. *Inf Organ.* 27:171–189.
10. Kimaro HC, Nhampossa JL 2005. Analyzing the problem of unsustainable health information systems in less-developed economies: Case studies from Tanzania and Mozambique. *Inf Technol Develop.* 11:273–298.
11. Kakarla S. 2019. Securing large datasets involving fast-performing key bunch matrix block cipher. Healthcare Data Analytics and Management, Advances in Ubiquitous Sensing Applications for Healthcare, Elsevier Publications, Paperback ISBN: 9780128153680, Vol 2, 111–132, https://doi.org/10.1016/C2017-0-03245-7.
12. Shirisha K, Sastry VUK. 2013. A block cipher involving the elements of a key bunch matrix as powers of the plaintext elements. *Int J Comput Netw Secur.* 23(2):1192–1197, Recent Science Publications, ISSN: 2051-6878, USA.
13. Resources.data.gov. 2017 Accessed. Demographic statistics by zip code, Retrieved from https://catalog.data.gov/data set/demographic-statistics-by-zip-code-cfc9/resource/e43f1938-3c4a-4501-9aaf- 46891bb21553.
14. Stallings W. 2003. *Cryptography and Network Security: Principle and Practices.* 3rd Edition. New Delhi: Springer, 29–30.
15. Hill L. 1929. Cryptography in an algebraic alphabet. *Am. Math. Mon.* 36(6): 306–312.
16. Albanesius C. 2011. LulzSec on Hacks: 'We Find it Entertaining'. *PC Magazine.* Available at: https://news.yahoo.com/lulzsec-hacks-entertaining-132426237.html.

7 Sentiment Analysis of Scraped Consumer Reviews (SASCR) Using Parallel and Distributed Analytics Approaches on Big Data in Cloud Environment

Mahboob Alam, Mohd. Amjad, and Mohd. Amjad

CONTENTS

7.1 Introduction: Background and Driving Forces 121
7.2 Big Data 122
7.3 Methodology 122
7.4 Tool and Techniques 123
7.5 System Design 125
7.6 Technology Used 128
7.7 Result 128
7.8 Conclusion 129

7.1 INTRODUCTION: BACKGROUND AND DRIVING FORCES

Sentiment analysis or opinion mining is the process of deriving the opinion or the attitude of the speaker with respect to an entity. All the topics and events may be represented as an entity. These topics are most probably covered by reviews. Sentiment analysis determines whether a given review is positive, negative, or neutral. In practice, artificial intelligence is used to convert every word into a specific point in the feature space, and the distance between these points is used to identify the context of messages, which are similar to the concept we are exploring. The data sets used in implementation of sentiment analysis are important. Consumer reviews are extracted as data sets for analysis. These reviews serve as guidance to the manufacturers and

DOI: 10.1201/9781003323426-7

the merchants with respect to the customer's choices and their views related to the product. If the new customers get a glimpse of what and how the customers who had reviewed the products earlier had felt about the product, they can easily make better choices. As a result, the mistakes made by previous customers can be avoided adding to the fact that a lot of confusion and doubts regarding a product could be cleared.

Nowadays, all the leading e-commerce websites show results biased toward the products and their features rather than considering what users actually expect and how they react when the actual product is delivered to them. The focus of this paper is to semantically analyze all such customer reviews and evaluate the product thereafter. On the contrary, with respect to comparing a product against its features, it considers the reviews of the users for this purpose. Therefore, a global reputation score can be assigned to every product, thus emphasizing its degree of trust. One of the many ways to achieve this can be providing the user with a number of reviews and asking him/her to select one as per his/her preference. Thus, the review provided by the user can be analyzed on the basis of the user's sentiments associated with the product.

The history of paper is in the growing craze of sentiment analysis and online shopping. The motive behind this project was to combine both the technologies and generate a third one where online shopping can be even more simplified by using the aspects of the products from top websites. The technologies used earlier were not able to show the aspects of small products, which do not have a huge crowd of buyers. There were different websites, which focused on different products but there did not exist even a single platform for a subtle product, having a small number of buyers. This chapter not only focuses on small products but also considers those rare products which do not have a lot of online buyers. This initiative will decrease the time wasted by a consumer while focusing on small details such as price variation on different websites of a product or reviews found while reading [5][6]. Since everything is available online from crawling to collecting reviews, the processing speed also depends mostly on the internet connection and nothing else.

7.2 BIG DATA

Big data refers to endless sized data whose size varies from terabytes to petabytes. Most of the data generated today is big data. The five important V's of big data are shown in Figure 7.1. The five V's include velocity, volume, veracity, value, and variety. Here, we deal with a huge amount of unstructured data. Weather forecasting is one such field, which is very important in many areas such as agriculture, flight delay system, railways, disaster management, etc.

7.3 METHODOLOGY

The methods that are applied to collect the reviews for aspect-oriented opinion mining are broadly classified into two parts. In the first part, the server gets the information from the websites to get links to the related products available on different platforms. After getting access to all such products, we collect the links to our NoSQL database and then come to the second step where we go to each link and collect the reviews in the form of different files for each product. Our main task is to

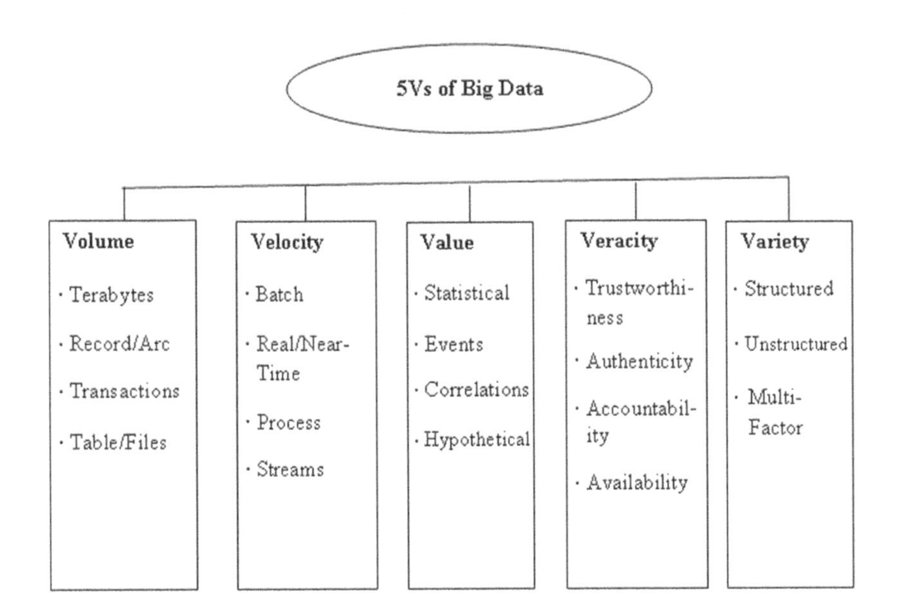

FIGURE 7.1 5V of Big Data.

collect features of a product entered by our user and according to our features sort the products and even sort our features according to the most prominent one, i.e., which has the maximum number of reviews related to it.

The methodology is rather complex for getting access to reviews on websites, as Flipkart requires technologies like phantoms that are headless browsers, which will help in our automated tasks. The methods for getting aspects require usage of natural language toolkits, which is a great choice for sentiment analysis and aspect mining of a given data set of reviews as shown in Figure 7.2, which gives the steps for aspect-oriented opinion mining. We classify each product on the basis of their aspects and further classify them on the basis of sentiment analysis of that particular aspect. Basically, we have four tuples, positive, negative, neutral, and total. The main motive is to promote the aspect that has the maximum total number of products in it and then focus on promoting the products that have the most positive reviews in it. The main purpose of our website is to ease shopping. Different products with subtle aspects are not discussed on the internet. This way we get aspects of almost anything with reviews without actually reading any reviews at all.

7.4 TOOL AND TECHNIQUES

We have used various tools like Hadoop, Hadoop Distributed File System (HDFS), and MapReduce for implementation of the given objective.

Hadoop: Apache developed Hadoop as an open-source framework and can be used to accumulate, practice, and evaluate a very vast volume of data. Its modules give an easy way to use languages, graphical interfaces, and organization tools for handling petabytes of data on thousands of computers.

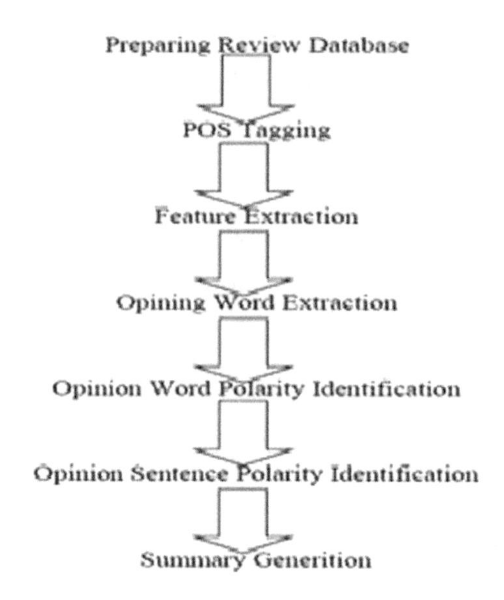

FIGURE 7.2 Steps for aspect-oriented opinion mining.

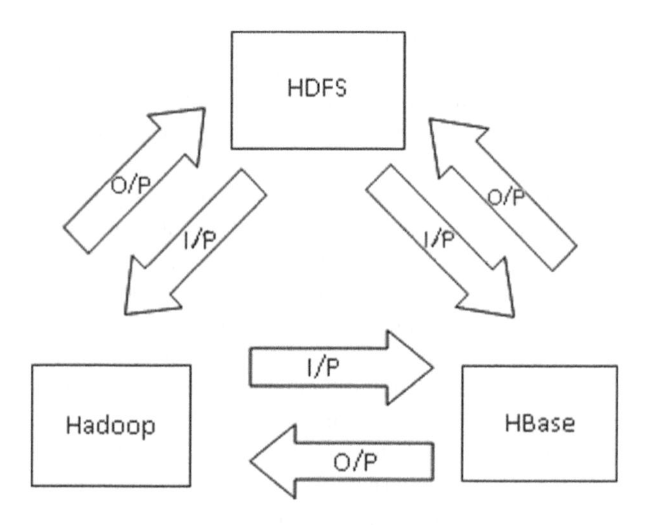

FIGURE 7.3 Relationship between Hadoop and HDFS.

HDFS: A file converts into blocks by using the Hadoop distributed file system. The distributed architecture carried these blocks in nodes. As HDFS has a master/slave architecture and we have a single name node in HDFS cluster, the file system namespace manages by a master server and also regulates access to files by clients. We can see in Figure 7.3 the relationship between Hadoop and HDFS.

FIGURE 7.4 MapReduce architecture.

MapReduce: Google has a patent for MapReduce, which it describes as a "system and method for efficient large-scale data processing" and is a real-world example of an innovation in collecting and analyzing data having been successfully patented. MapReduce is a framework and it supports the parallel computation in Java programs by using key-value pairs on huge data. The Map assignment converts input data into a data set and this data set can be computed by a key-value pair. To get the required result, the output of Map assignment is given to reduce assignment. It splits huge chunks of data. We can see the architecture of MapReduce in Figure 7.4.

Table 7.1 gives a comparative study of the available cloud providers with the type of service provided by each one of them.

Table 7.2 gives a brief description of the product types offered by various cloud service providers used for computing through cloud with the monthly cost of using them.

7.5 SYSTEM DESIGN

We have our basic workflow that we have followed to get the opinion mining of our test case of iPhone 5 reviews on amazon.com. Figure 7.5 gives the work flow diagram that our test case follows wherein first data scraping takes place, then data is cleaned, and processed in the form of a CSV file.

The processed data is stored in the database, which is then visualized. Figure 7.6 gives the detailed block diagram showing each task in detail.

TABLE 7.1
Comparison of Analytical Tools in Big Data Analysis (BDA)

Available cloud providers	Amazon AWS	Cloudera	Microsoft Azure
Analytic Data Base	Relational DBMS Service	Hbase	SQL Server 2012
Memory Data Base	No-Third Parties Options	Apache Spark	SQL Server Memory OLTP
Hadoop Distribution	Elastic MapReduce	CDH, Cloudera Standard	HDInsight
Software and H/W Systems	Not Applicable	Appliance Providers: Cisco and Dell	Dell Parallel DW, HP Enterprise Parallel D W

TABLE 7.2
Cost Comparison among Amazon, Cloudera, and MS Azure

Product Types	Amazon	Cloudera	Microsoft Azure
SQL Server License	0	Not available	0
	On-Demand SQL Server RDS: $495	Not available	P1 Instance (125 DTUS): $465
	10 EC2 Small Instances: $264	Not available	10 STD A1 Instances
	2 EC2 Medium: $105	Not available	2 STD A2 Instances $268
	500GB Tx Outbound to Internet: $45	Not available	500GB Tx Outbound to Internet: $30
	Elastic Loader Balancer: $19	Not available	Load Balancer: $0
	SQL Server 2012 license: $0		SQL Server 2012 license: $0
	1 EC2 Small Instance: $ 26		1 Standard A1 Instance: $15
	1 GB Tx Outbound to Internet: $0		1 GB Tx Outbound to Internet: $0
Total Monthly Cost	$954	$2600 Yearly or $650 Monthly	$928

The steps are as follows:

Input from user: Input is taken from the user on our website.
Check if input in the database: If there is data in the database related to the mentioned input, then it is accessed otherwise it is scraped and posted in the database.
If not then perform the following tasks:

- Scraping products from websites
- Storing in NoSQL database

Thus, data is collected in the database.

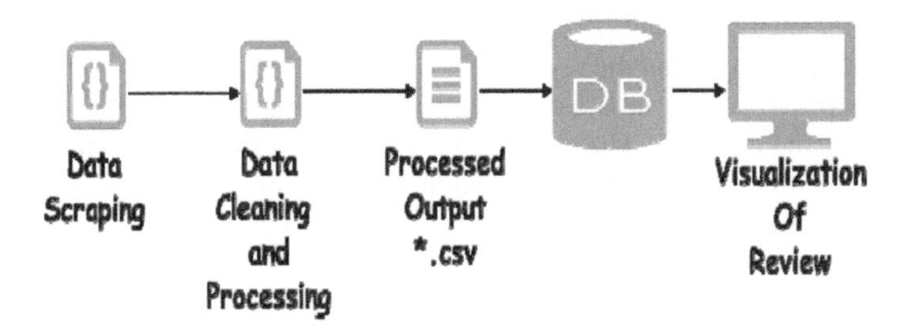

FIGURE 7.5 Basic computations.

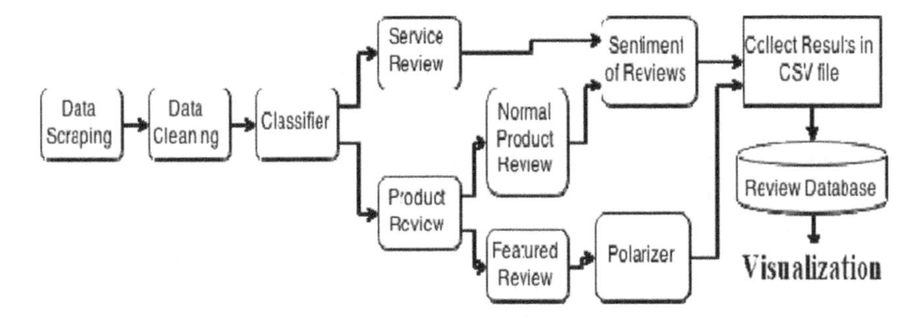

FIGURE 7.6 Detailed block diagram of the system.

Get data of related products from the database: Data from the database is collected, which is required to access each product. This data includes from names of products to their races and their links.

Web scraping: In this step, each website is scraped for reviews. This is the most complex step as when scraping is performed, in some cases the reviews are not readily available. Hence, we have to use various technologies like phantoms to automate the process. Crawling of websites for each product and then scraping their names, links, images, reviews, and prices are done. For each product, these are stored in JSON key-value format. Scraping is performed using multiple technologies in this project and in an asynchronous form.

This step requires that the crawling should be done in the fastest time span so that the result is retrieved fast. Generation of sentiment for each aspect is done in tuple format. The tuple consists of positive, negative, neutral, and sum of all aspects.

Analysis of the data: This step involves the aspect-oriented opinion mining of our reviews. Here, we majorly use python libraries for this purpose. The data that is received is cleaned and reprocessed before performing sentiment analysis and aspect mining of them. While cleaning it is ensured that all the stop words are removed and crucial aspects are given more priority while sorting the products.

Sentiment analysis is done by creating two functions majorly. The first function is to tokenize words and find aspects and each aspect then calls a function on its wordnet created from its reviews. This function is to practice sentiment analysis on the reviews.

Send output: *The* output is sent to the server and thus displayed in the form of rows and columns. The output is sent to users in the JSON format so that all the arrays can be easily displayed in a tabular form. The server sends the output to the page using angular.

7.6 TECHNOLOGY USED

AngularJS: We have used angular to link our server backend with the front end. Angular is a very useful tool which can be used for building dynamic websites.

Cascading Style Sheets (CSS): We use it to add more styling to our website. It enhances the user's experience by adding different graphical features in our website.

Node.js: Node.js is used to build the server of our website. It is used for its very useful CORS feature. CORS, which stands for cross origin resource sharing, helps in building our website on a single controller and we basically use get operation multiple times for the same post operation. Once the data is stored in the database only get operation is performed on it.

Python: We use packages like NLTK and WordNet for sentiment analysis of the aspects from product reviews. Python dictionary feature has been utilized to ensure that the data which is in tuple format is easily accessible and understandable.

JSON: The data that is stored in the database is the key-value format provided by JSON primarily.

PhantomJS: It is a headless browser which helps in automating the process even under a secure environment of websites. It helps in getting reviews from websites with a lot of security.

7.7 RESULT

While implementing the idea as described in this paper, we aimed at providing the customer with the most beneficial results on the basis of reviews of the customers and the aspects of the products as highlighted by them in their reviews.

We have implemented the proposed system and provided below the screenshots, which show the outputs of the system. The user can add a product to cart and buy it. Our system presents the results in the form of bar charts and pie charts and it classifies the reviews by performing a sentiment analysis on it.

Users can view the result and directly understand the sentiments extracted.

In Figure 7.7, the value 372, indicates that original sentiments are not only of the products but also of the service.

In Figure 7.8, we can see the overall sentiment that is in the form of negative, positive, and mixed for all reviews. Every column represents a numerical value to tell us the extent to which a review is positive or negative.

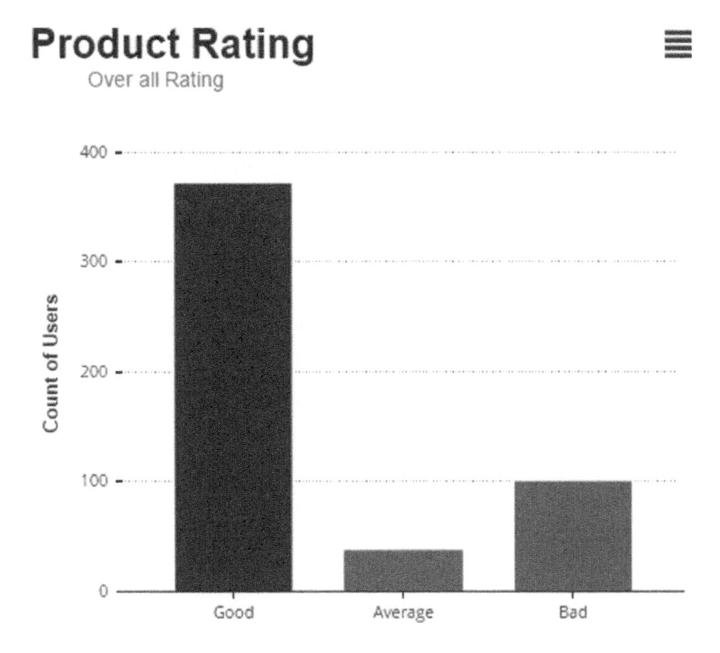

FIGURE 7.7 Product rating.

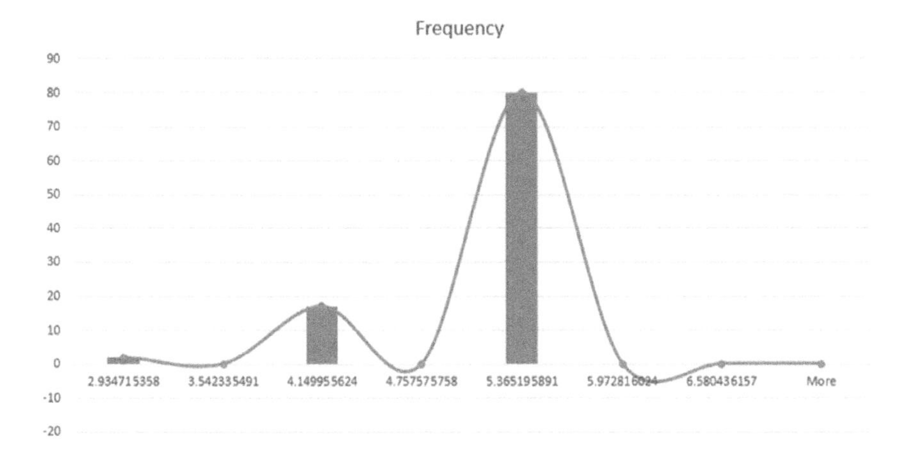

FIGURE 7.8 Overall sentiment analysis.

7.8 CONCLUSION

This chapter, gives a description, shows the implementation done, and puts forward a working model for extraction of customer reviews related to the product. This will further help them in buying the best product. Classification of products on the basis of polarity is also done in this paper. With respect to sentiment analysis, our methodology correctly integrated with the existing approaches of sentiment analysis.

Furthermore, classifying reviews on the basis of a sentiment approach further adds to the users' ease of selecting the right product and the place to buy it from.

Sentiment analysis empowers all types of competitive analysis and market research. Sentiment analysis will build all the distinctions, when you are anticipating future trends, exploring a replacement market, or keeping a position on the competition.

Future prospects include the following things:

(1) Research of the segmentation analysis using parallel and distributed analytics approaches on big data using the cloud environment.
(2) The structure optimization of the segmentation analysis using parallel and distributed analytics approaches on big data.

8 The UAV-Assisted Wireless Ad hoc Network

Mohd Asim Sayeed, Raj Shree, and Mohd Waris Khan

CONTENTS

8.1 Introduction 131
8.2 Dynamic Mobility Inclusion in a Routing Protocol 132
8.3 UAV Placement and Ground Node Grouping 133
8.4 Data Dissemination 133
8.5 Trajectory Optimization 133
8.6 Discussion 146
8.7 Summary 154
References 154

8.1 INTRODUCTION

The unmanned aerial vehicle (UAV)-assisted wireless ad hoc network has opened a new paradigm for enhanced wireless connectivity and coverage. The collaborative network formations between ground and aerial nodes opens a wide dimension for versatile and economical solutions for wireless network deployment. The degree of maneuverability, fine control, and the ability to be placed as and when required has helped researchers to design optimized capacity, coverage, service availability, and scheduling solutions. A UAV can be used as aerial base stations for sensor networks and Internet of things (IoT) devices, user equipment's and evolved NodeBs for cellular networks, wireless relays, and gateways for wireless ad hoc networks, providing coverage solutions when infrastructure is not available [1–4]. The UAV can take care of the data transmission needs of a wireless network by boosting the capacity and enhancing the coverage. The fundamental aspects in the design of a collaborative wireless network are data scheduling, UAV placement, trajectory mapping, and making the ground nodes aware of UAV's presence [1].

One of the benefits of being a wireless network is that nodes can move around in the topology, the nodes will fail to be localized around a specific point or coordinate. Using a UAV to provide wireless network services requires careful topological decisions including aerial node placements, aerial node mobility and association with the ground nodes, and data dissemination strategies. The most basic of the mobility patterns that an aerial node can adopt are static and dynamic mobility schemes. This paper discusses the strategies of using UAV-assisted wireless networks including their usage as service providers, mobile relays, mobile base stations, and routing

DOI: 10.1201/9781003323426-8

paradigms that offer inclusion of high-speed UAV nodes into a topology. The strategies of using a UAV-assisted wireless network can be broadly classified into:

- Routing strategies.
- UAV placement and node grouping strategies.
- Data dissemination and scheduling strategies.
- Trajectory optimization.

8.2 DYNAMIC MOBILITY INCLUSION IN A ROUTING PROTOCOL

Wireless ad hoc networks are made up of wireless nodes capable of calculating routes in a topology like a router. The route calculation by the node is done using a routing protocol. Routing protocols help a node in neighbor discovery, topology formations, and data packet transmission. The most common of the routing protocols are AODV [5], OLSR [6], DSDV [7], ZRP [8], DSR [9], etc. A routing protocol initializes the wireless node and transmits control packets (for example, an OLSR hello packet) for neighbor discovery. When a neighboring node responds to the initialization packet, the source recognizes the neighbor and adds an entry to its routing table. The routing table is used for topology formation and packet forwarding. Neighbor discovery and routing table calculations are perhaps the most important steps for a wireless node before any kind of data transmission can commence. The routing protocol can be proactive, that is, it can calculate all the routes from a node to every other node in the topology in advance, and it can be reactive in the sense that a route is calculated whenever data packets need to be sent. Some literature studies differentiate between the two in the sense that it is table-driven or not. However, every protocol needs to maintain the routing table for faster route lookup. A reactive protocol, however, updates its table on demand. A third kind of routing protocol exists for this reason. The hybrid protocols imitate the functioning of both proactive and reactive routing protocols.

A wireless ad hoc network routing provides for the mechanism for link discovery and maintenance of routes to and from all the nodes in the topology. When the speed of the participating node increases, the wireless links are broken frequently and new links are formed. A wireless ad hoc network deals with these fluctuations by transmitting control packets such as route request and hello packets. When new links are discovered the routing table is promptly updated and broken links are removed. With a high degree of mobility, the traditional routing barely copes, as an efficient handover mechanism does not exist. A recent trend in predictive routing protocols tries to handle this situation by using a predictive scheme. The wireless ad hoc network nodes take into account the position and velocity of a destination node to predict the route from source to destination. Predictive routing techniques are most suited routing techniques for using a UAV as a mobile base station (MB) or a mobile relay (MR).

Dynamic and predictive routing has been proposed in the literature for efficiently using UAV as a MB and a MR to enhance the overall network performance. These protocols map the UAV positioning and mobility as a function of time for seamless packet forwarding. Link quality and stability with strategic placement and deployment of UAV are the focal point of these approaches. Existing routing-based approaches for UAV-incorporated wireless ad hoc networks are listed in Table 8.1.

8.3 UAV PLACEMENT AND GROUND NODE GROUPING

The UAV can provide service to ground nodes in a topology. The topology must have techniques for assimilating the overhead aerial node. The topology is divided into smaller sections, which a UAV can serve efficiently. Unlike cluster-based routing protocols, the sections of a topology are formed on the basis of the number of nodes that can be served by a UAV in an area. The ground topology sections are primarily formed on the basis of the antenna beam width of the UAV. In a single UAV architecture, the UAV moves from one cluster/area to another. In a multi-UAV architecture, joint path formulation is required.

Communication between ground nodes and aerial nodes requires nodes to be assigned to the service-providing aerial node or relay. An aerial node placement and ground topological division serve the following purposes:

- Ground node coverage.
- Optimal multi-route data flow solutions.
- Conceptual view of the topology.
- Resource allocation.
- Ground network lifetime enhancement.

The existing approaches for ground node positioning and clustering for the UAV-assisted wireless ad hoc network are listed in Table 8.2.

8.4 DATA DISSEMINATION

A UAV-assisted ground network can utilize the UAV's line-of-sight connectivity and versatile mobility for increasing its capacity for data communication. A UAV can act as a mobile base station to relay the data flow from the ground network to a backbone or an ISP. Using a UAV as a mobile base station or a relay comes with a set of tasks cutout for the designer. A scheme of data dissemination works with the collaboration of the two networks, namely, aerial and ground. This collaboration generates multi-route solutions for data flow. Nodes are associated with the UAVs that service them and a trajectory or flight path designed for performance goals. In some work, priorities are assigned to different areas of the topology and or nodes. Data flows are prioritized for a node or set of nodes with time-sensitive data and energy efficiency [45–71]. The existing techniques for data dissemination in UAV-assisted ground networks are given in Table 8.3.

8.5 TRAJECTORY OPTIMIZATION

A UAV is free to move in a three-dimensional plane. Designing a path in the 3D plane with reference to ground nodes requires that the UAV can move parallel to

TABLE 8.1
Routing-Based Approach

Author, Year	Focus	Features/Findings, Advantages and Limitations
Xie, 2019 [10]	Ocean FANETs. Enhanced OLSR.	Incorporates rapid changes in the network topology, utilizing GPS co-ordinates and link expiration constructs. Network performance enhancement with low routing overhead. Adaptive solution for rapid changes in the topology. Scalability and coverage issues.
Chen et al., 2018 [11]	Multi-UAV-aided wireless network.	UAVs as mobile relays or base stations. Mutli-UAV placement optimization technique. LOS coverage with multi-UAVs. Quality of service and interference management. Limited experimental setting of four nodes.
Choi et al., 2018 [12]	Flying ad hoc networks. Position-based routing protocol.	Geolocation for calculating and maintaining routes using neighboring node information. Low routing overhead. Large node count in a vicinity is overlooked. Multi-hop routing supports limited QoS.
Khelifi et al., 2018 [13]	UAV ad hoc network. Data dissemination and energy efficiency.	Weighted centroid cluster-based routing protocol. UAV co-ordinates are predicted using fuzzy logic. Network performance gains in terms of data transmission and energy consumption. Unoptimized coverage and data dissemination.
Pu, 2018 [14]	UAV ad hoc network. Wireless link quality and load aware OLSR (LTA-OLSR).	Routing protocol with wireless link and data traffic-based optimizations in OLSR. Wireless link quality estimates using the radio range of the nodes. Data traffic latency estimations for load balancing and higher throughput. Multi-hop paths introduce delays. High routing overhead. Coverage issues.

(*Continued*)

TABLE 8.1 (CONTINUED)
Routing-Based Approach

Author, Year	Focus	Features/Findings, Advantages and Limitations
Song et al., 2018 [15]	UAV ad hoc network Mobility and packet latency prediction enhanced OLSR (OLSR-PMD).	Kalman filter for mobility prediction and Multiple Provider Router (MPR) selection based on stability. Data traffic load balancing using routing metrics. Performance gain in terms of reduced latency and higher packet delivery ratio (PDR). Multi-hop paths introduce delays. High routing overhead. Coverage issues.
Dong, 2017 [16]	UAV ad hoc network. MPEAOLSR.	MPR selection procedures consider link congestion and node energy levels. Reduces packet loss and end-to-end delay over traditional OLSR. High routing overhead. Coverage issues.
Gankhuyag et al., 2017 [17]	Flying ad hoc network.	Combined use of omni-directional and directional antenna with node position prediction to enable long range transmission using directional antenna. Network performance improvements. Topological changes are monitored. High routing overhead. Coverage issues. QoS not insured.
Li and Yan, 2017 [18]	Flying ad hoc network Link stability estimation preemptive routing.	Based on AODV and uses UAV co-ordinates to compute link stability and mobility predictions. Multiple paths are calculated and switched in case of broken links toward more reliable routes. Performance gains over AODV and DSR. High routing overhead. Multi-hop and preemptive routing introduce end-to-end delays in data flows.
Oubbati et al., 2017 [19]	UAV-assisted VANET Routing Protocol (UVAR).	Grid location and node association, node density estimates and link detections. Packet forwarding using the greedy algorithm. Low latency, high PDR, and high throughput. Unoptimized coverage and interference management. Multi-hop and preemptive routing.

(Continued)

TABLE 8.1 (CONTINUED)
Routing-Based Approach

Author, Year	Focus	Features/Findings, Advantages and Limitations
Rovira-Sugranes and Razi, 2017 [20]	UAV ad hoc networks. Predictive routing.	Predictive routing scheme with estimation of location for destination and next hops. Reduced packet delay in large networks of higher speeds. Multi-hop and predictive routing.
Lee et al., 2016 [21]	UAV network.	Multi-UAV routing protocol GCS-routing. Relies on geographical and topological information. Throughput maximization and centralized management. Multi-hop routing. Unoptimized congestion management.
Rosati et al., 2016 [22]	UAV ad hoc networks P-OLSR as an extension for OLSR.	Prediction of link quality evolutions with expected transmission count metric. Hello messages of OLSR are used to share node co-ordinates for three-dimensional position vectors. Reliable routing in a highly dynamic topology. High throughput. Multi-hop predictive routing introduces end-to-end delays and packet drops in data flows.
Biomo et al., 2014 [23]	UAV ad hoc network routing. Reactive-greedy-reactive protocol.	Reactive-greedy-reactive routing protocol with route reliability and stability restrictions. Network performance improvements. Congestion management. Multi-hop routing introduces end-to-end delays and packet drops in data flows.
Zheng et al., 2014 [24]	UAV ad hoc networks Mobility and load aware OLSR.	Link stability analysis for stable node calculation and reachability of other nodes from stable nodes, and packet queue are taken into consideration for multipoint relay node calculations. Higher PDR. Network performance improvement and congestion management. Multi-hop routing introduces end-to-end delays and packet drops in data flows.

(Continued)

TABLE 8.1 (CONTINUED)
Routing-Based Approach

Author, Year	Focus	Features/Findings, Advantages and Limitations
Rosati et al., 2013 [25]	UAV ad hoc networks. Reliable Multi-hop communications.	Link quality estimation with the use of GPS and relative speed of the nodes. Prediction of link quality and broken links to improve overall routing capability. Reduced packet loss rate. Multi-hop routing introduces end-to-end delays and packet drops in data flows. Unoptimized coverage.
Lin et al., 2012 [26]	UAV ad hoc network.	Probability density function for prediction of UAV positioning and mobility for finding routes. Minimized control traffic and reliable routing. Multi-hop routing. Unoptimized coverage and congestion management.
Alshabtat and Dong, 2011 [27]	UAV ad hoc networks. Low latency routing. Directional Antenna.	New routing techniques DOLSR based on OLSR. MPR count minimization in a topology. Reduced control packets and end-to-end delay. High throughput. Coverage issues and QoS is not guaranteed.
Hunjet et al., 2010 [28]	Mobile ad hoc network capacity and lifetime enhancement.	Strategic node placement and control node operations using swarm optimization. Interference reduction and reduced power consumptions. Coverage not insured for all nodes. Unoptimized congestion handling.
Cheng et al., 2007 [29]	UAV-relay network for scarce ground wireless network.	Optimized for delay tolerant applications. Data packets are picked up by a UAV and then delivered to their destinations. Delay tolerant, higher coverage, and PDR. QoS not insured and lower net throughput.
Rubin and Zhang, 2007 [30]	UAV-aided mobile ad hoc wireless network.	UAVs are placed as a relay to serve ground wireless ad hoc networks. Optimized three-dimensional placements yield significant gains in network capacity. Unoptimized congestion handling.

TABLE 8.2
Node Association and Cluster-based Approach

Author, Year	Focus	Features/Findings, Advantages and Limitations
Bouhamed et al. 2020 [31]	UAV-assisted WSN Trajectory and scheduling.	Learning algorithms to train the UAV about topology. Energy efficiency and throughput maximization. Minimum congestion and coverage optimization.
L. Wang, Hu, and Chen 2020 [32]	UAV-BS energy efficiency	UAV placement based on convex approximation techniques. Transmit power minimization. Unoptimized data dissemination.
Lai, Chen, and Wang 2019 [33]	UAV placement in a 3D plane.	On-demand UAV base station placement. The algorithm considers the circle placement problem and optimizes UAV placement as a knapsack problem. Maximized ground node coverage. Unoptimized congestion control and uncertain fairness.
Qi et al., 2019 [34]	Flying ad hoc sensor network. Cluster-based routing.	Clustering using a software-defined networking (SDN) controller for hierarchy management. Flows are assigned weight in terms of latency, reliability, and QoS. Higher throughput and PDR, lower latency. Uncertain fairness and higher routing overhead.
Aadil et al., 2018 [35]	Flying ad hoc networks. Energy-efficient clustering.	Optimized wireless coverage. Density-based clustering and cluster head selection. Enhanced network life time and packet delivery ratio. Mobility model puts constraint on the cluster head. Incurs higher cluster maintenance cost with high mobility.
Alzenad et al. 2017 [36]	UAV-assisted ground wireless network 3D UAV placement.	UAVs acting as a base station. Optimal placement with altitude and position adjustment of the UAV. Circle placement algorithms to maximize nodes served and minimizes energy requirements. Energy efficiency and optimal coverage. Unoptimized congestion control.
J. Wang et al. 2017 [37]	UAV-assisted WSN clustering	Cluster division using PSO. Nodes are clustered based on distance and energy. Network performance enhancements. Cluster head incurs the cost of higher mobility patterns.

(*Continued*)

TABLE 8.2 (CONTINUED)
Node Association and Cluster-based Approach

Author, Year	Focus	Features/Findings, Advantages and Limitations
Lyu et al. 2017 [38]	UAV-MBS for ad hoc network.	Algorithm for placement of mobile base stations. Minimizing the number of UAVs required to service an area. Coverage maximization Unoptimized congestion control. Latency.
Mozaffari et al. 2016 [39]	UAV-assisted ground wireless network.	UAV placement determined using signal width. Circle packing for UAV placement and altitude adjustment. Studied minimum UAV count requirements for coverage of a topology. Optimal maximized coverage and lifetime. Unoptimized congestion control and higher latency.
Kalantari, Yanikomeroglu, and Yongacoglu 2016 [40]	UAV base stations for cellular network QoS	Optimal minimal count of UAV for topology coverage. PSO-based 3D placement of UAV-BS. Coverage optimization. Unoptimized congestion control and higher latency.
Bor-Yaliniz, El-Keyi, and Yanikomeroglu 2016 [41]	UAV-BS for cellular networks.	UAV placement to maximize the number of users served while keeping the search area to a minimum. QoS. Unoptimized coverage solutions.
Shu et al., 2011 [42]	UAV ad hoc network. Predictive cluster-based routing.	Well-connected UAV nodes are deemed cluster heads for cluster stabilization. Ferry nodes provide coverage to isolated sections of the topology. Higher throughput and PDR. No provision of congestion control. Uncertainty in fairness.
Zang and Zang, 2011 [43]	UAV ad hoc network. Clustering.	Clustering based on mobility predictions and duration of transmission link. On-demand clustering. Network stability and performance improvements. Unoptimized congestion control. Latency issues. Uncertainty in fairness.
Fu and DaSilva, 2007 [44]	UAV wireless mesh network (WMN). Cluster-based routing protocol.	UAV form an airborne WMN and serve as cluster heads. Self-organizing hierarchical routing structure. Disruption tolerant wireless link stabilization. Unoptimized congestion control. Higher latency. Uncertainty in fairness.

TABLE 8.3
Data Dissemination Techniques

Author, Year	Focus	Features/Findings, Advantages and Limitations
Sharma et al., 2018 [45]	UAV guided ad hoc network.	Routing protocol designed for congestion-free efficient routing. Congestion management. Multi-hop.
Wu et al., 2017 [46]	Data processing and dissemination technique for drone swarms.	Reinforcement learning-based data dissemination. Knowledge sharing and management. Adaptive processing and dissemination. Congestion management issues.
Chandhar, Danev and Larsson, 2017 [47]	Drone swarm data communication with the ground station. Multi-antenna base station.	MIMO systems at the base station with a line-of-sight connectivity with UAV swarm. Increased network capacity. Congestion management issues.
Sanchez-Garcia et al. 2016 [48]	Data dissemination. Deployment	Wireless network optimization for providing services to rescuers and victims in a disaster scenario. The Jaccard distance for checking service efficiency. AI algorithms, hill-climbing, and simulated annealing for optimizing nodes served. UAVs act as a service provider. Mobility model for ground node movements. Congestion management and resource allocation issues.
G Reina et al. 2016 [49]	Data dissemination Delay-tolerant network	A multi-objective genetic algorithm for optimization and tree-based decision-making for performance improvement. Probabilistic data dissemination algorithm. Optimizations in terms of packet delivery and latency. Delay tolerant wireless network only.
Sharma, Beenis, and Kumar 2016 [50]	Capacity enhancements Heterogeneous network.	On-demand UAV assignment over a topology. Neural network-based model for UAV association. Gains in spectral efficiency and delays. Congestion management and resource allocation issues.

(*Continued*)

TABLE 8.3 (CONTINUED)
Data Dissemination Techniques

Author, Year	Focus	Features/Findings, Advantages and Limitations
Say et al. 2016 [51]	Data acquisition in WSN. Throughput maximization.	Topology is divided into frames. Priorities are assigned to nodes in frames. Modified IEEE 802.11 MAC priorities, nodes with higher priorities get preference in data transmission. Minimized packet loss and transmission distance. Optimized energy consumption. Uncertain fairness in priority assignment.
Zheng et al., 2016 [52]	Data dissemination in WSN.	Bulk data dissemination with negotiation and flooding. Survival of the fittest algorithm for time efficiency and selection of the fittest between negotiation and flooding for data dissemination. Maximized data rate. Multi-hop.
Kim, Lim, and Krishnamachari, 2016 [53]	Data dissemination in a vehicular cloud system.	Data prefetching technique. Based on a greedy and online learning algorithm. Data dissemination from cloud to RSU to vehicle. Efficient data dissemination rates. Multi-hop.
Rosati et al., 2016 [22]	FANET routing protocol.	OLSR-based predictive OLSR. It relies on GPS information metrics. Improved network performance. Multi-hop.
Sharma, You, and Kumar, 2016 [54]	Data dissemination in a UAV-coordinated WSN. Energy efficiency.	Attraction factor is assigned to nodes based on node's energy levels. The technique shows performance gains in terms of throughput, coverage, and network lifetime. Congestion management issues. Multi-hop.
Wichmann and Korkmaz, 2015 [55]	Wireless multimedia sensor network. Latency minimization. Path formulation.	UAVs as relays and gateways. Path formulation based on the traveling salesman problem. Maximized throughput and lower energy consumption. Further improvements in the data rate using multiple UAVs. Time slotted data collection. Uncertain resource allocation.

(Continued)

TABLE 8.3 (CONTINUED)
Data Dissemination Techniques

Author, Year	Focus	Features/Findings, Advantages and Limitations
Tunca et al., 2015 [56]	WSN routing protocol. Mobile relays.	Energy-efficient routing protocol utilizing mobile sinks. Data acquisition is made with the help of mobile sinks. Energy efficiency. Multi-hop. Not extensible for more than one sink.
Zhu et al., 2015 [57]	Data dissemination in the WSN. Energy efficiency.	Target-specified bulk data dissemination. Low latency and contention-free. Uncertain congestion management. Multi-hop.
Temel and Bekmezci, 2015 [58]	MAC protocol for FANET.	Directional antennas. Probing and data transmission phases. Gains in throughput, lower delays, and fairness. Directional antenna.
Yan, Zhang, and Wang, 2014 [59]	Data dissemination in VANET. Latency minimization	It aims at controlling a set of receivers in an area. The algorithm is based on process scheduling algorithms. Delay and communication overhead minimization. Higher latency. Multi-hop.
Shen et al., 2014 [60]	Data dissemination in VANET.	Data dissemination scheduling. Reduced latency. Multi-hop.
Ciobanu et al., 2014 [61]	Throughput maximization in delay tolerant network.	Probabilistic forwarding. The algorithm is based on the Jaccard distance and history. High reachability throughput. Only for delay tolerant network.
Sharma and Kumar, 2014 [62]	Cooperative ad hoc network.	Maps of topological area for searching and tracking. Bayesian Kalman filters for position estimates. Maps provide waypoints for UAV mobility. Multi-hop.
Ros and Ruiz, 2013 [63]	Data dissemination in VANET.	Modeled data dissemination to the number of transmissions. Network performance gains. Multi-hop

(*Continued*)

TABLE 8.3 (CONTINUED)
Data Dissemination Techniques

Author, Year	Focus	Features/Findings, Advantages and Limitations
Cai et al., 2013 [64]	QoS for MAC in FANET. Full-duplex radio and multi-packet reception.	MPR and full-duplex radio MAC scheme for dynamic mobility of FANET. Optimization for channel state information. Network updates are done with token passing. Better performance with a half-duplex radio and significant gains with a full-duplex radio. Multi-hop
Alasmary and Zhuang, 2012 [65]	Impact of mobility VANET	Performance analysis of IEEE802.11p MAC. Contention window method for boosting network performance. Priority-based service. Multi-hop.
Ho and Shimamoto, 2011 [66]	WSN-UAV communication protocol.	Sensors are divided into groups and assigned priority. Sensors communicate directly with the UAVs. Priorities are assigned using PFS. Uncertain resource allocation.
Ding and Xiao, 2010 [67]	Data dissemination in VANET.	Multi-path routing scheme with static nodes as relays. Static nodes are placed at the intersection for relaying data packets and link formation. Latency minimization. Static nodes can store the packets for the time when a route is made available. Multi-hop.
Zhao, Zhang, and Cao, 2007 [68]	Data dissemination in VANET.	Periodic data broadcasting, buffering, and rebroadcasting. Data pouring method for capacity maximization. Multi-hop.
Eichler, 2007 [69]	Performance of IEEE802.11p WAVE.	Packet collision is minimized by increasing the contention window, which results in some throughput degradation. Shape the traffic to reduce priority messages. Congestion management issues.

(*Continued*)

TABLE 8.3 (CONTINUED)
Data Dissemination Techniques

Author, Year	Focus	Features/Findings, Advantages and Limitations
Chen et al., 2006 [70]	QoS IEEE802.11e.	Rate control based on channel estimates and average delay. Maximizing channel utilization. Traffic regulation.
Palat, Annamalai and Reed, 2005 [71]	UAV swarm WSN MIMO	Studied MIMO antennas and beamforming communication with UAV-assisted networks. Performance increases in range and reliability. Congestion management issues.

the ground for waypoint selections and adjust its height for various application scenarios. Figure 8.1 describes the static mobility pattern when using a single UAV to provide coverage to the ground ad hoc network. The UAV represented by the grid in a sequential path. The grid spacing is essentially defined by the antenna coverage of the UAV on ground. The circle formed by the antenna coverage can be viewed as moving from one grid section to another.

All the nodes that fall inside these overlapping circles will essentially have UAV assistance and can transmit their data via the UAV to a backbone network, base station, or other nodes on the ground topology using multiple UAVs. Ground nodes require some form of control packet dissemination to form new routes using UAVs.

Another such approach of providing UAV assistance to the ground ad hoc network is by a closed geometric flight plan such as a hexagon or a circle. Figure 8.2 describes the UAV flight path in a circular fashion. Following a closed geometric flight path enables the UAV to provide coverage to the topology with some major benefits:

i. **Recurring coverage.** A closed geometric trajectory plan ensures that the UAV will provide services at regular intervals. This ensures fairness. A multi-UAV system can use multiple UAVs to shorten the waiting periods for the ground nodes and guarantee access to the services provided by the UAV.
ii. **Multi-UAV usage for data dissemination.** A multi-UAV system can provide direct routes from one side of the topology to another using an aerial route via multiple UAV from source to destination, thereby reducing the hop count. This mechanism also saves node energy spent in packet hopping.
iii. **Means to implement sleep timers more effectively in a wireless sensor network (WSN).** For a UAV-assisted WSN, the UAV will be in range at regular intervals. This is more feasible for designing sleep timers and efficient medium access control (MAC) layer designs for energy efficiency.

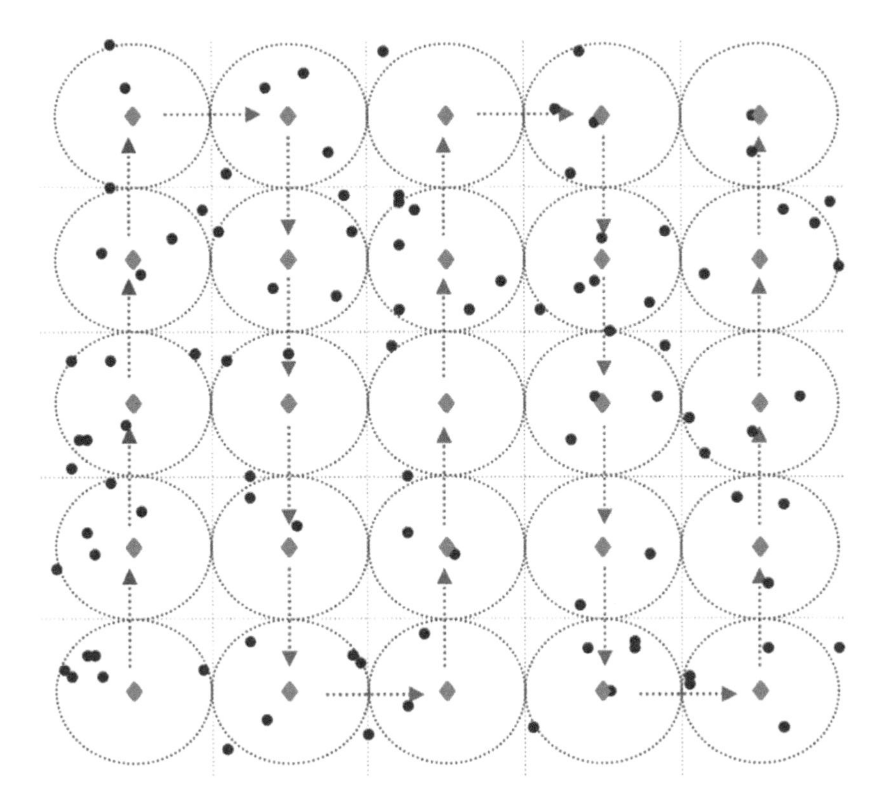

FIGURE 8.1 UAV-Assisted Wireless Ad hoc Network with a Static UAV Path in a Grid.

Figure 8.3 describes a scenario where multiple UAVs can provide services over a large topology using multiple closed geometric shaped flight paths. Another placement and path formulation technique employs calculating the UAV placement and path selection dynamically at runtime. Dynamic UAV placement and path selection techniques make the path selection in real time. After an initial placement and path formulation, the system allows for rapid changes in the UAV trajectory to serve the ground nodes more efficiently. This model makes path decisions based on mission parameters. The model can be designed to optimize ground wireless ad hoc network operations and help alleviate inherent problems in the wireless ad hoc network.

Figure 8.4 describes the dynamic flight plan selection scenario. Figure 8.2 with dots nodes are UAVs, black dots are ground wireless ad hoc nodes, and the blue circle defines the area the UAV antenna beam covers. It is clearly evident that multiple decisions can be made in this scenario as the UAV is freely deployable in all directions.

The ease of mobility makes it ideal for providing service to the ground wireless network. The ease of mobility in a 3D plane makes a UAV suitable for line-of-sight communications. UAV path formulation to service ground nodes requires multi-parameter considerations. Optimizing a UAV path is done for the following reasons:

i. **Throughput fairness and maximization.** Giving an equal opportunity to all nodes in the topology.

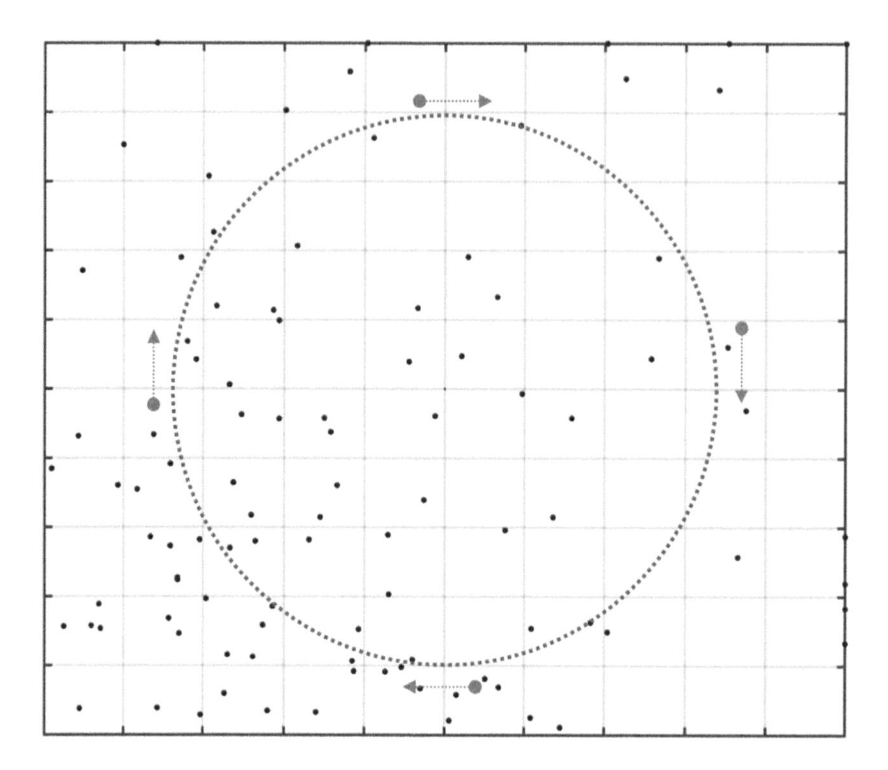

FIGURE 8.2 UAV-Assisted Wireless Ad hoc Network with a Static UAV Path in a Geometric Pattern.

ii. **Reducing latency.** Packet delays can be significantly reduced by providing UAV coverage to a node in a timely fashion.
iii. **Energy Efficiency.** The energy efficiency of the drone in service of the ground nodes and the ground nodes.
iv. **Resource Allocation.**

The existing techniques for path formulation and maintenance are given in Table 8.4.

8.6 DISCUSSION

For data dissemination solution in UAV-assisted wireless ad hoc networks, perhaps the most critical design choice is node association. A node must get associated with the UAV before using the multi-path data flow solution offered by the UAV or UAVs. In the literature, node association is done by dividing the topology into smaller areas. These areas can be called a sector, sub-group, and clusters. A clustering algorithm tightly couples a set of a node in a given area into clusters. Nodes in a cluster get associated with a single overhead UAV. These associated nodes must make changes in their routing information to reflect the available multi-path solution. Priorities can be assigned to areas with higher data flow, congestion, or power requirements.

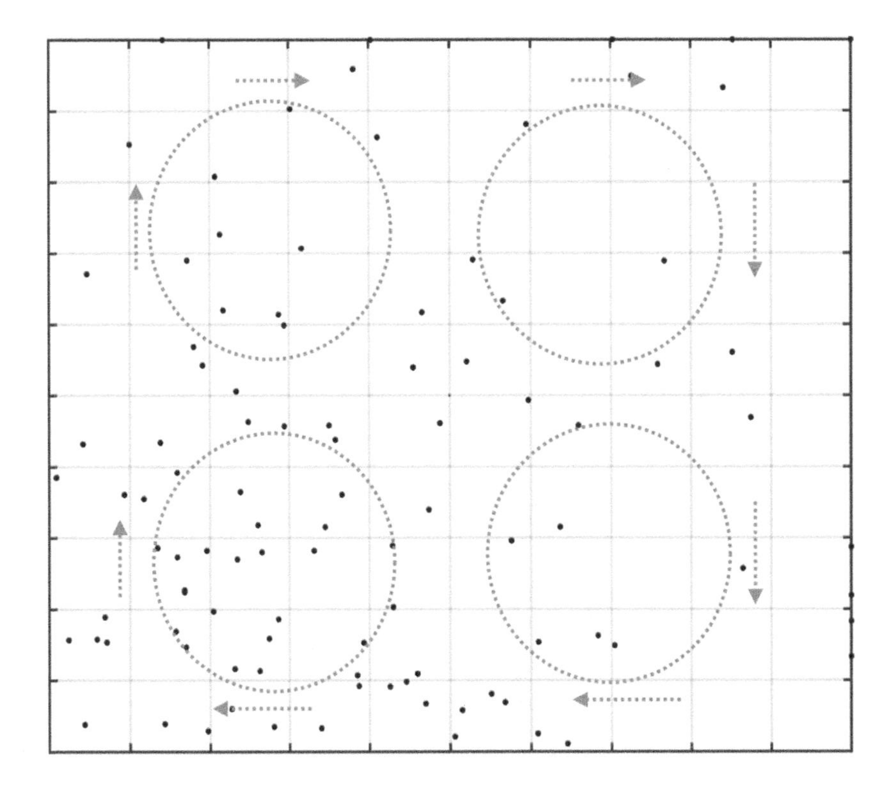

FIGURE 8.3 Static Multi-UAV Paths in a Geometric Pattern.

Reducing end-to-end delays increases the throughput. UAV assistance reduces delays in ground ad hoc network traffic. A packet can be picked up by one UAV and transmitted to another UAV at the destination on the ground, completely bypassing the ground network. The number of packets transmitted from one end of the ground topology to another using a UAV reduces the ground nodes' load. They can thus remove congestion in the ground link from source to destination. UAV-assisted wireless ad hoc networks may use adaptive and predictive techniques to monitor the ground ad hoc network for network state and provide assistance when and where required.

A random mobility model can be of no good when optimized ground wireless network coverage is required. The flight path should be adaptive and should be welcoming to changes in the flight path. With a set of waypoints, a UAV can adjust the flight path according to the required needs. These waypoints define the UAV position with respect to time in a three-dimensional plane and the UAV's velocity at that time. Waypoints form the basics of all the mobility models. A UAV can use two states to serve the ground network, namely, fly-by and hovering. In a fly-by state, the UAV can fly past the nodes without stopping mid-air, also called hovering. This type of mobility pattern is taxing on the ground nodes as the nodes need to continuously reorganize the routing information or they can predict the location of UAV with

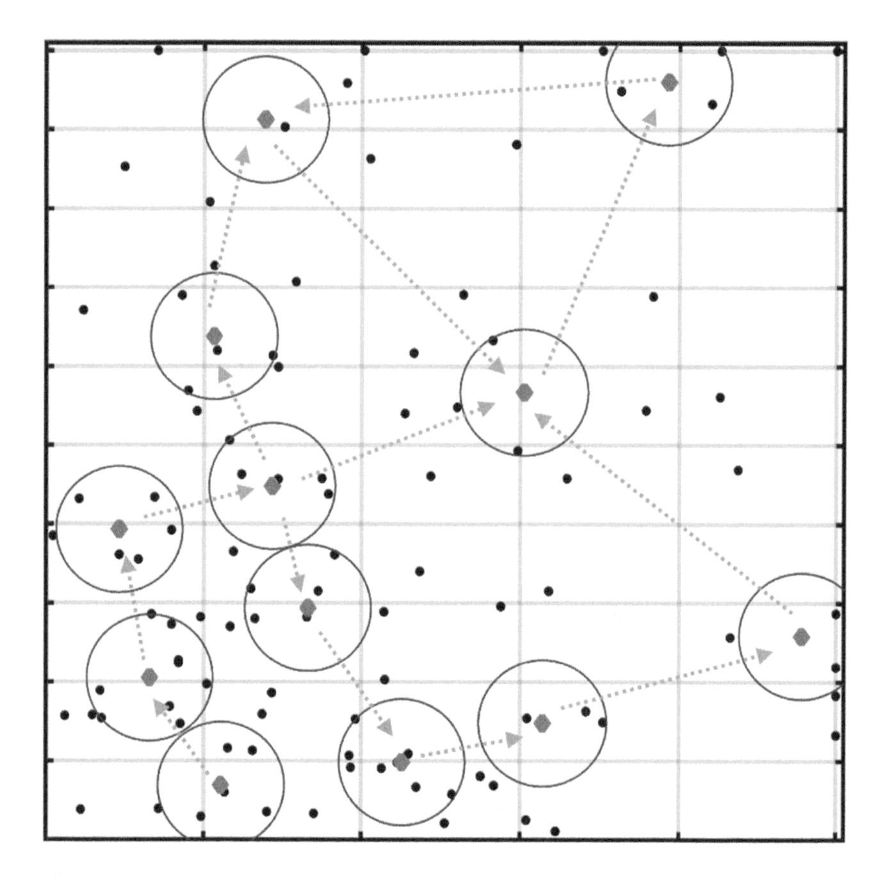

FIGURE 8.4 Dynamic UAV Flight Paths.

respect to time for packet forwarding. However, a fly-by mechanism does save UAV energy and thus increases the number of nodes that can be served in a single flight.

As the UAV speed and the data traffic play an important role in this scenario. In the literature, different data flow scheduling is applied at UAVs head and tail. A head means the heading direction; nodes are made aware that a relay is present in advance. A tail means the UAV is moving off the area, so any consecutive data flowing toward the UAV will not be supported by an aerial network. While data flow at the head can be managed accurately because a present UAV will announce its presence on the wireless channel, the data flow at the rear side of the UAV cannot be managed just by a beacon announcing that it has left the area. The ongoing transmission will start dropping packets until the ground nodes reconfigure it by changing routing information. When the data traffic and UAV speed are low, a fly-by pattern will generally be of little consequence. A slow-moving UAV can accept and forward data flows of a lower rate, and this is because the number of packets transmitted is low. As the UAV speed increases, ground nodes will quickly lose connection to the UAV as it flies past. Even at moderate speeds, if the data traffic is high, the ground nodes will get penalized for latency.

TABLE 8.4
Trajectory Optimization Techniques

Authors, Years	Focus	Features/findings, Advantages and Limitations
Xie et al., 2020 [72]	Wireless powered communication system. Uplink throughput maximization.	Trajectory formulations. UAV path and resource allocation for links. Throughput maximization for IoT devices. One-to-one node association.
Ahmed et al., 2020 [73]	Throughput maximization. Energy-efficient trajectory design.	2D terrestrial UAV deployment with fixed UAV altitudes. Two-phase UAV scheduling. Considerations for non-line of site users. Resource allocation. Optimized throughput and energy consumption. No congestion management and interference management.
G. Tang et al., 2020 [74]	Path planning	Minimum snap trajectory methods for optimized UAV paths with minimum deviations. Optimized trajectory. Mission objective.
J. Tang et al., 2020 [75]	Trajectory optimization. Resource allocation.	Throughput maximization through path optimization and resource allocation. Multi-agent deep learning strategy. 3D UAV path and channel resource optimization in wireless powered communication networks (WPCNs) serving IoT. No congestion management.
Qian et al., 2019 [76]	Mobile edge computing application.	User association. UAVs deployed as an edge server. Maximized offloading of the task. Trajectory optimization. One UAV can service one user at a time.
Mardani et al., 2019 [77]	UAV flight planning. Throughput maximization	Maximized throughput for cellular video streaming. Deployed A* algorithm for distance and throughput optimization. Distance and throughput optimization. Uncertain fairness in resource location.
Liu et al., 2019 [78]	Throughput maximization.	Full-duplex links to UAVs. 5G networks. UAV path and power optimization. No congestion management.
Hua et al., 2019 [79]	Throughput maximization. Small cell wireless system.	Optimizations for UAV trajectories, User scheduling and Transmit power. Throughput maximization. No congestion management.

(Continued)

TABLE 8.4 (CONTINUED) Trajectory Optimization Techniques

Authors, Years	Focus	Features/findings, Advantages and Limitations
B. Liu & Zhu, 2019 [80]	Data gathering. Energy efficiency. Path formulations	Transmission policies and trajectory optimization. Recursive random search for UAV route formulation. Energy efficiency. Timely service for ground nodes. No congestion management and non-adaptive path selection.
Zeng et al., 2018 [81]	Energy efficiency.	Position and path optimization using the traveling salesman problem. Time slot allocation for ground nodes. Optimized energy requirements for minimum throughput requirements of ground nodes. No Congestion management. Resource allocation issues.
F. Wu et al., 2019 [82]	Minimum throughput maximization.	Multi-UAV WPCN. Throughput maximization with trajectory and resource allocation. Throughput maximization. Trajectory optimization for fair resource allocation. No congestion management and non-adaptive path selection.
R. Zhang et al., 2017 [83]	Average throughput maximization per user.	Ground node and aerial node association and scheduling. Trajectory optimization using convex optimization techniques. Trajectory optimization. Algorithm primarily for the downlink.
Xie, Xu and Zhang., 2018 [84]	WPCN.	UAV transfers power (WPT) using a radio frequency in the downlink. Ground nodes use harvested power for uplink data transmission. UAV trajectory designs with hovering and move on. Throughput maximization. Only downlink considerations. Resource allocation issues.
Wu and Zhang., 2018 [85]	OFDMA systems with UAV.	Network delay affects UAV mobility. UAVs serve as a base station for ground nodes. UAV trajectory optimization and OFDMA resource allocation. The algorithm uses the block coordinate descent method. Throughput maximization and reduced latency. No congestion management. Resource allocation issues.

(*Continued*)

TABLE 8.4 (CONTINUED)
Trajectory Optimization Techniques

Authors, Years	Focus	Features/findings, Advantages and Limitations
Sallouha et al., 2018 [86]	Path formulation. Node localization.	Path formulation for the dense environment. Waypoint, flight vectors, and altitude considerations. Fast and accurate node localization. No congestion management and resource allocation issues.
Sayeed et al., 2018 [87]	Path formulation.	Mobility model with waypoint selection. Attraction factors-based path formulation for throughput maximization. Capacity enhancement. No congestion management. Resource allocation issues.
Xu et al., 2018 [88]	Path formulation. Throughput Maximization.	UAVs are deployed as sinks where ground nodes have subtle energy. Fair resource allocation and optimized for minimum throughput for each ground node. Optimized path formulation, communication, and scheduling. Resource allocation issues and no congestion management.
Zeng et al., 2017 [89]	Trajectory design for UAVs in multicast ground networks.	Optimal path formulation. Minimum latency. Higher coverage for ground networks Improved network performance. Fair resource allocation. No congestion management.
Cheng et al., 2018 [90]	Trajectory optimization.	For data offloading at the base station. Throughput maximization for edge users. Throughput maximization. Trajectory optimization for edge user. No congestion management.
Ouyang et al., 2018 [91]	Trajectory optimization.	Downlink throughput maximization by optimizing path and transmission power. Throughput maximization. Downlink only.
Bulut & Guevenc, 2018 [92]	Trajectory optimization. Cellular network.	UAV path is optimized to keep it in contact with the ground base station at all times. Network performance gains. Coverage enhancements. No congestion management.

(*Continued*)

TABLE 8.4 (CONTINUED)
Trajectory Optimization Techniques

Authors, Years	Focus	Features/findings, Advantages and Limitations
ur Rahman et al., 2018 [93]	Trajectory optimization. Disaster area UAV communication networks.	Network monitors for disaster areas maintain data flow and Maximizes throughput using path formulations. Throughput maximization. No congestion management. Uncertain fairness.
Jiang et al., 2018 [94]	Power and trajectory optimization. UAV-enabled amplify and forward relay networks.	Transceiver power and UAV trajectory optimizations. UAVs for data transmission between isolated ground nodes. Throughput maximization. No congestion management.
Zhang et al., 2018 [95]	Energy efficiency. Trajectory optimization.	Multi-hop UAV network for data dissemination. Throughput maximization using trajectory and transmission power optimization. Energy efficiency. Throughput maximization No congestion management. Multi-hop attracts end-to-end delays.
J. Liu et al. 2018 [96]	WSN-UAV Path planning.	Age-based data categorization for path formulation. A genetic algorithm for optimal trajectory formulation. Energy efficiency. Optimized UAV trajectories. Congestion management. Uncertain fairness.
Lin and Saripalli., 2017 [97]	Path planning. Collision avoidance.	Collision prediction and waypoint sampling. Closed-loop is rapidly exploring random tree algorithm for path generation. Collision-free paths for UAVs Mission objectivity.
Kumar et al. 2019 [98]	SDN-based secure mobility model. Path formulation.	SDN controller as certifying authority. Path formulations. Density-based waypoint selection metrics. Network performance improvements No congestion management. Uncertain fairness.
Zeng & Zhang, 2017 [99]	Energy efficiency. Trajectory optimization.	Circular UAV trajectories. Path radius, direction, speed, and acceleration optimization for energy efficiency. Optimal throughput with UAV flight planning. No congestion management.

Another critical design decision is calculating the flight path beforehand or letting the system adapt and allow for dynamic changes to the flight path. On the one hand, a static path such as a UAV flying in circular fashion guarantees service availability at regular intervals. On the other, it may be too slow for a dying node or a node in need of more resources. A dynamic path formulation is adaptive, that is, it will allow a node in need of network resources more time to finish its data transmission. A dynamic path formulation must have a degree of fairness to its design. It should not let a node starve or wait for more extended periods.

Timely service is of paramount importance. A UAV-assisted ground ad hoc wireless network will look to the aerial network for the much-needed service; a node left waiting by the aerial network will increase the packet flow and increase latency around that node. A UAV trajectory plan must incorporate design choices to timely service a node or a set of nodes.

As ad hoc networks are self-organizing, nodes have the ability and responsibility to forward data packets. Sometimes a node cluster or a single node may become isolated from the network topology. A UAV flight plan must adapt to such a situation and provide services to the isolated cluster of nodes. Another design perspective is how to associate the ground wireless ad hoc network with the aerial network. This integration failure is because a UAV might come in a range of ad hoc nodes such that a set of nodes update their routing table to reflect this. Furthermore, routing tables are updated again when the UAV leaves. These updates will have to be frequent to accommodate the speed of the UAV.

A multi-UAV wireless ad hoc network makes the trajectory formulation a more complex task. A multi-UAV communication network can serve the ground nodes with higher efficiency. A multi-UAV-assisted wireless ad hoc network must make design choices for signal interference, non-redundant deployment, seamless aerial connectivity (handover mechanisms), service reliability, and maximized coverage. The UAV deployment must not have signal interference between neighboring UAVs. The ground wireless network will experience packet collision if two UAVs serve a common subset of nodes in two areas. Minimizing the number of nodes in direct contact with multiple UAVs will have dual benefits; first reduced packet loss and higher data rates. Second, adjusting the UAV will lead to coverage of more ground area. The system must stay adaptive to changes happening due to other UAVs. Most importantly, a multi-UAV path formulation should avoid collisions between the UAVs. Following deductions are derived from the literature review:

- Ground wireless ad hoc networks will benefit from using UAVs as mobile base stations, relays, or multi-path solutions for congested links.
- Topological subdivisions are essential for data dissemination, priorities, and path formulation.
- Topological subdivisions effectively create smaller visions of more extensive topology. Larger topologies create a larger control flow in the network.
- Topological subdivisions can yield performance gains by limiting the number of nodes serviced by a single overhead UAV.
- The static flight plan can provide services to the ground ad hoc network in a time-bound manner.

- Adaptive flight plans are most suited for the ad hoc network as they can reflect a change when required.
- Waypoint selection and UAV flight patterns play an essential role in ground network performance improvement.
- Interference management in a multi-UAV environment and enhanced coverage for isolated nodes is some of the least researched areas.
- Predictive algorithms in the literature generally rely more on mobility situations than link and channel states in a node locality.
- Topological subdivisions can be combined with predictive algorithms to yield better results and assess the network state at a given instance.

8.7 SUMMARY

In this paper, the design consideration for the UAV-assisted wireless ad hoc network was discussed. A comprehensive literature review is presented along with some state-of-the-art solutions for UAV-assisted wireless ad hoc networks. The literature review establishes that topological subdivisions and their utilization, coverage, node isolation, congestion, and link state are some of the least researched areas of UAV-assisted wireless networks and require further analysis.

REFERENCES

1. M. Mozaffari, W. Saad, M. Bennis, Y. H. Nam, and M. Debbah, "A tutorial on UAVs for wireless networks: Applications, challenges, and open problems," arXiv, vol. 21, no. 3, pp. 2334–2360, 2018.
2. L. Gupta, R. Jain, and G. Vaszkun, "Survey of important issues in UAV communication networks," *IEEE Commun. Surv. Tutorials*, vol. 18, no. 2, pp. 1123–1152, 2016, doi: 10.1109/COMST.2015.2495297.
3. Y. Chen, H. Zhang, and M. Xu, "The coverage problem in UAV network: A survey," in Fifth International Conference on Computing, Communications and Networking Technologies (ICCCNT), New Delhi, Jul. 2014, pp. 1–5, doi: 10.1109/ICCCNT.2014.6963085.
4. R. A. Nazib and S. Moh, "Routing protocols for unmanned aerial vehicle-aided vehicular Ad Hoc Networks: A survey," *IEEE Access*, vol. 8, pp. 77535–77560, 2020, doi: 10.1109/ACCESS.2020.2989790.
5. C. Perkings, E. Belding-Royer, and S. Das, "Ad hoc on-demand distance vector (AODV) routing," *IETF RFC*, 3561, pp. 1–37, July 2003, [Online]. Available: http://tools.ietf.org/pdf/rfc3561.pdf.
6. T. Clausen and P. Jacquet, "Optimized link state routing protocol (OLSR)," RFC Editor, 2003, [Online]. Available: https://www.rfc-editor.org/info/rfc3626.
7. C. E. Perkins and P. Bhagwat, "Highly dynamic destination-sequenced distance-vector routing (DSDV) for mobile computers," *SIGCOMM Comput. Commun. Rev.*, vol. 24, no. 4, pp. 234–244, Oct. 1994, doi: 10.1145/190809.190336.
8. N. Beijar, "Zone routing protocol (ZRP)," *Netw. Lab. Helsinki Univ. Technol. Finl.*, vol. 9, pp. 1–12, 2002.
9. D. B. Johnson and D. A. Maltz, "DSR: The dynamic source routing protocol for multi-hop wireless ad hoc networks," *Comput. Sci. Dep. Carnegie Mellon Univ. Addison-Wesley*, vol. 5, no. 1, pp. 139–172, 1996, [Online]. Available: http://www.monarch.cs.cmu.edu/.

10. P. Xie, "An enhanced OLSR routing protocol based on node link expiration time and residual energy in ocean FANETS," in 2018 24th Asia-Pacific Conference on Communications, APCC 2018, Chicago, 2019, pp. 598–603, doi: 10.1109/APCC.2018.8633484.
11. Y. Chen, X. Liu, N. Zhao, and Z. Ding, "Using multiple UAVs as relays for reliable communications," in 2018 IEEE 87th Vehicular Technology Conference (VTC Spring), Omaha, June 2018, pp. 1–5, doi: 10.1109/VTCSpring.2018.8417733.
12. S. C. Choi, H. R. Hussen, J. H. Park, and J. Kim, "Geolocation-based routing protocol for flying ad hoc networks (FANETs)," in International Conference on Ubiquitous and Future Networks, ICUFN, Washington, DC, 2018, vol. 2018-July, pp. 50–52, doi: 10.1109/ICUFN.2018.8436724.
13. F. Khelifi, A. Bradai, K. Singh, and M. Atri, "Localization and energy-efficient data routing for unmanned aerial vehicles: Fuzzy-logic-based approach," *IEEE Commun. Mag.*, vol. 56, no. 4, pp. 129–133, 2018, doi: 10.1109/MCOM.2018.1700453.
14. C. Pu, "Link-quality and traffic-load aware routing for UAV ad hoc networks," in Proceedings: 4th IEEE International Conference on Collaboration and Internet Computing, CIC 2018, Phoenix, 2018, pp. 71–79, doi: 10.1109/CIC.2018.00-38.
15. M. Song, J. Liu, and S. Yang, "A mobility prediction and delay prediction routing protocol for UAV networks," in 2018 10th International Conference on Wireless Communications and Signal Processing, WCSP 2018, Plymouth, 2018, pp. 1–6, doi: 10.1109/WCSP.2018.8555927.
16. S. Y. Dong, "Optimization of OLSR routing protocol in UAV ad HOC network," in 2016 13th International Computer Conference on Wavelet Active Media Technology and Information Processing, ICCWAMTIP 2017, Oakland, 2017, pp. 90–94, doi: 10.1109/ICCWAMTIP.2016.8079811.
17. G. Gankhuyag, A. P. Shrestha, and S. J. Yoo, "Robust and reliable predictive routing strategy for flying ad-hoc networks," *IEEE Access*, vol. 5, pp. 643–654, 2017, doi: 10.1109/ACCESS.2017.2647817.
18. X. Li and J. Yan, "LEPR: Link stability estimation-based preemptive routing protocol for flying ad hoc networks," in Proceedings: IEEE Symposium on Computers and Communications, Atlanta, 2017, pp. 1079–1084, doi: 10.1109/ISCC.2017.8024669.
19. O. S. Oubbati, A. Lakas, F. Zhou, M. Güneş, N. Lagraa, and M. B. Yagoubi, "Intelligent UAV-assisted routing protocol for urban VANETs," *Comput. Commun.*, vol. 107, pp. 93–111, 2017, doi: 10.1016/j.comcom.2017.04.001.
20. A. Rovira-Sugranes and A. Razi, "Predictive routing for dynamic UAV networks," in 2017 IEEE International Conference on Wireless for Space and Extreme Environments, WiSEE 2017, New York, 2017, pp. 43–47, doi: 10.1109/WiSEE.2017.8124890.
21. J. Lee et al., "Constructing a reliable and fast recoverable network for drones," in 2016 IEEE International Conference on Communications, ICC 2016, Tacoma, 2016, pp. 1–6, doi: 10.1109/ICC.2016.7511317.
22. S. Rosati, K. Kruzelecki, G. Heitz, D. Floreano, and B. Rimoldi, "Dynamic routing for flying ad hoc networks," *IEEE Trans. Veh. Technol.*, vol. 65, no. 3, pp. 1690–1700, 2016, doi: 10.1109/TVT.2015.2414819.
23. J. D. M. M. Biomo, T. Kunz, and M. St-Hilaire, "Routing in unmanned aerial ad hoc networks: Introducing a route reliability criterion," in 2014 7th IFIP Wireless and Mobile Networking Conference, WMNC 2014, Vail, May 2014, pp. 1–7, doi: 10.1109/WMNC.2014.6878853.
24. Y. Zheng, Y. Wang, Z. Li, L. Dong, Y. Jiang, and H. Zhang, "A mobility and load aware olsr routing protocol for uav mobile AD-HOC networks," in IET Conference Publications, San Diego, May 2014, vol. 2014, no. 650 CP, pp. 1–7, doi: 10.1049/cp.2014.0575.

25. S. Rosati, K. Kruzelecki, L. Traynard, and B. Rimoldi, "Speed-aware routing for UAV ad-hoc networks," in 2013 IEEE Globecom Workshops, GC Workshops, San Ramon, 2013, pp. 1367–1373, doi: 10.1109/GLOCOMW.2013.6825185.
26. L. Lin, Q. Sun, S. Wang, and F. Yang, "A geographic mobility prediction routing protocol for ad hoc UAV network," in 2012 IEEE Globecom Workshops, GC Workshops, Shipshewana, 2012, pp. 1597–1602, doi: 10.1109/GLOCOMW.2012.6477824.
27. A. I. Alshabtat and L. Dong, "Low latency routing algorithm for unmanned aerial vehicles ad-hoc networks," *World Acad. Sci. Eng. Technol.*, vol. 80, no. 1, pp. 705–711, 2011, doi: 10.5281/zenodo.1061573.
28. R. A. Hunjet, A. Coyle, and M. Sorell, "Enhancing mobile adhoc networks through node placement and topology control," in 2010 7th International Symposium on Wireless Communication Systems, Boca Raton, 2010, pp. 536–540, doi: 10.1109/ISWCS.2010.5624347.
29. C. M. Cheng, P. H. Hsiao, H. T. Kung, and D. Vlah, "Maximizing throughput of UAV-relaying networks with the load-carry-and-deliver paradigm," in IEEE Wireless Communications and Networking Conference, WCNC, Boston, Mar. 2007, pp. 4420–4427, doi: 10.1109/WCNC.2007.805.
30. I. Rubin and R. Zhang, "Placement of UAVs as communication relays aiding mobile ad hoc wireless networks," in MILCOM 2007: IEEE Military Communications Conference, Oct. 2007, pp. 1–7, doi: 10.1109/MILCOM.2007.4455114.
31. O. Bouhamed, H. Ghazzai, H. Besbes, and Y. Massoud, "A UAV-assisted data collection for wireless sensor networks: Autonomous navigation and scheduling," *IEEE Access*, vol. 8, pp. 110446–110460, 2020, doi: 10.1109/ACCESS.2020.3002538.
32. L. Wang, B. Hu, and S. Chen, "Energy efficient placement of a drone base station for minimum required transmit power," *IEEE Wirel. Commun. Lett.*, vol. 9, no. 12, pp. 2010–2014, 2020, doi: 10.1109/LWC.2018.2808957.
33. C. C. Lai, C. T. Chen, and L. C. Wang, "On-demand density-aware UAV base station 3D placement for arbitrarily distributed users with guaranteed data rates," *IEEE Wirel. Commun. Lett.*, vol. 8, no. 3, pp. 913–916, Jun. 2019, doi: 10.1109/LWC.2019.2899599.
34. W. Qi, Q. Song, X. Kong, and L. Guo, "A traffic-differentiated routing algorithm in Flying Ad Hoc Sensor Networks with SDN cluster controllers," *J. Franklin Inst.*, vol. 356, no. 2, pp. 766–790, 2019, doi: 10.1016/j.jfranklin.2017.11.012.
35. F. Aadil, A. Raza, M. F. Khan, M. Maqsood, I. Mehmood, and S. Rho, "Energy aware cluster-based routing in flying ad-hoc networks," *Sensors (Switzerland)*, vol. 18, no. 5, pp. 1589–1597, 2018, doi: 10.3390/s18051413.
36. M. Alzenad, A. El-Keyi, F. Lagum, and H. Yanikomeroglu, "3-D placement of an unmanned aerial vehicle base station (UAV-BS) for energy-efficient maximal coverage," *IEEE Wirel. Commun. Lett.*, vol. 6, no. 4, pp. 434–437, Aug. 2017, doi: 10.1109/LWC.2017.2700840.
37. J. Wang, Y. Cao, B. Li, H. jin Kim, and S. Lee, "Particle swarm optimization based clustering algorithm with mobile sink for WSNs," *Futur. Gener. Comput. Syst.*, vol. 76, pp. 452–457, Nov. 2017, doi: 10.1016/j.future.2016.08.004.
38. J. Lyu, Y. Zeng, R. Zhang, and T. J. Lim, "Placement optimization of UAV-mounted mobile base stations," *IEEE Commun. Lett.*, vol. 21, no. 3, pp. 604–607, 2017, doi: 10.1109/LCOMM.2016.2633248.
39. M. Mozaffari, W. Saad, M. Bennis, and M. Debbah, "Efficient deployment of multiple unmanned aerial vehicles for optimal wireless coverage," *IEEE Commun. Lett.*, vol. 20, no. 8, pp. 1647–1650, Aug. 2016, doi: 10.1109/LCOMM.2016.2578312.
40. E. Kalantari, H. Yanikomeroglu, and A. Yongacoglu, "On the number and 3D placement of drone base stations in wireless cellular networks," in IEEE Vehicular Technology Conference, Boston, 2016, doi: 10.1109/VTCFall.2016.7881122.

41. R. I. Bor-Yaliniz, A. El-Keyi, and H. Yanikomeroglu, "Efficient 3-D placement of an aerial base station in next generation cellular networks," *Journal of Computers*, vol. 78, pp. 156–171. 2016, doi: 10.1109/ICC.2016.7510820.
42. J. Shu, Y. Ge, L. Liu, and L. Sun, "Mobility prediciton clustering routing in UAVs," in Proceedings of 2011 International Conference on Computer Science and Network Technology, ICCSNT 2011, 2011, vol. 3, pp. 1983–1987, doi: 10.1109/ICCSNT.2011.6182360.
43. C. Zang and S. Zang, "Mobility prediction clustering algorithm for UAV networking," in *2011 IEEE GLOBECOM Workshops, GC Workshops*, 2011, pp. 1158–1161, doi: 10.1109/GLOCOMW.2011.6162360.
44. B. Fu and L. A. DaSilva, "A Mesh in the sky: A routing protocol for airborne networks," in Proceedings: IEEE Military Communications Conference MILCOM, New York, 2007, pp. 1–7, doi: 10.1109/MILCOM.2007.4454819.
45. V. Sharma, I. You, R. Kumar, and V. Chauhan, "OFFRP: Optimised fruit fly based routing protocol with congestion control for UAVs guided ad hoc networks," *Int. J. Ad Hoc Ubiquitous Comput.*, vol. 27, no. 4, pp. 233–255, 2018, doi: 10.1504/IJAHUC.2018.090596.
46. D. Wu et al., "ADDSEN: Adaptive data processing and dissemination for drone swarms in urban sensing," *IEEE Trans. Comput.*, vol. 66, no. 2, pp. 183–198, 2017, doi: 10.1109/TC.2016.2584061.
47. P. Chandhar, D. Danev, and E. G. Larsson, "Massive MIMO for communications with drone swarms," in arXiv, pp. 347–354, 2017.
48. J. Sanchez-Garcia, J. M. Garcia-Campos, S. L. Toral, D. G. Reina, and F. Barrero, "An intelligent strategy for tactical movements of UAVs in disaster scenarios," *Int. J. Distrib. Sens. Networks*, vol. 2016, no. 3, p. 8132812, 2016, doi: 10.1155/2016/8132812.
49. D. G Reina, R. I. Ciobanu, S. L. Toral, and C. Dobre, "A multi-objective optimization of data dissemination in delay tolerant networks," *Expert Syst. Appl.*, vol. 57, pp. 178–191, Sep. 2016, doi: 10.1016/j.eswa.2016.03.038.
50. V. Sharma, M. Bennis, and R. Kumar, "UAV-assisted heterogeneous networks for capacity enhancement," *IEEE Commun. Lett.*, vol. 20, no. 6, pp. 1207–1210, 2016, doi: 10.1109/LCOMM.2016.2553103.
51. S. Say, H. Inata, J. Liu, and S. Shimamoto, "Priority-based data gathering framework in UAV-assisted wireless sensor networks," *IEEE Sens. J.*, vol. 16, no. 14, pp. 5785–5794, 2016, doi: 10.1109/JSEN.2016.2568260.
52. X. Zheng, J. Wang, W. Dong, Y. He, and Y. Liu, "Bulk data dissemination in wireless sensor networks: Analysis, implications and improvement," *IEEE Trans. Comput.*, vol. 65, no. 5, pp. 1428–1439, 2016, doi: 10.1109/TC.2015.2435778.
53. R. Kim, H. Lim, and B. Krishnamachari, "Prefetching-based data dissemination in vehicular cloud systems," *IEEE Trans. Veh. Technol.*, vol. 65, no. 1, pp. 292–306, 2016, doi: 10.1109/TVT.2015.2388851.
54. V. Sharma, I. You, and R. Kumar, "Energy efficient data dissemination in multi-UAV coordinated wireless sensor networks," *Mob. Inf. Syst.*, vol. 2016, 2016, doi: 10.1155/2016/8475820.
55. A. Wichmann and T. Korkmaz, "Smooth path construction and adjustment for multiple mobile sinks in wireless sensor networks," *Comput. Commun.*, vol. 72, pp. 93–106, 2015, doi: 10.1016/j.comcom.2015.06.001.
56. C. Tunca, S. Isik, M. Y. Donmez, and C. Ersoy, "Ring routing: An energy-efficient routing protocol for wireless sensor networks with a mobile sink," *IEEE Transactions on Mobile Computing*, 2015, vol. 14, no. 9, pp. 1947–1960, doi: 10.1109/TMC.2014.2366776.

57. X. Zhu, X. Tao, T. Gu, and J. Lu, "Target-aware, transmission power-adaptive, and collision-free data dissemination in wireless sensor networks," *IEEE Trans. Wirel. Commun.*, vol. 14, no. 12, pp. 6911–6925, 2015.
58. S. Temel and I. Bekmezci, "LODMAC: Location oriented directional MAC protocol for FANETs," *Comput. Networks*, vol. 83, pp. 76–84, 2015, doi: 10.1016/j.comnet.2015.03.001.
59. T. Yan, W. Zhang, and G. Wang, "DOVE: Data dissemination to a desired number of receivers in VANET," *IEEE Trans. Veh. Technol.*, vol. 63, no. 4, pp. 1903–1916, 2014, doi: 10.1109/TVT.2013.2287692.
60. X. Shen, X. Cheng, L. Yang, R. Zhang, and B. Jiao, "Data dissemination in VANETs: A scheduling approach," *IEEE Trans. Intell. Transp. Syst.*, vol. 15, no. 5, pp. 2213–2223, 2014, doi: 10.1109/TITS.2014.2313631.
61. R. I. Ciobanu, D. G. Reina, C. Dobre, S. L. Toral, and P. Johnson, "JDER: A history-based forwarding scheme for delay tolerant networks using Jaccard distance and encountered ration," *J. Netw. Comput. Appl.*, vol. 40, no. 1, pp. 279–291, 2014, doi: 10.1016/j.jnca.2013.09.012.
62. V. Sharma and R. Kumar, "A cooperative network framework for multi-UAV guided ground ad hoc networks," *J. Intell. Robot. Syst. Theory Appl.*, vol. 77, no. 3–4, pp. 629–652, 2015, doi: 10.1007/s10846-014-0091-0.
63. F. J. Ros and P. M. Ruiz, "Minimum broadcasting structure for optimal data dissemination in vehicular networks," *IEEE Trans. Veh. Technol.*, vol. 62, no. 8, pp. 3964–3973, 2013, doi: 10.1109/TVT.2013.2244107.
64. Y. Cai, F. R. Yu, J. Li, Y. Zhou, and L. Lamont, "Medium access control for Unmanned Aerial Vehicle (UAV) ad-hoc networks with full-duplex radios and multipacket reception capability," *IEEE Trans. Veh. Technol.*, vol. 62, no. 1, pp. 390–394, 2013, doi: 10.1109/TVT.2012.2211905.
65. W. Alasmary and W. Zhuang, "Mobility impact in IEEE 802.11p infrastructureless vehicular networks," *Ad Hoc Networks*, vol. 10, no. 2, pp. 222–230, 2012, doi: 10.1016/j.adhoc.2010.06.006.
66. D. T. Ho and S. Shimamoto, "Highly reliable communication protocol for WSN-UAV system employing TDMA and PFS scheme," in 2011 IEEE GLOBECOM Workshops, GC Workshops, New Delhi, 2011, pp. 1320–1324, doi: 10.1109/GLOCOMW.2011.6162401.
67. Y. Ding and L. Xiao, "SADV: Static-node-assisted adaptive data dissemination in vehicular networks," *IEEE Trans. Veh. Technol.*, vol. 59, no. 5, pp. 2445–2455, 2010, doi: 10.1109/TVT.2010.2045234.
68. J. Zhao, Y. Zhang, and G. Cao, "Data pouring and buffering on the road: A new data dissemination paradigm for vehicular ad hoc networks," *IEEE Trans. Veh. Technol.*, vol. 56, no. 6 I, pp. 3266–3277, 2007, doi: 10.1109/TVT.2007.906412.
69. S. Eichler, "Performance evaluation of the IEEE 802.11p WAVE communication standard," in IEEE Vehicular Technology Conference, 2007, pp. 2199–2203, doi: 10.1109/VETECF.2007.461.
70. X. Chen, H. Zhai, X. Tian, and Y. Fang, "Supporting QoS in IEEE 802.11e wireless LANs," *IEEE Trans. Wirel. Commun.*, vol. 5, no. 8, pp. 2217–2227, 2006, doi: 10.1109/TWC.2006.1687738.
71. R. C. Palat, A. Annamalai, and J. H. Reed, "Cooperative relaying for ad-hoc ground networks using swarm UAVs," in *Proceedings - IEEE Military Communications Conference MILCOM*, vol. 2005, pp. 1588–1594, 2005, doi: 10.1109/MILCOM.2005.1605902.
72. L. Xie, J. Xu, and Y. Zeng, "Common throughput maximization for UAV-enabled interference channel with wireless powered communications," *Network Security*, vol. 68, no. 5, pp. 3197–3212, 2019.

73. S. Ahmed, M. Z. Chowdhury, and Y. M. Jang, "Energy-efficient UAV-to-user scheduling to maximize throughput in wireless networks," *IEEE Access*, vol. 8, pp. 21215–21225, 2020, doi: 10.1109/ACCESS.2020.2969357.
74. G. Tang, Z. Hou, C. Claramunt, and X. Hu, "UAV trajectory planning in a port environment," *J. Mar. Sci. Eng.*, vol. 8, no. 8, p. 592, 2020, doi: 10.3390/JMSE8080592.
75. J. Tang, J. Song, J. Ou, J. Luo, X. Zhang, and K. K. Wong, "Minimum throughput maximization for multi-UAV enabled WPCN: A deep reinforcement learning method," *IEEE Access*, vol. 8, pp. 9124–9132, 2020, doi: 10.1109/ACCESS.2020.2964042.
76. Y. Qian, F. Wang, J. Li, L. Shi, K. Cai, and F. Shu, "User association and path planning for UAV-aided mobile edge computing with energy restriction," *Journal of Network Security*, vol. 8, no. 5, pp. 1312–1315, 2019.
77. A. Mardani, M. Chiaberge, and P. Giaccone, "Communication-aware UAV path planning," *IEEE Access*, vol. 7, pp. 52609–52621, 2019, doi: 10.1109/ACCESS.2019.2911018.
78. X. Liu, D. He, and H. Ding, "Throughput maximization for UAV-enabled full-duplex relay system in 5G communications," *Phys. Commun.*, vol. 32, pp. 104–111, 2019, doi: 10.1016/j.phycom.2018.11.014.
79. M. Hua, L. Yang, C. Pan, and A. Nallanathan, "Throughput maximization for full-duplex uav aided small cell wireless systems," arXiv, vol. 9, no. 4, pp. 475–479, 2019.
80. B. Liu and H. Zhu, "Energy-effective data gathering for uav-aided wireless sensor networks," *Sensors*, vol. 19, no. 11, p. 2506, 2019, doi: 10.3390/s19112506.
81. Y. Zeng, J. Xu, and R. Zhang, "Energy minimization for wireless communication with rotary-wing UAV," *International Journal of Network Security & Its Applications*, vol. 18, no. 4, pp. 2329–2345, 2018.
82. F. Wu, D. Yang, L. Xiao, and L. Cuthbert, "Minimum-throughput maximization for multi-UAV-enabled wireless-powered communication networks," *Sensors*, vol. 19, no. 7, p. 1491, 2019, doi: 10.3390/s19071491.
83. R. Zhang, Y. Zeng, and Q. Wu, "Joint Trajectory and Communication Design for Multi-UAV Enabled Wireless Networks," *International Journal of Security and Networks*, vol. 17, no. 3, pp. 2109–2121, 2017.
84. L. Xie, J. Xu, and R. Zhang, "Throughput maximization for UAV-enabled wireless powered communication networks," *International Journal of Information Security*, vol. 6, no. 2, pp. 1690–1703, 2018.
85. Q. Wu and R. Zhang, "Common throughput maximization in UAV-enabled OFDMA systems with delay consideration," *Journal of Network Security Computer Networks*, vol. 66, no. 12, pp. 6614–6627, 2018.
86. H. Sallouha, M. M. Azari, and S. Pollin, "Energy-constrained UAV trajectory design for ground node localization," *Security and Communication Networks*, vol. 56, pp. 1–7, 2018.
87. M. Abuzar Sayeed and R. Kumar, "An efficient mobility model for improving transmissions in multi-UAVs enabled WSNs," *Drones*, vol. 2, no. 3, pp. 1–23, 2018, doi: 10.3390/drones2030031.
88. Y. Xu, L. Xiao, D. Yang, Q. Wu, and L. Cuthbert, "Throughput maximization in multi-UAV enabled communication systems with difference consideration," *IEEE Access*, vol. 6, pp. 55291–55301, 2018, doi: 10.1109/ACCESS.2018.2872736.
89. Y. Zeng, X. Xu, and R. Zhang, "Trajectory optimization for completion time minimization in UAV-enabled multicasting," *Journal of Cybersecurity*, vol. 17, no. 4, pp. 2233–2246, 2017.
90. F. Cheng et al., "UAV trajectory optimization for data offloading at the edge of multiple cells," *IEEE Trans. Veh. Technol.*, vol. 67, no. 7, pp. 6732–6736, 2018, doi: 10.1109/TVT.2018.2811942.

91. J. Ouyang, Y. Che, J. Xu, and K. Wu, "Throughput maximization for laser-powered UAV wireless communication systems," *International Journal of Computer Science and Network Security*, vol. 56, pp. 1–6, 2018.
92. E. Bulut and I. Guevenc, "Trajectory optimization for cellular-connected UAVs with disconnectivity constraint," in 2018 IEEE International Conference on Communications Workshops, ICC Workshops 2018 - Proceedings, 2018, pp. 1–6, doi: 10.1109/ICCW.2018.8403623.
93. S. Ur Rahman, G. H. Kim, Y. Z. Cho, and A. Khan, "Positioning of UAVs for throughput maximization in software-defined disaster area UAV communication networks," *J. Commun. Networks*, vol. 20, no. 5, pp. 452–463, 2018, doi: 10.1109/JCN.2018.000070.
94. X. Jiang, Z. Wu, Z. Yin, and Z. Yang, "Power and trajectory optimization for UAV-enabled amplify-and-forward relay networks," *IEEE Access*, vol. 6, pp. 48688–48696, 2018, doi: 10.1109/ACCESS.2018.2867849.
95. G. Zhang, H. Yan, Y. Zeng, M. Cui, and Y. Liu, "Trajectory optimization and power allocation for multi-hop UAV relaying communications," *IEEE Access*, vol. 6, pp. 48566–48576, 2018, doi: 10.1109/ACCESS.2018.2868117.
96. J. Liu, X. Wang, B. Bai, and H. Dai, "Age-optimal trajectory planning for UAV-assisted data collection," *International Journal of Computer Network and Information Security*, vol. 56, 2018, pp. 553–558.
97. Y. Lin and S. Saripalli, "Sampling-based path planning for UAV collision avoidance," *IEEE Trans. Intell. Transp. Syst.*, vol. 18, no. 11, pp. 3179–3192, 2017, doi: 10.1109/TITS.2017.2673778.
98. R. Kumar, M. A. Sayeed, V. Sharma, and I. You, "An SDN-based secure mobility model for UAV-ground communications," in *Communications in Computer and Information Science*, 2019, vol. 971, pp. 169–179, doi: 10.1007/978-981-13-3732-1_14.
99. Y. Zeng and R. Zhang, "Energy-efficient UAV communication with trajectory optimization," *IEEE Trans. Wirel. Commun.*, vol. 16, no. 6, pp. 3747–3760, Jun. 2017, doi: 10.1109/TWC.2017.2688328.

9 Integrating Cybernetics into Healthcare Systems

Security Perspective

Saquib Ali, Jalaluddin Khan, Jian Ping Li, Masood Ahmad, Kanika Sharma, Amal Krishna Sarkar, Alka Agrawal, and Ranjit Rajak

CONTENTS

9.1 Introduction 161
9.2 Data Breach in Healthcare 163
9.3 Previous Similar Research Initiatives 165
9.4 Needs and Importance of Cybernetics in Healthcare Security 167
9.5 Challenges 168
9.6 Future Works 169
9.7 Conclusion 171
References 171

9.1 INTRODUCTION

In the 1940s, the discipline of cybernetics was defined as "the scientific study of interconnection or the scientific analysis of control and communication in animal and machine" by its founders. In more recent usage, cybernetics has come to mean the study of human–machine interaction through the use of technology-based processes and control structures [1]. In another sense, cybernetics is the study of how humans and machines govern and interact with one another in sophisticated ways to achieve goals.

In more recent usage, cybernetics has come to mean the study of human–machine interaction through the use of technology-based process and control mechanisms. In other words, cybernetics is the study of how humans and machines govern and communicate with one another in sophisticated ways to achieve goals. This can be seen on a variety of levels. From pharmaceuticals to robots, technology is now pervasive in how we diagnose, treat, and give care. Technology influences how people engage with their own health on a personal level. The processes that pervade our system – care, administration, communication, and collaboration – are all soaked with, built on, and facilitated by technology. Health, on the other hand, is intrinsically human. Care is inextricably linked to the human condition. The human aspects of imbued

DOI: 10.1201/9781003323426-9

health, interactional connection, and personal approach – of people caring for other people – should and will remain at the forefront of how we presume about healthcare and health systems, despite all of the methodologies, apps, medical devices, analytics, and communication technologies.

Cybernetics is a philosophy or a way of thinking for designing and defining structures [2]. The name "cybernetics" comes from the Greek word "kybernetike," which meaning "administration." Many scholars and experts, on the other hand, believe that the word's true meaning is "steering." The terminology is defined as "the art of steering" in more detail. The entire concept of second-order cybernetics revolves upon this premise.

Healthcare cybernetics has the possibility of assisting us in better understanding and optimizing how humans and technology interact in complicated ways to accomplish desired outcomes in our health system. It may bring the role of each technology, process, and interaction in the system, as well as their futures, into sharp light. In some ways, it offers a chance to view the big picture and pursue it in radical new ways.

The healthcare industry is regarded as an essential infrastructure. Healthcare and the public health sector serve the most vulnerable citizens of our communities and have become one of the most targeted areas by cybercriminals. Attacks on public health infrastructure have been steadily increasing. When compared to other types of personal or financial data, patient history data has a far larger monetary worth [3]. The upsurge in criminality is fueled by a lack of awareness among technology users, as well as an upsurge in personal healthcare devices like monitors, personalized medical devices, the digitization of paper-based patient data, and the increasing existence of multiple healthcare systems on the Internet. In this paper, we explore the objectives, implications, and limitations of cybernetics in the healthcare security industry to raise awareness of cyber threats and countermeasures to battle the industry's expanding security issues.

This concept is to manipulate a system's output so that it follows the desired control signal known as the reference. To do this, a (typical feedback) controller is built that determines which outputs should be watched, how to evaluate them to the standard, which system behaviors should be changed, and how to modify them. The error signal, which represents the error between the measured and desired output, is sent back into the system input in order to bring the real output nearer to the standard. The controlled feedback loop is depicted in Figure 9.1. The most protection in healthcare industry is needed for data integrity and security of data, for protecting data breaches, cybernetics could be the technique to secure data because of its feedback loop.

It has been discovered that data integrity rifts are frequently difficult to spot. When the bigger repercussions of a data integrity breach are often unclear, and attackers exploit the leaked data to execute subsequent attacks; thus, data integrity protection becomes even more important. Rather than removing or disrupting access to digital information, cyber operations will use it in the near future to compromise its integrity [4]. Information tampering will have an impact on people's lives. Given the repercussions of data tampering, this rising type of cybercrime constitutes

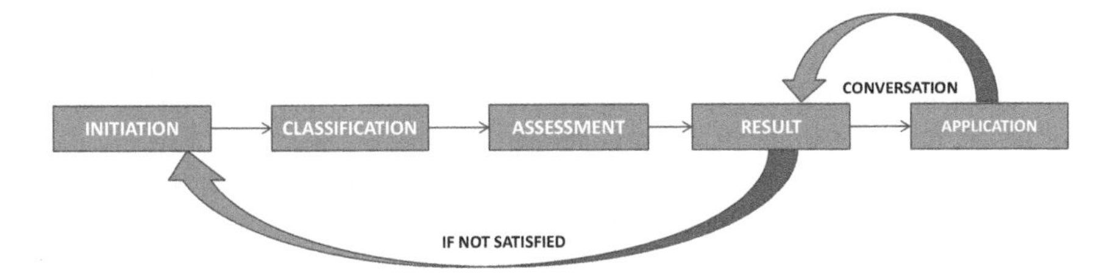

FIGURE 9.1 Feedback loop.

a massive threat that must be handled immediately [5]. As a result, security professionals and researchers must be aware of the dangers of data manipulation. In order to secure information from manipulation attacks, a constant and stringent data protection solution is critically needed. Data integrity is also one of the most pressing challenges in the healthcare industry. A breach of data integrity in a healthcare facility could have a variety of potentially serious effects. Cyber security breaches are now often regarded as the most serious threat to hospitals. In the healthcare industry, maintaining data integrity is critical.

9.2 DATA BREACH IN HEALTHCARE

In the digital age, data is the most precious asset. Every digitalized enterprise generates a massive amount of information. For any security professional, managing such a vast number of data effectively is a tough and time-consuming task. Each type of data has its own meaning and use. Data's worth is totally determined by its nature. For example, healthcare data provides value in people's lives. Threat actors are directly targeting this store of data, as evidenced by the recent increase in cyberattacks, to profit from the monetary rewards that hoarded or manipulated data can yield.

A breach of data integrity in a healthcare facility could have a variety of potentially serious effects. Cyber security breaches are now often regarded as the most serious threat to hospitals. In the healthcare industry, maintaining data integrity is critical. Due to the organizational structure of healthcare facilities, which includes high-end point intricacy and regulatory limitations, the healthcare industry has become a serious challenge. Many security breaches have demonstrated that the healthcare industry continues to fall behind other industries in terms of safeguarding the information integrity of its stakeholders. The company's brand image and consumer trust are both protected by data integrity. Any data breach can result in significant revenue loss as well as a loss of consumer faith in the company's legitimacy. This form of threat is more dangerous to businesses than attacks on secrecy and availability. When frequent data integrity attacks go undiscovered or unrecognized, and attackers use flawed information or data in various types of attacks, the importance of the data integrity issue becomes even more pressing.

The issue of data integrity is one of the most pressing concerns in the global healthcare business. A healthcare organization's integrity breach might have severe consequences. A patient whose data has been tampered with may be given the incorrect drugs, resulting in death. Most healthcare businesses now have insecure data storage practices and lack robust malware-defeating tools. All of these concerns present a slew of challenges for healthcare businesses when it comes to data integrity according to multiple reports, and the amount of data breaches affecting the healthcare business is on the rise. During 2009–2020, HIPPA, an online survey magazine, conducted a study on data breach attacks on healthcare businesses. According to this analysis, the healthcare industry is currently experiencing its greatest data breach attack since 2009 [6]. To maintain the integrity, confidentiality, and availability of data, the healthcare industry requires strong defenses against malware attacks, as shown in Figure 9.2. According to HIPPA's study, the healthcare industry has had 25 major data breaches in the last 12 years.

We were able to categorize the percentage ratio of the type of attack that was used more frequently in healthcare companies using this data. Events alone are responsible for 62% of significant healthcare assaults, according to Figure 9.3. This is a substantial ratio in any industry [6]. The requirement for a systematic and watertight package for controlling data integrity and smart hospital security is demonstrated by a critical examination of this sort of categorization. According to one analysis, 94% of healthcare organizations have experienced cyberattacks [7]. According to an annual research study on healthcare data breaches, the number of leaked records increased in 2018 compared to 2017 [8]. According to an internet post, on the dark web, any healthcare document may cost from $1 to $1000.

On the dark web, this is the second-highest cost for any asset [9]. Cancer Treatment Centers of America (CTCA), Southeastern Regional Medical Center, revealed 16,819 records of cancer patients in 2019 by targeting their emails [10]. According to an internet news outlet, the American Medical Collection Agency (AMCA) was hacked

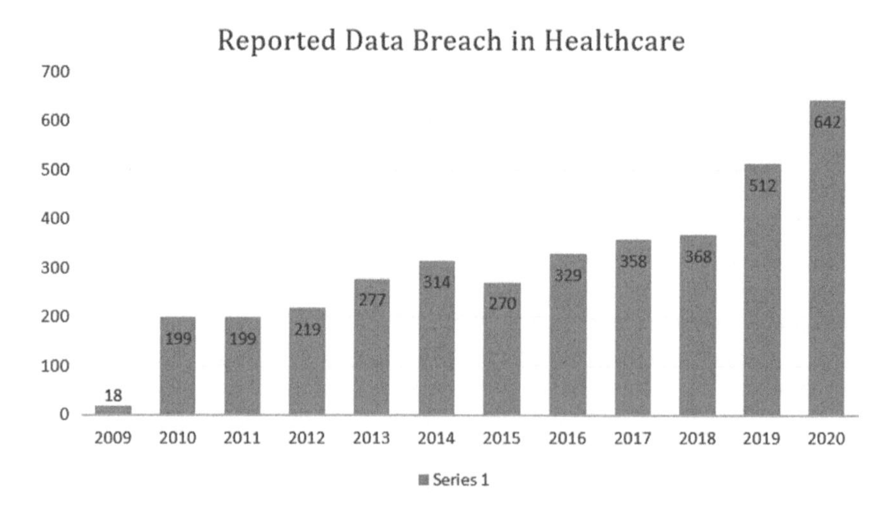

FIGURE 9.2 Data breaches in the healthcare industry in the last 12 years.

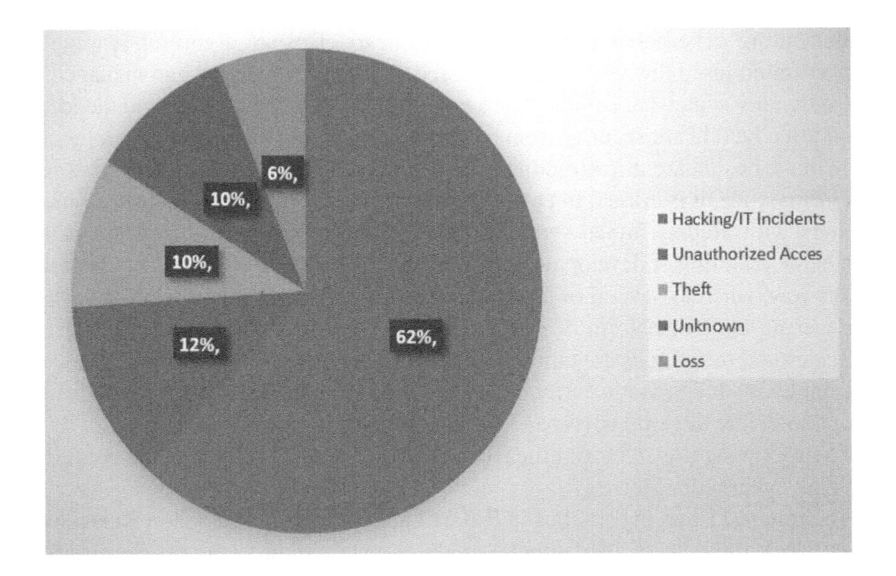

FIGURE 9.3 Percentage ratio of the type of breach in the healthcare industry.

for eight months in early May 2019, compromising the data of 25 million patients. During this attack, classified and sensitive data such as billing records and patient prescriptions were compromised [11].

Recent data breaches disclosed by two large healthcare companies, Quest Diagnostics and Labcorp, have affected the data of almost 19 million patients via a service provider they shared [12]. The worldwide healthcare cyber security market is expected to reach USD 27 billion by 2025, according to a new study by Global Market Insights. Another shocking incident from 2019 is the breach of 10,993 availers' data at the American Baptist Homes of the Midwest [13–15] due to hacked emails and network servers. The statistics presented in this section of the report provide a clear picture of attack trends and a review of assaults on healthcare services in past years. A thorough examination of these incidents reveals the current state of data security and cyberattacks in healthcare. Uncertainty is also bred through data tampering. The ramifications of ambiguity in today's data-driven world are terrifying [16–19]. A data breach may put commerce, health, infrastructure, national security, and political processes at jeopardy. Data tampering is more sophisticated, putting into question not just an industry's capacity to secure its data but also the data's validity. Consider the ramifications if terrorists tamper with or doctor sensitive military and government information. Manipulation of highly confidential data might have disastrous results. This example highlights the critical need for understanding the present state of data security and integrity in healthcare organizations.

9.3 PREVIOUS SIMILAR RESEARCH INITIATIVES

The authors reviewed important research initiatives conducted in this subject to grasp the scenario of healthcare security challenges for a healthcare system by implementing cybernetics principles in healthcare. Despite a careful search, the authors found just a few works that discussed healthcare modeling and management that were only somewhat similar. The suggested work's main concept and the idea is to examine healthcare security from a cybernetics perspective, allowing professionals to make healthcare infrastructure safer against a variety of difficulties [20–23]. In this perspective, it is critical to emphasize that cybernetics is not a method or methodology with predetermined symmetric stages. An ideology of thought processes aids in the planning of development milestones. In our study, we applied cybernetics in this way, whereas several of the works we cite simply associate cybernetics with engineering and some form of system development. In addition, several different strategies and techniques are tailored to the healthcare industry. Nonetheless, due to its complexity and breadth, entire healthcare data management must be redefined. It is vital to define the total governance of healthcare data.

The following are some significant studies that might be cited in the context of applying cybernetics ideology:

The role of IT and cybernetics in Russian healthcare is discussed in Korotkova's paper. With the help of fundamental analysis and examples, the article discusses the current state and extent of cybernetics in healthcare [24–27]. The main focus of the article is definitely on the evolution of information technology in the country's healthcare infrastructure. According to Korotkova, health is a domain that will never end since as long as humans exist on the planet, they will become ill or suffer from diseases due to the basic constitution of the human body. The researcher goes on to say that because computers are used in almost every part of daily life and business, the information technology association in healthcare will be a major revolution. The author goes on to say that striking a balance between digitization and healthcare can be achieved by employing cybernetics ideology.

Furthermore, Faggini et al. offered a methodology for ensuring healthcare sustainability using multiple digital infrastructures. The research suggested DocBox24, a theoretical model based on online sustainable healthcare service delivery [28–33]. The writers also depicted real-life cases and presented a comparative analysis based on various facts to corroborate their study's work. The study did an excellent job of demonstrating the importance and power of digital infrastructure in healthcare. Another study [14] explores blockchain-based safe data management and travel in an Internet of things (IoT) context to handle the digital infrastructure and data management of healthcare. This type of methodology and work is in great demand and importance in the current era of digitalization. The paper depicts a blockchain-based approach for secure communication in healthcare that works on all types of data layers.

Yang et al., for example, worked on data security and validation in the healthcare field. The paper evaluates a data validation scenario before proposing a secure and effective approach for dealing with data validation concerns in healthcare [15]. The study's findings make a substantial contribution to the current research in this field.

The use of sociocybernetics in health management is discussed in a technical report. Although the application of cybernetics presented in this study is from a social standpoint, we were able to connect it to the scope of cybernetics in healthcare.

9.4 NEEDS AND IMPORTANCE OF CYBERNETICS IN HEALTHCARE SECURITY

The growing digitalization of information in the healthcare sector has improved services, but it has also resulted in a negative side-effect: a security risk. Healthcare providers must have a robust and trustworthy information security service in place due to the sensitivity of healthcare data and the rising threat of information security. Not only should the strategies be able to respond to and safeguard healthcare data, but they should also be able to predict and avert cybercriminal threats.

Protecting patient information is more important and difficult than cutting costs. To safeguard your healthcare information, you must have a thorough data security framework and plan in place. Instead of being reactive, your cyber security service should be proactive. It should be capable of detecting and preventing an attack before it occurs.

The healthcare information should be inventoried and monitored by a professional information security firm. Figure 9.4 depicts how data is gathered, stored, utilized, and communicated between departments, on the cloud, in devices, in data centers, and across the network. It will then provide custom-tailored data security solutions which help. Because it employs a feedback looping system, the notion of cybernetics could be effective in preventing data breaches. Every communication step provides input on their position, making the system more dependable, secure, and authenticated.

Control theory, or cybernetics (Wiener, 1948), is a cross-disciplinary field of engineering and computer logic. It is concerned with the behavior of dynamical systems with inputs and how feedback affects that behavior. In 1948, Norbert Wiener coined the concept "the study of communication and control in the human and machine." Nowadays, the word is frequently misused to mean "technology-assisted control of any system."

The controlled object and the controller are the two parts of a control system. The managed object, which is the element that implements the business's functionality,

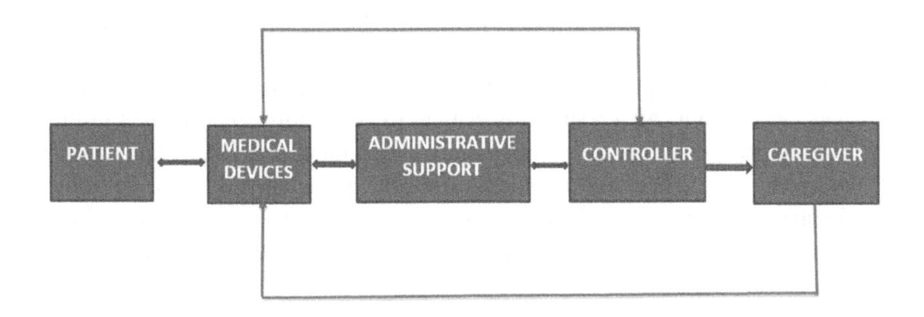

FIGURE 9.4 Object connectivity in healthcare systems.

is what is provided. The controller [16] is what is produced based on the model of the controlled item. The system's purpose is to keep the specified properties of the controlled object's output at, or suitably close to, the enabling factors in the presence of various disruptions to the controlled item (also called the setpoint). To achieve the system's goal, the control system must carry out certain actions that have an influence on the controlled object via the manipulated variables. This is a list of the controller's controls.

9.5 CHALLENGES

1. Validation of data: The authentication service offers permission, which is required by both medical and non-medical online services. Authentication is necessary for each medical sensor and base station in a cybernetics-based healthcare system to ensure that data is transmitted by a trustworthy sensor.
2. Key distribution: When two parties communicate information, they must share a session key, which must be kept secret. The secure session key aids in the security of subsequent communications and protects data against a variety of security threats. As a result, in a cybernetics-based healthcare system, an efficient key distribution system is a critical necessity to safeguard patient privacy [17].
3. Strong user authentication: Because the vulnerability of wireless messages to unauthorized users is a major issue in a wireless healthcare environment, it is desirable to consider a strong user authentication force in which each user must prove their authenticity before accessing the patient's physiological information. Strong user authentication, often known as two-factor authentication in cybernetics, improves the security of healthcare applications that employ wireless medical sensor networks [18].
4. Data integrity: Data integrity services guarantee that data has not been changed during transit by an adversary on the receiving end. Patient information can be tampered with by an adversary due to the sensor network's broadcast nature; this can be extremely harmful in the case of critical life events. To ensure data integrity, one must be able to spot any data alteration carried out by unauthorized parties. As a result, adequate data integrity procedures guarantee that the information received is not tampered with.
5. Confidentiality of patient health data: Patient health data is typically subject to legal and ethical confidentiality responsibilities. This medical information must be kept private and only approved physicians and nurses should have access to it. As a result, it is critical to keep individual medical information private so that an enemy cannot snoop on patient data. Data eavesdropping can be harmful to patients since the adversary might use the data for any illegal reason, infringing on the patient's privacy. As a result, data confidentiality is a critical necessity in the cybernetics-based healthcare system.
6. Data refresh: In a cybernetics healthcare system, data security and integrity are insufficient without consideration of data freshness. Because the data is

fresh, it suggests that the patient's physiological signals are new or discernible, and therefore, an adversary did not repeat previous messages. There are two sorts of refresh: low novelty, which gives a complete ordering on a request-response pair and permits latency estimate, and high novelty, which provides a total ordering on a request-response pair but does not contain delay information.

7. Access control: Because doctors, nurses, pharmacists, insurance companies, laboratory staff, social workers, and others are directly involved in a patient's physiological data in a healthcare system, a role-based access control mechanism that can restrict access to physiological information, such as user roles, should be implemented in a real-time cybernetics healthcare system.
8. Data availability: Data availability guarantees that services and information are available when needed. As a result, the medical sensor node's availability assures that caregivers have access to patient data at all times. Because data availability will be lost if a sensor node is seized by an enemy, it is critical to maintain the cybernetics healthcare system up and operating in the case of a loss of availability.
9. Patient permission: When a healthcare practitioner exposes a patient's medical records to another healthcare specialist, the patient must give their approval.
10. Secure localization: Estimating the patient's position is critical in a cybernetics healthcare system. The absence of intelligence in patient monitoring in a real-time cybernetics healthcare system allows attackers to communicate erroneous patient positions via fake signals.
11. Bottom-up and top-down privacy: Because new medical sensors are frequently deployed when old sensors fail in a real-time network health system, bottom-up and top-down privacy is critical. A medical sensor in secure mode or above cannot read future transmitted messages after leaving the network, but a sensor in secure mode or below cannot read any previously transmitted messages [12].
12. Communication and computation costs: Because wireless medical sensors are restricted in resources and medical system functions require space to function, security systems must be cost-effective in terms of communication and computation.

9.6 FUTURE WORKS

Healthcare solutions that are both functional and cost-effective. Hospitals and healthcare systems may increase production, profitability, and efficiency by converting humans into human labor and providing them with an artificial intelligence (AI)-powered technician to assist them around the clock to the influence of human labor on investment.

An AI "curator" that lives on the system of a healthcare practitioner aids the cybernetics solution. It collaborates with the user to give useful information at precisely the correct time. Cybernetics helps to minimize the time it takes to accomplish important functions like patient check-ins, payments, and more by continuously depending on signals from its surroundings. By offloading frequent and heavy data input, cybernetics may enhance your security and workforce. Hospitals and healthcare systems are always seeking for innovative methods to increase efficiency and reduce friction in their operations. Healthcare employees, on the other hand, are currently drowning in a sea of data and programs. Burnout is at an all-time high, and COVID19 has only added to the need for efficiency and capacity. Healthcare organizations that want to accomplish more with less are turning to technology to aid their workers in completing their tasks swiftly and effectively.

Executives in the healthcare business are turning to artificial intelligence, including automation and the emerging science of cybernetics, to boost productivity and decision-making while decreasing fatigue. At a time when time is more valuable than ever, AI is giving tailored information to workers to drive better and quicker outcomes in hospitals and healthcare systems.

When healthcare organizations think about investing in AI, they usually start by automating processes with a robot or an intelligent automation solution. We can unload laborious and repetitive data entry into machines with smart automation, allowing our team to focus on higher-level projects that require the touch of a consumer, people. The way hospitals and healthcare systems handle healthcare is changing fundamentally, allowing our employees to focus on what matters most: providing a high-quality patient experience.

However, increasing the workload of intelligent automation robots is not the sole useful application of AI. Once time-consuming and repetitive jobs are automated, there is a great potential to enhance processes that still require human intervention. This is why cybernetics is being used to enhance labor that cannot be mechanized in healthcare systems. Cybernetics is a fast-evolving field that will play a critical role in assisting companies in maximizing worker unload time while also increasing employee happiness and work-life balance.

Cybernetics increases the habit of offloading people and laborious data entry by decreasing mistakes, boosting efficiency, and improving decision-making. Cybernetics is being used in the planning department of The Ohio State University Wexner Medical Center to assist staff plan patients faster and with fewer mistakes. When an X-ray or imaging scan is planned, a function that automatically raises the patient's weight is provided. This information is required, but it may be too humiliating for the patient to ask directly, or it may need the patient to open another program and conduct more research. When a patient is scheduled, critical information about the appointment may be given to the scheduler and then double-checked for correctness before the appointment is booked.

Cybernetics is a method of delivering focused and correct information to employees. Employees are more productive when they do not have to switch programs to get information. You enhance decision-making every time a recommendation is provided utilizing sophisticated analysis. You save money by reducing costly mistakes every time a typo is spotted. When cybernetics and automation are integrated, the healthcare system will be able to accomplish more with less resources.

Artificial intelligence is already altering the way we work and will continue to do so in the not-too-distant future. However, this is not the AI described in science fiction in which machines take over our jobs. Artificial intelligence, intelligent automation (IA), and cybernetics (Cybernetics) are now affordable.

9.7 CONCLUSION

Running the healthcare system smoothly in this era is a difficult task. Every new update and patch in the system creates a vulnerability or possibility of failure, thus posing a huge risk to the system as well as to the user. This scoping review identified the need and importance of cybernetics in healthcare security to mitigate cyberattacks targeting the healthcare sector, as well as the challenges of cybercriminals. Studies in healthcare security and areas for improvement suggest future work with AI-based cybernetics in healthcare security. We have provided useful information for cybernetics in the healthcare sector on the issue of cyber security.

REFERENCES

1. http://healthlab.edu.au/cybernetics/
2. Glanville, R. A (cybernetic) musing: Design and cybernetics. *Cybern. Hum. Knowing* 2009, 16, 175–186.
3. Alhakami, W.; Baz, A.; Alhakami, H.; Pandey, A.K.; Khan, R.A.Symmetrical model of smart healthcare data management: A cybernetics perspective. *Symmetry*2020, 12, 2089.
4. Agrawal, A.and Alharbe, N. R., Need and importance of healthcare data integrity. *Int. J. Eng. Technol.*, Aug. 2019, 11, no. 4, 854–859.
5. Chakraborty, R., Mathew, J., and Vasilakos, A., Eds., Security and fault tolerance in Internet of things. *Signal and Communication*. Springer,2019, doi: 10.1007/978-3-030-02807-7.
6. *Healthcare Data Breach Statistics*. 2019. Accessed: Oct. 21, 2019. [Online]. Available: https://www.hipaajournal.com/healthcare-data-breach-statistics/
7. Filkins, B.*SANS Health Care Cyber-threat Report: Widespread Compromises Detected, Compliance Nightmare on Horizon*. Norse, 2014. Accessed: Oct. 21, 2019. [Online]. Available: https://www.sans.org/reading-room/whitepapers/_rewalls/paper/34735
8. *Breached Patient Records Tripled in 2018 vs 2017, as Health Data Security Challenges Worsen*. 2018. Accessed: Oct. 23, 2019. https://www.protenus.com/press/press-release/breached-patient-records- tripled-in-2018-vs-2017-as-health-data-security challenges-worsen
9. *Here's How Much Your Personal Information Is Selling for on the Dark Web*. 2017. Accessed: Oct. 23, 2019. https://www.experian.com/blogs/ask-experian/heres-how-much-your-personal-information-is-selling-for-on-the-dark-web/
10. *Healthcare Data Breaches Reach Record High in April*. 2019. Accessed: Oct. 27, 2019. https://www.modernhealthcare.com/ cybersecurity/healthcare-data-breaches-reach-record-high-april
11. *The 10 Biggest Healthcare Data Breaches of 2019*. 2019. Accessed: Nov. 4, 2019. [Online]. Available: https://healthitsecurity.com/news/the-10-biggest-healthcare-data-breaches-of-2019-so-far
12. WangY., AtteburyG., RamamurthyB.A survey of security issues in wireless sensor networks. *IEEE Commun. Surv. Tutor.*2006;8, 2–23.
13. Korotkova, O.M.; Korotkova, O.M.; Belokoneva, I.V.; Belokoneva, I.V.Development of it and cybernetics in Russian healthcare: Past, present, future. *Ìîëîäåæíûéèííîâàöèîííûéâåñòíèê*2019, 8, 107–108.

14. Faggini, M.; Cosimato, S.; Nota, F.D.; Nota, G.Pursuing sustainability for healthcare through digital platforms. *Sustainability*2019, 11, 165. [CrossRef]
15. Yang, P.; Stankevicius, D.; Marozas, V.; Deng, Z.; Liu, E.; Lukosevicius, A.; Min, G.Lifelogging data validation model for internet of things enabled personalized healthcare. *IEEE Trans. Syst. Man Cybern. Syst.*2016, 48, 50–64. [CrossRef]
16. https://www.sciencedirect.com/topics/computer-science/cybernetics?__cf_chl_captcha_tk__=pmd_jxGS6UioOnd8USXGiV4bmksldK_y_fSvhoosFycd62o-1632890525-0-gqNtZGzNAuWjcnBszQi9
17. MisicJ., MisicV.Enforcing Patient Privacy in Healthcare WSNs Through Key Distribution Algorithms. *Secur. Commun. Network.* 2008;1, 417–429.
18. http://www.techrepublic.com/whitepapers/strong-user-authentication-and-hipaa-cost-effective-compliance-with-federal-security-mandates/2345053.
19. Kumar, R.; Khan, S. A.; Khan, R. A. Revisiting software security: durability perspective. *Int. J. Hybrid Inf. Technol.* 2015, 8, no. 2, 311–322.
20. Kumar, R.; Khan, S. A.; Khan, R. A. Durability challenges in software engineering. *Crosstalk-J. Defense Software Eng.* 2016, 29–31.
21. Sahu, K.; Shree, R.; Kumar, R. Risk management perspective in SDLC. *Int. J. Adv. Res. Comput. Sci. Software Eng.*, 2014, 4, no. 3, 1–15.
22. Sahu, K.; Alzahrani, F. A.; Srivastava, R. K.; Kumar, R. Hesitant fuzzy sets based symmetrical model of decision-making for estimating the durability of Web application. *Symmetry*, 2020, 12, no. 11, 1770.
23. Kumar, R.; Khan, S. A.; Khan, R. A. Analytical network process for software security: a design perspective. *CSI Trans. ICT*, 2016, 4, no. 2, 255–258.
24. Sahu, K.; Alzahrani, F. A.; Srivastava, R. K.; Kumar, R. Evaluating the impact of prediction techniques: Software reliability perspective. *Comput. Mater. Continua*, 2021, 67, no. 2, 1471–1488.
25. Kumar, R.; Khan, S. A.; Khan, R. A. Durable security in software development: Needs and importance. *CSI Commun.*, 2015, 39(7), 34–36.
26. Ansari, M. T. J.; Baz, A.; Alhakami, H.; Alhakami, W.; Kumar, R.; Khan, R. A. P-STORE: Extension of STORE methodology to elicit privacy requirements. *Arabian J. Sci. Eng.*, 2021, 46, no. 9, 8287–8310.
27. Kumar, R.; Khan, S. A.; Khan, R. A. Software security testing: A pertinent framework. *J. Global Res. Comput. Sci.*, 2014, 5(3), 23–27.
28. Attaallah, A.; Alsuhabi, H.; Shukla, S.; Kumar, R.; Gupta, B. K.; Khan, R. A. Analyzing the big data security through a unified decision-making approach. *Intell. Autom. Soft Comput.*, 2022, 32, no. 2, 1071–1088.
29. Almulihi, A. H.; Alassery, F.; Khan, A. I.; Shukla, S.; Gupta, B. K.; Kumar, R. Analyzing the Implications of Healthcare Data Breaches through Computational Technique. *Intell. Autom. Soft Comput.*, 2022, 1763–1779.
30. Pandey, A. K.; Al-Amri, J. F.; Subahi, A. F.; Kumar, R.; Khan, R. A. Analyzing the implications of COVID-19 pandemic through an intelligent-computing technique. *Comput. Syst. Sci. Eng.*, 2022, 959–974.
31. Kumar, R.; Khan, A. I.; Abushark, Y. B.; Alam, M. M.; Agrawal, A.; Khan, R. A. An integrated approach of fuzzy logic, AHP and TOPSIS for estimating usable-security of web applications. *IEEE Access*, 2020, 8, 50944–50957.
32. Kumar, R.; Zarour, M.; Alenezi, M.; Agrawal, A.; Khan, R. A. Measuring security durability of software through fuzzy-based decision-making process. *Int. J. Comput. Intell. Syst.*, 2019, 12, no. 2, 627.
33. Kumar, R.; Khan, A. I.; Abushark, Y. B.; Alam, M. M.; Agrawal, A.; Khan, R. A. A knowledge-based integrated system of hesitant fuzzy set, ahp and topsis for evaluating security-durability of web applications. *IEEE Access*, 2020, 8, 48870–48885.

10 Threats and Countermeasures in Digital Crime and Cyberterrorism

Mohit Kumar, Ram Shringar Raw, and Bharti Nagpal

CONTENTS

10.1 Introduction 173
10.2 Literature Review 179
10.3 Proposed Framework 181
10.4 Data Flow Diagram 182
10.5 Countermeasures against Cybercrime and Cyberterrorism 183
10.6 Comparative Analysis 184
10.7 Application Work 185
10.8 Research Implications 185
10.9 Research Limitations 187
10.10 Conclusion and Future Scope 187
References 188

10.1 INTRODUCTION

Cyber threat is more threatful for the information system security and cyber security. A threat is defined as the ability to do serious harm to the computer system, leading to a cyberattack, which can compromise our information system and communication system in the network [1–3].

In the above Figure 10.1, various types of threats are shown, which are discussed as follows (Table 10.1):

Vulnerability is defined as the weakness in a computer system or in the network system that an attacker can exploit and can gain unauthorized access to the information or can perform a cyberterrorism activity by using it [4, 5].

DOI: 10.1201/9781003323426-10

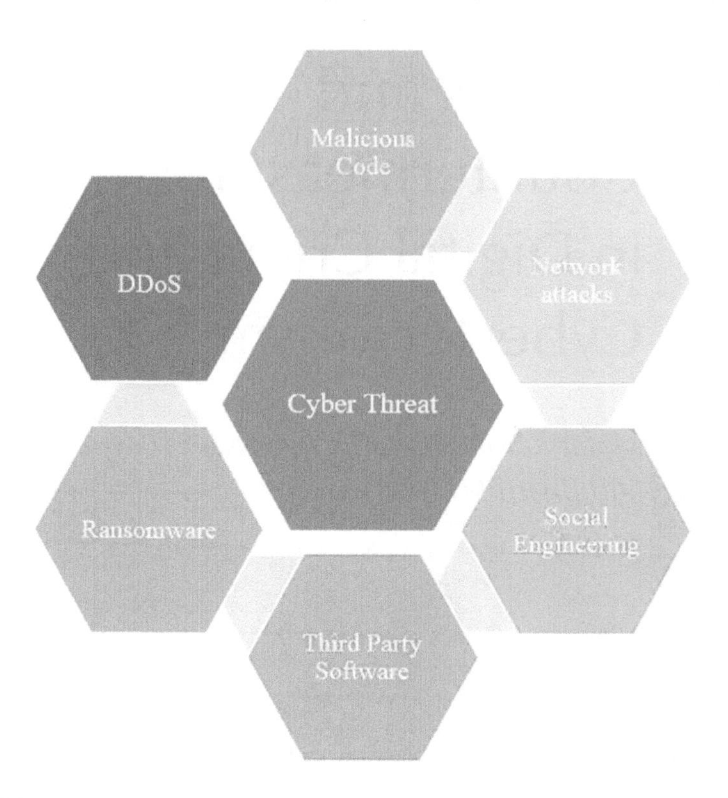

FIGURE 10.1 Types of Cyber Threats.

There are different types of vulnerabilities in a computer system, which are as follows:

- **Type 1 Vulnerability:** In this type of vulnerability, we can check suspicious inputs for malicious activities in a website with the help of input validation. Suspicious inputs may permit a malicious code to be executed many times without proper and exact verification on the original intention [4, 5].
- **Type 2 Vulnerability:** In this type of vulnerability, there is the difficulty in the characterization of different data types which we use in the programming language for web development.
- **Type 3 Vulnerability:** This type of vulnerability is defined as any process delay in the analysis stage until the runtime stage as the present variables are measured despite the source code using an expression to achieve the attack.
- **Type 4 Vulnerability:** In this type of vulnerability, there is improper definition of the datatype while designing [4, 5].

In a digital crime or cybercrime, a computing device is used to execute the cybercriminal activity. It can harm confidential information, cause security breaches,

TABLE 10.1
Types of Threats [1–3]

S. No.	Threats	Description
1.	Malicious Code	It is a type of code, which is embedded in any application and automatically executes itself.
2.	Network Attacks	In this, an attacker can compromise the networks running in the organizations, governments, etc. to make the whole network vulnerable to this attack.
3.	Social Engineering	In this, the attacker manipulates the people to extract confidential information, breaking the communication network for cyberattack.
4.	Third-Party Software	When we download third-party software from the internet, it can make our computer system vulnerable to cyberattack.
5.	Ransomware	In this threat, an attacker applies encryption on user confidential information and wants ransom for giving the confidential information.
6.	DDoS Attacks	In this attack, different attackers in remote locations can increase the traffic of a particular network or server to make the server inaccessible to the user.

financial harm, privacy violations, and steal government, country, and military secret information to fulfill the criminal objective [6–8].

In the above Figure 10.2, classification of cybercrimes are discussed as follows:

1. **Crime against individual:** In this type of cybercrime, an attacker does criminal activity against individual users to gain access to confidential information, commit financial frauds, etc. For example, phishing, spoofing, sniffing [6–8].
2. **Crime against property:** In this type of cybercrime, an attacker does unauthorized trespassing with the use of computers, stealing someone's copyright information, causing computer vandalism, etc. [6–8].
3. **Crime against organization:** In this type of cybercrime, an attacker does the crime against the organization, governments, country, etc. to gain access to confidential information, cause security breaches, violate privacy, etc. For example, trojan horse, logic bomb, etc. [6–8].
4. **Crime against society:** In this type of attack, an attacker can gain access to confidential information, malware, network intrusions, etc. against any society and compromise the information security [6–8].

There are various kinds of cybercrimes in the digital world, which are as follows (Table 10.2):

Cyber terrorism is the most harmful word in the cyber space. It can create a havoc or cause violent actions such as making threats or causing serious bodily harm or loss of life with the help of a computer system and the internet. The main focus of

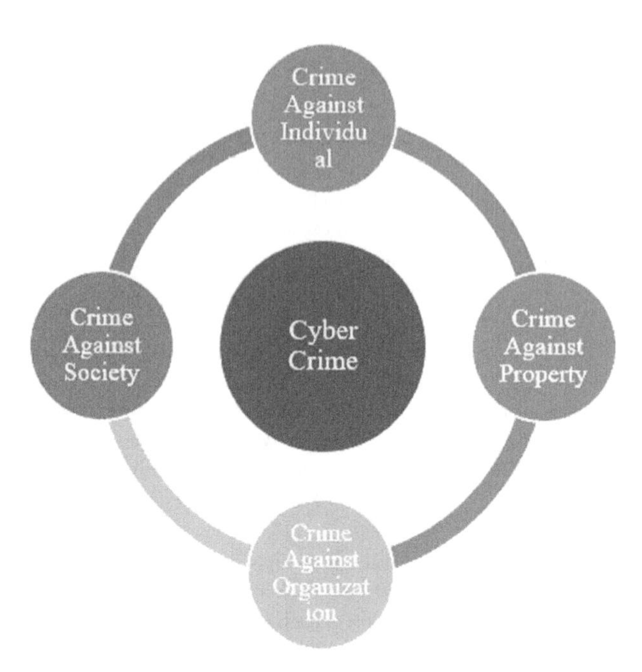

FIGURE 10.2 Classification of Cybercrime.

cyberterrorism is to gain political and ideological advantages by making a more fearful or threatful environment in the mind [9–11].

The above Figure 10.3 of cyberterrorism shows how a cyberattack occurs to execute terrorism against any country, organizations, etc., which creates a fear in the mind of the government and organization of any country and also in the normal people about the cyberattack in their system. Cyberterrorism is harmful for the cyber security of countries and organizations, but it is more destructive in nature as it destroys and harms the information system, and to resolve that cyber terror the cyber terrorist demands ransom from the government of the country or from the people earnings [9–11].

Five types of cyberattacks are as follows:

1. **Incursion:** In this type of attack, an attacker can gain unauthorized access to information for the purpose of stealing the information or modifying the information [9–11].
2. **Destruction:** In this type of attack, the main motive of the attacker is to intrude into someone's computer system and intentionally harm the computer system or destroy the information from the computer system.
3. **Disinformation:** In this type of attack, an attacker spreads rumors across the internet to harm a particular nation and can create threats and fears in the minds of the various developing or developed nations.
4. **Denial of service:** In this attack, an attacker can make any website or servers unavailable by overloading the network traffic on that website and make the web application not accessible for the normal user.

TABLE 10.2
Various Kinds of Cybercrimes [6–8]

S. No.	Digital Crime	Description
1.	Phishing	An attacker sends the malicious URLs to the user to gain the access of the user information and user computer to do cybercrime.
2	Malware	This type of crime can be done with the help of virus, worms, malicious codes and compromise the user computer system and harm the computer system.
3.	Ransomware	In this attack, an attacker applies encryption on user confidential information and wants ransom for giving the confidential information.
4.	Identity Theft	This type of cybercrime occurs when an attacker uses the identity of another person for stealing personal information, money, etc.
5.	Cyber Bullying	In this crime, an attacker harasses, bullies the normal user with the help of electronic or digital communication for personal benefit.
6.	Virus	It is an executable file embedded in an application and when we install or click the application, it harms the computer system.
7.	DoS Attack	In this attack, an attacker can increase the traffic of a particular network or server to make the server inaccessible to the normal user.
8.	Trojan Horse	It is an auto executable virus or malware which embeds itself in the application or in the file which is downloaded from the internet by the user and can harm the user's computer system.
9.	Cyberstalking	In this crime, an attacker harasses or frightens the normal user with the help of electronic or digital communication for personal benefit.
10.	Spoofing	In this an attacker impersonating himself as a legitimate user and stealing confidential information, performing financial fraud, etc.
11.	Network Sniffing	In this cybercrime, the attacker sniffs the data packets in between the communication channel of the sender and the receiver or the interception between the communication channel and steals the confidential information.

5. **Defacements of websites:** In this attack, an attacker changes the websites partially or totally for sharing threatful messages in the public domain and also to other websites and creates a threatful and fearful environment for the public [9–11].

This section gives you the knowledge about the threats, digital crime or cybercrime, and cyberterrorism, and cyber laws are needed to compensate and put the cyber terrorist behind bars. A cyber law is defined as the rules and regulations defined by government bodies to protect and prevent the public and nation from cyberterrorism, and in India, the cyber law came into existence as Information Technology Act, 2000 (IT Act), which needs to be followed by every person in the nation strictly as a protective and preventive measure because existing laws are not sufficient when it comes to the cyber space activity; hence, the need for the cyber law to remind us to strictly follow the rules [12, 13].

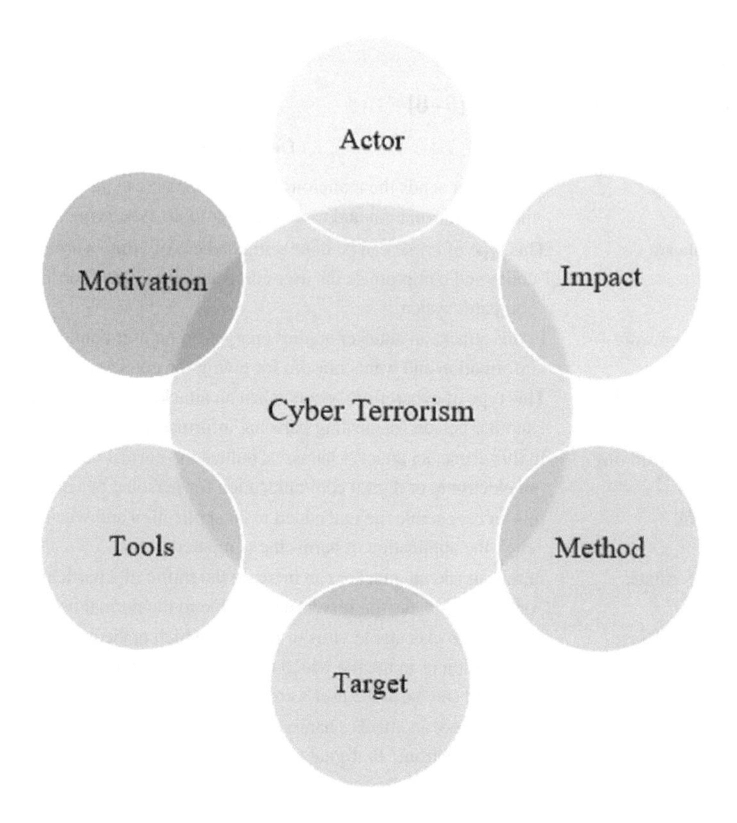

FIGURE 10.3 Cyberterrorism.

There are different sections in the cyber sections, which we need to strictly follow; some of them are as follows:

1. Section 43 punishment to the attacker in which the computer device has been damaged by the attacker intentionally [12, 13].
2. Section 66 imprisonment of 3 years to the attacker for unethical hacking and compensation of Rs. 1 crore and a fine of Rs. 2 lakhs.
3. Section 67 for publishing of the obscene material information and for this a fine of Rs. 1 lakh and imprisonment of 5 years and also double charge for the 2nd attempt.
4. Section 68 not fulfilling the guidelines of the controller.
5. Section 72 punishment for breaching the confidentiality of the computer system.
6. Section 73 punishment for publishing a false virtual signature.
7. Section 74 punishment for publishing the digital signature for fulfilling their fraudulent motive and under this section imprisonment of 10 years and a fine of Rs. 1 lakh [12, 13].

10.2 LITERATURE REVIEW

In the paper, authors [1] give an overview about the threats and vulnerabilities in cloud computing and how the organization decision maker responds after detecting the threat and vulnerability in the cloud and makes cloud computing more secure.

In the paper, authors [2] give an overview about the threats and vulnerabilities in blockchain technology and categorize the threats and vulnerabilities as per the last 10 years of real cyber security breaches. This paper also concerns about the future research work in developing countermeasures against the threats and vulnerabilities.

In the paper, authors [6] investigate the threats and vulnerabilities in the smart grid system. This paper also proposed solutions as countermeasures for the theft of electricity in the smart grid system.

In the paper, authors [14] identify and analyze the different types of threats and vulnerabilities in a web-based wallet application to give us the understanding of the threats which affect a mobile wallet application. This paper also provides countermeasures against these threats and vulnerabilities.

In the paper, authors [15] surveyed the reason for vulnerabilities in IoT technologies and also the limitations in the existing research. This paper also discussed the different methodologies for the IoT system. This paper also presented the classification of existing IoT protocols and did the comparison between them and also discussed the comparative analysis of the different IoT-based simulation tools.

In the paper, authors [7] proposed the security reference architecture for the blockchain for studying the threats, vulnerabilities, and defenses or countermeasures against the threats. This paper gives the understanding of the security and privacy aspects of the blockchain.

In the paper, authors [9] showed how we can assist the network operators to understand the need of IoT in network security and also the network attacks, threats, vulnerabilities, and their countermeasures.

In the paper, authors [10] give an overview about the IoT and its applications and also present the challenges of the IoT. This paper surveyed the state-of-the-art data security solutions for the IoT.

In the paper, authors [16] presented a systematic review of the threats, vulnerabilities, and their countermeasures to mitigate the security problem that happened in the SaaS environment in cloud security with the help of the SALSA framework.

In the paper, authors [17] presented the cyber threat in the industrial IoT and also surveyed the attacks and countermeasures in industrial IoT. This paper also provides a thorough analysis of the solutions for the cyberattack in the industrial IoT.

In the paper, authors [18] presented the cyberattacks, threats, and vulnerabilities, and this paper also proposed the framework for the countermeasures in the applications of the IoT.

In the paper, authors [19] discussed the different security issues such as cyberattacks, threats, vulnerabilities, and countermeasures for the IoT system. This paper also provides the increase in the awareness about the cyberattacks and also improvements in the security system of the IoT devices.

In the paper, authors [20] showed the identification of the security vulnerabilities, attacks, and countermeasures of the C3I (command, control, communication, and intelligence) system.

In the paper, authors [21] presented the security aspects of the private data centers and clouds in the software-defined network and also provided the information about the threat, vulnerabilities, and the countermeasures in the software-defined network.

In the paper, authors [22] discussed the various types of vulnerabilities that can be exploited by the attackers in the neural network-based system. This paper also presented the different challenges in the implementation of the attacks. This paper overviews the different attacks on deep neural networks.

In the paper, authors [23] discussed the various types of cybercrime and the different types of cyberterrorism against the state or the country.

In the paper, authors [24] discussed blockchain technology to counter the risk of cyberterrorism and increase the information security system for the country or the state and make the information system less vulnerable to cyberterrorism.

In the paper, authors [25] presented the development of cyberterrorism against the various developed and developing nations and also the challenges and threats it poses to global security.

In the paper, authors [26] discussed the reasons for the worldwide increase of cyberterrorism in the current time and in the upcoming time and how to eliminate or reduce cyberterrorism worldwide.

In the paper, authors [27] focused on the role of the attacker of these cybercrimes in India and what was the impact of the current cyber laws on the cyber terrorists' intentions.

In the paper, authors [28] presented the understanding of the level of cyber terrorism and cybercrime in India and how to prevent our country from cyberterrorism and with the prevention we have to take initiative to create awareness about cyberterrorism among common people.

In the paper, authors [29] discussed the investigative process of computer forensics and the response of the judicial system toward digital evidence for providing a better judgment against cybercrime.

In the paper, authors [12] discussed the different types of cyber forensics tools and techniques which help in the investigation against cyberterrorism, cyber stalking, spams, etc. This paper also discussed the five standards steps for the investigative process and also the challenges faced in the investigative process of cyber forensics.

In the paper, authors [30] reviewed the cybercrimes in India. This paper also analyzes the study that fraud and victims of cybercrime are mostly between the age of 20 and 29 years and how it can affect children and women and how we can create awareness in the people of India against cybercrime.

In the paper, authors [31] presented the analysis of cybercrime prevention awareness by using the nearing neighboring algorithm in India and also discussed the countermeasures against cybercrime.

In the paper, authors [32] discussed the different types of cybercrimes and threats common people face on the internet and also in the cyber world, which can affect the information system security. This paper also classifies cyber threat and the impact of cyber threat in India.

In the paper, authors [33] studied the cyber laws and the use of the cyber laws to counter cyber threat, cybercrime, vulnerabilities, and cyberterrorism.

In the paper, authors [34] discussed the emergence and evolution of the main cyber terrorists' threats and the inabilities of countries to differentiate between two different types of threats.

In the paper, authors [35] give a brief overview of digital forensics and also about the investigative process of digital forensics and the different types of tools used in digital forensics and how it provides the evidence to the forensic person.

In the paper, authors [36] showed the comparative study of analysis and investigation using digital forensics in which the author conducted the comparative study of the different digital forensics' tools for analysis and investigation purpose and the use of these digital forensics' tools for different purposes at different instances of time.

In the paper, author [4] implemented the detection of the SQL injection attack using the NIST method in network forensics in which first it created SQL injection scenarios and after that it created the log file using the snort tool rule and after that the snort tool mitigated the SQL injection attack by alerting the system using email and analysis of the result with the help of user acceptance testing.

In the paper, author [37] showed the study of the digital forensics' branches and the digital forensics' tools and also the proper use of the digital forensics' tools in the digital forensics process to deal with cybercrimes.

In the paper, author [38] presented the network analysis using the AI-powered packet analysis methodology for network traffic classification and pattern identification in cybercrime, which can give us both hardware and network properties.

10.3 PROPOSED FRAMEWORK

Cyber security is the most important aspect for the cyber world. For providing cyber security to our information system, computer system, and network system, we have to take different countermeasures against cyberterrorism. Now, in this section we are proposing the framework for countering cyberterrorism in which it gives the understanding of which countermeasures we have to take at different scenarios or different times.

In the above Figure 10.4, the proposed framework provides an architecture of how we can counter cyberterrorism and cybercrime activities in the cyber world and also provide more security functionalities to the information system as well as to the computer system and also increase the efficiency of the cyber secure system in the cyber world.

Steps for the working of the proposed framework in Figure 10.4 are as follows:

1. In this system first, it identifies the activities or the number of occurrences of cyberterrorism activity in the information system.
2. After that it analyzes the severity of the cyberterrorism activity in which it analyzes the effectiveness of the cyberterrorism in the information system.
3. Then, it alerts the expert system by using the IDS in the information system because it provides the proactiveness against the cybercrime and cyberterrorism attack in the information system.

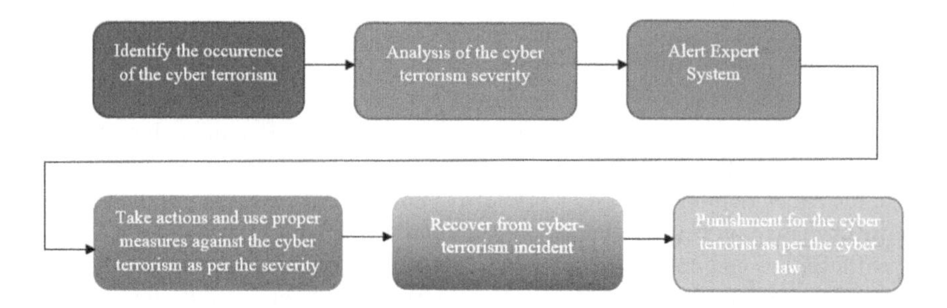

FIGURE 10.4 Proposed Framework.

4. Then, after alerting the expert system, the cyber secure system takes the proper actions and also the necessary countermeasures against the cyberterrorism activity as per the severity, if the severity of the cyberterrorism activity is high then it will take the customized measures or we can say the multiple and complex measures to counter the cyberterrorism activity and if the severity is low or medium then it will take general measures that must be taken and provide the cyber security to the information system.
5. Then, after taking the necessary actions, the information system or the cyber system needs to recover from the cyberterrorism activity and acquire the consistent state the information system had before the cyberterrorism attack. To recover and acquire the consistent state, there are multiple tools available in the cyber world, which can be used such as ProDiscover, Magnetic RAM Capture, Autopsy, and many more.
6. Then, after recovery, the cyber secure system needs to identify and find out the person behind the cyberterrorism activity by using the proper cyber forensic process, and after finding the culprit of the cyberterrorism activity, strict punishment must be there as per the cyber law for the culprit, which can give the message to the cyber world that no one should do this cyberterrorism activity against the cyber world and if done so forgiveness will not be given to them as per the cyber law.

This proposed framework gives the better understanding of how the countermeasure has to be taken by the cyber security expert and also by the normal user and this proposed work can enhance the efficiency and working of the cyber secure system and also decrease the occurrence of the cybercrime and cyberterrorism activity in the cyber world.

10.4 DATA FLOW DIAGRAM

In this section, we discuss the data flow diagram for the countermeasures against the cyberterrorism and the digital crime in the cyber world which consist of the events that are linked among them and functioned, which usually shows the flow of the data and the countermeasures we need to take against cyberterrorism.

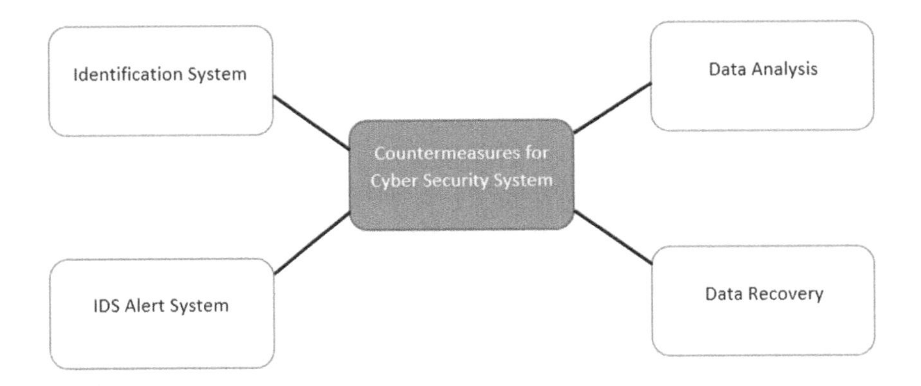

FIGURE 10.5 Data Flow Diagram.

The above data flow diagram in Figure 10.5 shows the flow of the event through the cyber security system and the countermeasures we have to take against cyberterrorism and digital crime. This data flow diagram consists of the following things:

- **Identification System:** To identify the occurrence of cybercrime activity in the cyber security system.
- **IDS Alert System:** This is used to alert the cyber system against the digital crime activity in the cyber security system.
- **Data Analysis:** This is used to analyze the data of the cybercrime activity in the cyber security system.
- **Data Recovery:** This is used to recover the cyber secure system after any cybercrime activity.

This data flow diagram for the cyber secure system can be helpful in reducing the cyber terrorism activity, safeguard the information system, streamline processes, and protect our cyber world. This system can be helpful in increasing the efficiency of the infrastructure of the cyber secure system. For protecting sensitive data, this structure or system can eliminate data breach failures against cyberterrorism activity and decrease the occurrence of cyber terrorism activity.

10.5 COUNTERMEASURES AGAINST CYBERCRIME AND CYBERTERRORISM

This paper discussed the threats, digital crime, and cyberterrorism, how they occur, and how they compromise the information system, and it also discussed the related research work in the literature survey. Now, in this section, how countermeasures against cybercrime or digital crime and cyberterrorism can help the user combat these activities and protect their information system from cyberattack is discussed [12, 39, 40].

Countermeasures against cybercrime and cyberterrorism are as follows [12, 39, 40]:

1. Need of proper awareness in the public about the information system security concerns in the cyber world to protect themselves from cyberterrorism by using antivirus, proper firewall functioning, IDS system, etc.
2. Government needs to encourage ethical hacking, which can be helpful in finding out the vulnerabilities in the computer and information system and prevent the normal user from these.
3. There is the need to update and modify rules and regulations as the technology emerges in the upcoming time, which can affect the computer system and most important information systems from cyberterrorism attack.
4. There is the need to create the combined and strong international rules which need to be followed by every country or nation in the whole world.
5. Use of the combination of the firewall and IDS in the computer system to alert against cyberattack, which provides the proactiveness against cyberterrorism attack in the information system.
6. Government needs to control the functioning of the social media sites and make the rules and regulations on the social media sites, which can prevent cyberterrorism. Nowadays, government follows these measures or rules strictly on the social media sites and the sites must follow them.
7. Government should provide training to the normal user for performing cyber space activity and increase the knowledge of the working in the cyber space, which can result in the reduction in the occurrence of cyberterrorism attack.
8. Government and cyber security agencies must develop strong firewall security rules that need to be implemented across the country and increase the functionality of the information system.
9. There should be proper conduction of the security audit from time to time for monitoring the cyber security in the country, which can prevent us from cyberterrorism activity and also decrease the occurrence of cyberterrorism activity in the information system.
10. Government should take strict actions against cyber terrorists for doing the cyberattack in the country as per the cyber law written by the constitution of the country and make our cyber defense system stronger and more effective and also increase the efficient working of the information system [12, 39, 40].

10.6 COMPARATIVE ANALYSIS

This paper discussed the different types of threats, vulnerabilities, cyberterrorism, and also the countermeasures against these threats and cyberterrorism. For taking the countermeasures against cyberterrorism, different types of cyber forensics or computer forensics tools are needed. Now, in this section, we are doing the comparative analysis of the different types of tools for cyber forensics. This comparative analysis gives the functionalities of cyber forensics tools and the use of these different tools in different scenarios [4, 36, 37].

In Table 10.3, we compare the functionalities of the different cyber forensics tools by using different parameters. This comparison provides a better understanding of these cyber forensics tools and provides the details for the appropriate use of these tools in appropriate situations, which gives a better understanding of the cyber forensics tools. This comparison will provide the identification, acquisition, and analysis of the cybercrime and cyber terror attacks in the cyber world, and working of these tools can be useful in the different architectures or in the different frameworks that the whole cyber world follows and also it is useful in the proposed framework, which is discussed in this research work [4, 36, 37].

10.7 APPLICATION WORK

This project has the different fields of applications in which it can be used to enhance the cyber security system, which are as follows:

- This research work can be used in the research area in which it enhances researcher's knowledge and proceeds with another research work in the field of countering cyberterrorism [16, 18].
- This research work can also enhance the cyber security in any organization by the efficient use of the digital forensics tools for the purpose of enhancing the cyber security functionalities [17, 19].
- This research work is also useful for the cyber security experts or cyber security organizations as it provides a better way to counter cyberterrorism and increase the efficiency of the cyber secure system by using the proper digital forensics tool at different instances of time [21, 23].
- This research work can also be used by the normal user as it gives a better understanding of the digital forensics tools to the user and security features in their personal information system or computer system [20, 22].

10.8 RESEARCH IMPLICATIONS

In the cyber world, cybercrime and cyberterrorism are the major issue of concern for the computer system as well as for the information system, and for that we have to take the countermeasures and the necessary steps to reduce the impact of cyberterrorism. This chapter discusses the different types of threats in cyberattack, different types of cyberterrorism, and different technologies that are used by the cyber terrorist. This chapter will also provide the discussion on the rules and regulations, policies, laws, and emerging technologies that are introduced in the cyber world, which can enhance and change the infrastructure of the cyber security, which can reduce and stop illegal activities in the cyber world and enhance the security infrastructure for the information system. This chapter gives an overview about the research work to follow, which can provide the review of network security, cyber security, internal threats, and different topics in cyber security. This chapter provides the comparative analysis of the different tools for countermeasuring cyberterrorism. This chapter proposed the framework of how we can counter cyberterrorism and what type of

TABLE 10.3
Comparative Analysis [4, 36, 37]

S. No.	Parameters	Wireshark	Autopsy	Nmap	Network Miner	Magnetic RAM Capture	Pro Discover
1.	Packet Sniffing	Yes	No	No	Yes	No	No
2.	Traceroute	Yes	No	Yes	Yes	No	No
3.	Identifying Anonymous Activity	Yes	Yes	No	Yes	Yes	Yes
4.	Recovery and Acquisition	No	Yes	No	No	Yes	Yes
5.	Analysis	Yes	No	Yes	Yes	Yes	No
6.	Evidence	Evidence for the network traffic and also the analysis for the network traffic.	Evidence for the deleted data from the digital devices.	Evidence for open and closed ports.	Evidence for the network traffic.	Evidence for the physical memory in the computer system.	Evidence for the deleted data from the digital devices.

countermeasure we have to take at a particular scenario to reduce the effect of cybercrime in the computer system, information system, and in the network system. This chapter also gives the preventive and detective measures against cybercrime and cyberterrorism and provides the security to the information system, which can be helpful in the enhancement of cyber security and decrease the occurrence of cyberterrorism in the cyber world.

10.9 RESEARCH LIMITATIONS

- Cyber criminals or cyber terrorists exploit the threat for fulfilling their own motives like personal, political, financial, and many other reasons because every person or country relies on the information and communication network [24, 28].
- Doing cyberterrorism is very easy and economical compared to the other terrorist attacks because doing a cyberterrorism attack is cheaper than the other terrorism in the country [25, 26].
- Cyber criminals execute their cybercrime by using guest accounts, malwares, and other activities that make them anonymous to the other person [29, 31].
- Cyberterrorism can achieve multiple goals such as government, individual, state bodies, etc. because of the vulnerabilities lying in the targeted system, which are easily detected by cyber terrorists [28, 30].
- Cybercrimes can be executed from a remote location and due to this, cyber experts have difficulties in tracking them [32, 33].
- Impact of cyberterrorism has a wide coverage because cyber terrorists can attack on one target system and develop the fear in a lot of people's minds in very short time and create the havoc of cyberterrorism [34, 35].

10.10 CONCLUSION AND FUTURE SCOPE

This research work discussed the different types of threats, vulnerabilities, cybercrimes, and cyberterrorism occurring in the cyber world and also discussed the cyber laws which are needed against cyberterrorism. This research work also discussed the proposed framework for taking action against cybercrime and cyberterrorism activities in the cyber world and also the countermeasures we need to take against cybercrime activities in the cyber world. This research work also discussed the use of the digital forensics tools in the process of the proposed framework and also gave the comparative analysis of some of the digital forensics tools, which can be useful to protect our information system. This research work provides the working of the proposed work in the different fields of the cyber world and enhances the functionality of the cyber secure system. In the future, more techniques and methodologies will be needed to decrease cyberterrorism activities in a more effective manner and also provide a cyber secure environment in the cyber world.

REFERENCES

1. Suryateja, P. S. "Threats and vulnerabilities of cloud computing: A review." *International Journal of Computer Sciences and Engineering* 6.3 (2018): 297–302.
2. Alkhalifah, A., Ng, A., Kayes, A. S. M., Chowdhury, J., Alazab, M., & Watters, P. A. "A taxonomy of blockchain threats and vulnerabilities." In *Blockchain for Cybersecurity and Privacy* (pp. 3–28). CRC Press, 2020.
3. Raw, R. S., Kumar, M., & Singh, N. "Software-defined vehicular adhoc network: A theoretical approach." *Cloud-Based Big Data Analytics in Vehicular Ad-Hoc Networks.* IGI Global, 2021. 141–164.
4. Caesarano, A. R., & Riadi, I. "Network forensics for detecting SQL injection attacks using NIST method." *International Journal Cyber-Security Digital Forensics* 7.4 (2018): 436–443.
5. Ambedkar, M. D., Ambedkar, N. S., & Raw, R. S. "A comprehensive inspection of cross site scripting attack." In 2016 International Conference on Computing, Communication and Automation (ICCCA). IEEE, 2016.
6. Al Yahmadi, F., & Ahmed, M. R. "Taxonomy of threats and vulnerabilities in smart grid networks." *International Journal of Energy and Power Engineering* 15.4 (2021): 168–171.
7. Homoliak, I., Venugopalan, S., Reijsbergen, D., Hum, Q., Schumi, R., & Szalachowski, P. The security reference architecture for blockchains: Toward a standardized model for studying vulnerabilities, threats, and defenses. *IEEE Communications Surveys & Tutorials* 23.1 (2020): 341–390.
8. Kamal, R., Raw, R. S., Saxena, N. G., & Kaushal, S. K. "Implementation of security & challenges on vehicular cloud networks." In Communication and Computing Systems: Proceedings of the International Conference on Communication and Computing Systems (ICCCS 2016), Gurgaon, India, 9-11 September, 2016 (p. 379). CRC Press, 2017, February.
9. Hamza, A., Gharakheili, H. H., & Sivaraman, V. *IoT Network Security: Requirements, Threats, and Countermeasures.* IEEE, 2020.
10. Reddy, A. M., Reddy, K. S., Prasad, M., & Obulesh, A. "Internet of things (IoT) security threats and countermeasures." *Network Security* 5.1 (2021): 12–26.
11. Raw, R. S. "The amalgamation of blockchain with smart and connected vehicles: Requirements, attacks, and possible solution." In 2020 2nd International Conference on Advances in Computing, Communication Control and Networking (ICACCCN), Lucknow. IEEE, 2020.
12. Maheshwari, S., & Sharma, N. "Cyber forensic: A new approach to combat cyber crime1, 2." *International Journal of Computer Network and Information Security* 56.3 (2021): 15–29.
13. Yadav, A. K., Bharti, R. K., & Raw, R. S. "Security solution to prevent data leakage over multitenant cloud infrastructure." *International Journal of Pure and Applied Mathematics* 118.7 (2018): 269–276.
14. Bosamia, M., & Patel, D. "Wallet payments recent potential threats and vulnerabilities with its possible security measures." *International Journal of Computer Sciences and Engineering* 7 (2019): 810–817.
15. Srivastava, A., et al. "Future IoT-enabled threats and vulnerabilities: State of the art, challenges, and future prospects." *International Journal of Communication Systems* 33.12 (2020): e4443.
16. Rocha, M., Manuel, V. "A systematic review of security threats and countermeasures in SaaS." *Instituto de Ingeniería y Tecnología* 45.6 (2020).

17. Tsiknas, K., Taketzis, D., Demertzis, K., & Skianis, C. "Cyber threats to industrial IoT: A survey on attacks and countermeasures." *IoT* 2.1 (2021): 163–188.
18. Ghazal, T. M., Hasan, M. K., Hassan, R., Islam, S., Abdullah, S. N. H. S., Afifi, M. A., & Kalra, D. (2020). "Security vulnerabilities, attacks, threats and the proposed countermeasures for the Internet of Things applications." *Solid State Technology* 63(1s): 2513–2521.
19. Choudhary, Y., Umamaheswari, B., & Kumawat, V. "A study of threats, vulnerabilities and countermeasures: An IoT perspective." *Humanities* 8.4 (2021): 39–45.
20. Ahmad, H., Dharmadasa, I., Ullah, F., & Babar, A. (2021). *A Review on C3I Systems' Security: Vulnerabilities, Attacks, and Countermeasures.* arXiv preprint arXiv:2104.11906.
21. Abdelrahman, A. M., Rodrigues, J. J., Mahmoud, M. M., Saleem, K., Das, A. K., Korotaev, V., & Kozlov, S. A. "Software-defined networking security for private data center networks and clouds: Vulnerabilities, attacks, countermeasures, and solutions." *International Journal of Communication Systems*, 34.4 (2021): e4706.
22. Khalid, F., Hanif, M. A., & Shafique, M. "Exploiting vulnerabilities in deep neural networks: Adversarial and fault-injection attacks." arXiv preprint arXiv:2105.03251 (2021).
23. Ambika, T., & K. Senthilvel. "Cyber crimes against the state: A study on cyber terrorism in India." *Webology* 17.2 (2020): 15–25.
24. Antonyan, E. A., & N. A. Grishko. "New technologies in cyber terrorism countering." In XVII International Research-to-Practice Conference Dedicated to the Memory of MI Kovalyov (ICK 2020), Boston. Atlantis Press, 2020.
25. Sebastian, J., & P. Sakthivel. "Cyber terrorism: A potential threat to global security." 2020.
26. Serebrennikova, A. V. "Cyber terrorism: Modern challenges." *Colloquium-Journal.* 19.71 (2020): 178–189.
27. Valsalan, K. *A Critical Analysis on Cyber Crimes and Security Issues in India.* IEEE, 2020.
28. Sebastian, J., & Sakthivel, P. *Cyber Terrorism: A Potential Threat to National Security in India.* IEEE, 2020.
29. Prakash, N., & Duhan, R. "Computer forensic investigation process and judicial response to the digital evidence in India in light of rule of best evidence." 8.05 Springer, (2020).
30. Datta, P., Panda, S. N., Tanwar, S., & Kaushal, R. K. "A technical review report on cyber crimes in India." in 2020 International Conference on Emerging Smart Computing and Informatics (ESCI) (pp. 269–275). IEEE, 2020, March.
31. Ravichandran, K. "Awareness of cyber crime prevention analyzed by the nearest neighbour analysis in India." *Network Security*, 45.6 (2020): 245–256.
32. Tanwar, S., Paul, T., Singh, K., Joshi, M., & Rana, A. "Classification and impact of cyber threats in India: A review." In 2020 8th International Conference on Reliability, Infocom Technologies and Optimization (Trends and Future Directions) (ICRITO) (pp. 129–135). IEEE, 2020, June.
33. Vaishnav, Ms N. T., & Barde, S. "Study on cyber laws of India." *Journal of Computer, Internet and Network Security* 1.2 (2020): 1–5.
34. Singh, R. "Counterterrorism in India: An ad hoc response to an enduring and variable threat." In *Non-Western Responses to Terrorism.* Manchester University Press, 2020.
35. Pawar, S., Bhusari, C., & Vaz, S. "*Survey on digital forensics investigation and their evidences.*" *Journal of Computer, Internet and Network Security* 12.6 (2020): 147–156.

36. Pansari, N. "A comparative study of analysis and investigation using digital forensics." *International Journal of Linguistics and Computational Applications (IJLCA)* 7.2 (2020): 147–162.
37. Zinge, P. A., & Chatterjeem M. "Comprehensive study of digital forensics branches and tools." *International Journal of Forensic Computer Science (IJoFCS)* 14.5 (2018): 556–569.
38. Sikos, L. F. "Packet analysis for network forensics: A comprehensive survey." *Forensic Science International: Digital Investigation* 32 (2020): 200892.
39. Singh, N., Dayal, M., Raw, R. S., & Kumar, S. "SQL injection: Types, methodology, attack queries and prevention." In 2016 3rd International Conference on Computing for Sustainable Global Development (INDIACom) (pp. 2872–2876). IEEE, 2016, March.
40. Aliyu, A., Abdullah, A. H., Kaiwartya, O., Cao, Y., Usman, M. J., Kumar, S., Lobiyal, D. K., & Raw, R. S. "Cloud computing in VANETs: Architecture, taxonomy, and challenges." *IETE Technical Review* 35.5 (2018): 523–547.

11 Cryptography Techniques for Information Security

A Review

Ganesh Chandra, Satya Bhushan Verma, and Abhay Kumar Yadav

CONTENTS

11.1 Introduction 191
11.2 Literature Review 193
11.3 Cryptography Techniques 195
11.3.1 Substitution Technique 195
11.3.1.1 Caesar Cipher 195
11.3.1.2 Mono Alphabetic Ciphers 196
11.3.1.3 Homophonic Cipher 196
11.3.1.4 Polyalphabetic Cipher 196
11.3.1.5 Playfair Cipher 196
11.3.2 Transposition Technique 197
11.3.2.1 Rail Fence Technique 199
11.3.2.2 Simple Columnar Technique 199
11.4 Discussion 199
11.5 Conclusion 199
References 199

11.1 INTRODUCTION

The word "cryptography" was coined by Leon Battista Alberti around AD 1467; it is a Greek word which means "secret writing" and it provides secure communication between the participants.

In the era of computer technology, a huge amount of data is available on the web or internet. People communicate with each other all over the world using the internet, so the need occurs for security of data (i.e., network security, web security). Hence, cryptography is used to secure information over the internet from hackers, enemies, and unauthorized users. The parameter used for security is authentication, confidentiality, integrity, non-repudiation.

DOI: 10.1201/9781003323426-11

Authentication is the process to verify the identity of the sender or confirming a proof of identities. Confidentiality secures the information by allowing only those who are authorized to access the information. Integrity confirms that the received information is achieved from the authorized party. Non-repudiation used to verify the sender sending the message. The sender cannot deny at a later stage that he or she did not send the message.

There are two processes used in cryptography, the first one is encryption and the second one is decryption. In encryption, the original message that is (plaintext) is converted into another form that is ciphertext. In decryption, the coded form of a message is converted again into an original format so as the receiver of the message can read it. Encryption and decryption are always performed using algorithms and secure keys.

Types of Cryptography:
Cryptography is a method used for information security, which is classified into the following categories as depicted in Figure 11.1.

1. Symmetric key cryptography.
2. Asymmetric key cryptography.

1. Symmetric Key Cryptography
Symmetric key cryptography or secret key cryptography uses a single/same key for both encryption and decryption as defined in Figure 11.2 [3]. In this approach, both the sender and the receiver should know about the secret key. Data Encryption Standard (DES), Triple DES, Advanced Encryption Standard (AES), and Blowfish are the symmetric key algorithms.

2. Asymmetric Key Cryptography
In this type of cryptography one key is used for the encryption of the data and the other is used for the decryption of the data. In public key cryptography (PKC), keys come in pairs of matched public and private keys, one of the private keys or secure keys should be protected and another key called as the public key can be given to anybody [5]. Every user in PKC has two keys, one is the enciphering key Ek (public

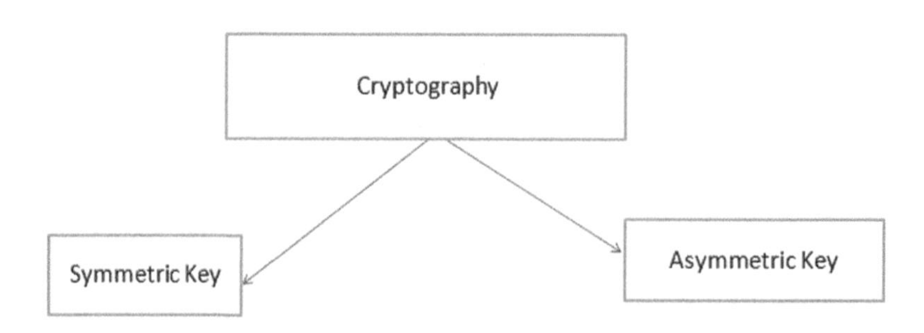

FIGURE 11.1 Types of cryptography.

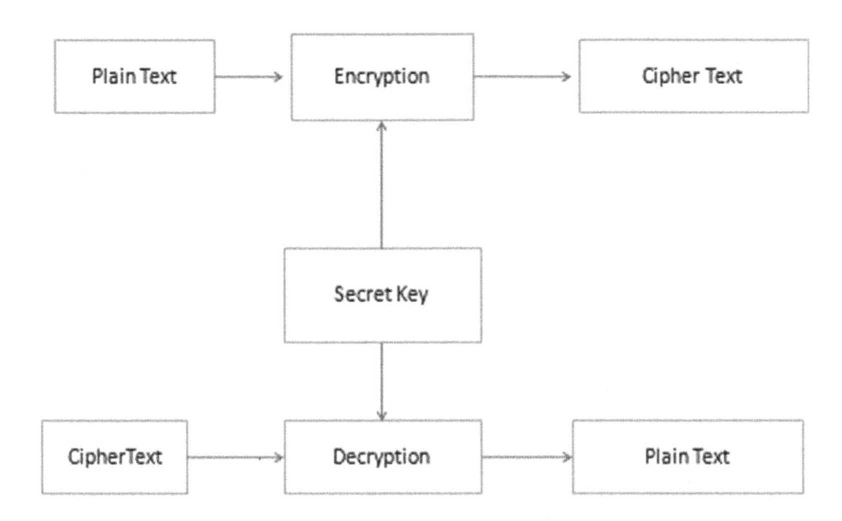

FIGURE 11.2 Symmetric key cryptography process [5].

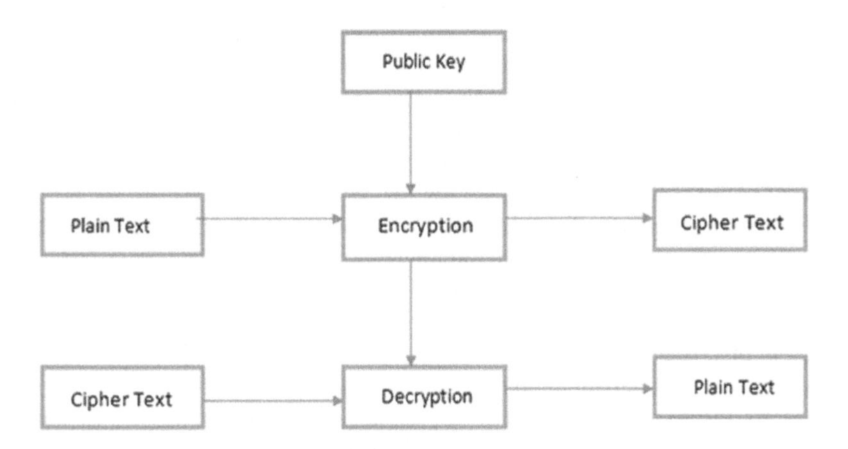

FIGURE 11.3 Asymmetric key cryptography process [5].

key) and the other is the deciphering key Dk (private key). The primary key feature is to remove the dependency on a single key for both encryption and decryption, as shown in Figure 11.3 [5].

11.2 LITERATURE REVIEW

There is a large amount of work done by the researchers in the field of cryptography techniques for data security, explained in this paper as follows:

In 2011, Ashwak M. AL-Abiachi [1] worked on "A Competitive Study of Cryptography Techniques over Block Cipher." This paper focused on different cryptography techniques for providing secure communication, and also, this paper reviews different research studies that have been done for encryption and decryption in block cipher.

In 2012, Ganesh Chandra et al. [3] worked on "ECC Public Key Cryptosystem for Security Services in Mobile Communication: A Study." Elliptic Curve Cryptography is a complex public key cryptosystem, where several parameters have to be selected carefully before its implementation for wireless communication systems. In this paper, we study the various applications of Elliptic Curve Cryptography (ECC) in open communication environments like cell phones, PDAs, sensor networks, etc. The major benefits of ECC in wireless communication are low bandwidth implementation, etc.

In 2012, Vinod Kumar Yadav et al. [2] worked on "Public Key Cryptosystem Technique Elliptic Curve Cryptography with Generator g for Image Encryption." In this paper, ECC points convert into cipher image pixels at the sender side and a decryption algorithm is used to get the original image within a very short time with a high level of security at the receiver side.

In 2014, Sourabh Chandra et al. [3] worked on "A comparative survey of symmetric and asymmetric key cryptography." Cryptography is the most important technique for secure transmission of data. Both symmetric and asymmetric key algorithms are essential for providing security of the data. This paper gives the comparative analysis of symmetric and asymmetric cryptography techniques.

In 2014, Bidisha Mandal et al. [4] worked on "A Comparative and Analytical Study on Symmetric Key Cryptography." This paper focused on a comparison study of symmetric cryptography techniques, e.g., AES, DES, 3DES, and Blowfish. After analyzing all algorithms, we came to know that Blowfish is a highly proposed algorithm for security purposes.

In 2015, Laiphrakpam Dolendro Singh [5] worked on "Image Encryption using Elliptic Curve Cryptography." Cryptography plays a very significant role in transferring images securely. With the help of Elliptic Curve Cryptography, it provides a high level of security with smaller key size compared to other cryptographic techniques. In this paper, we implement the Elliptic Curve Cryptography for encryption, decryption, and transferring of the image from one end to another end.

In 2016, Priyadarshini Patila et al. [6] worked on "A Comprehensive Evaluation of Cryptographic Algorithms: DES, 3DES, AES, RSA and Blowfish." In order to find the best cryptography algorithm, we have to analyze the performance, strength, and weakness of all the algorithms. We compare DES, AES, Blowfish, and RSA. Then, we find RSA consumes more time for encryption and decryption as compared to Blowfish.

In 2016, Payal Patel et al. [7] worked on "Integrated ECC and Blowfish for Smartphone Security." This paper gives the hybrid approach of ECC and Blowfish, which provides stronger security of data in the mobile cloud. To transmit the data more securely, a random number is used to increase the computational complexity for an adversary.

In 2017, Sarika Y. Bonde [8] worked on "Analysis of Encryption Algorithms (RSA, SRNN and 2 key pair) for Information Security." The encryption algorithm plays an essential role for secure communication, where the encryption time is the major issue of concern. For performance evaluation, the RSA algorithm, two key pair algorithm, and short-range natural number (SRNN) algorithm are used. Also, RSA consumes the least decryption time as compared to SRNN and two key pair algorithms.

In 2017, Vania Beatrice Liwandouw [9] worked on "The Existence of Cryptography: A Study on Instant Messaging." This paper studies and analyzes several cryptography applications running on android and iOS platforms so that privacy and confidential communication can be achieved. The results of this study shows the best recommended cryptography application.

In 2017, Sarita Kumari [10] worked on "A research Paper on Cryptography Encryption and Compression Techniques." This paper defines cryptography techniques as a very popular way of sending information secretly. There are many techniques available for achieving the goal of cryptography, e.g, confidentiality, authentication, integrity, and non-repudiation.

In 2018, Dimas Natanaelaet al. [11] worked on "Text Encryption in Android Chat Applications using Elliptical Curve Cryptography." In this paper, we implement the ECC algorithm to secure text messages in a smartphone. We also give the experimental result of our chat apps' performance such as time of accuracy of the received text message, average encryption, and decryption time.

In 2018, Marek R. Ogiela et al. [12] worked on "Cognitive cryptography techniques for intelligent information management." This paper discusses the cognitive cryptography techniques for secure information. In cognitive cryptography, it is particularly legitimate to use personal information contained in biometric information sets, as well as semantic information, which is unambiguously used to identify the individual features of all protocol users

In 2018, Naglaa F. Saudy [13] proposed "Error analysis and detection procedures for elliptic curve cryptography." As new applications are being developed and dependence on systems to offer new services is continually expanding, the requirement for enhanced security near-ideal models is in high demand. To meet this competitive need, the ECC public key cryptosystem has been created.

11.3 CRYPTOGRAPHY TECHNIQUES

There are two main cryptographic techniques, i.e., substitution and transposition, by which a plaintext is converted into a ciphertext.

11.3.1 Substitution Technique

Substitution swaps the part of the message with another part rendering to some mapping (e.g., replace one letter with the another one). In this technique, letters of plaintext are replaced by other letters or numbers [6].

11.3.1.1 Caesar Cipher

The drawback of this technique is that this method is not very secure [13].

Plain text- WORK IS WORSHIP

Cipher text- ZRUN LV ZRUVKLS

11.3.1.2 Mono Alphabetic Ciphers

Due to a small key space, the Caesar cipher is not secure, and to increase the key space, every letter is replaced by an arbitrary substitution, e.g., A is replaced by any letter from A to Z, such type of substitution is called as monoalphabetic substitution cipher. There is no relation between the replacement of A and B. This random substitution of characters makes this method difficult to crack.

11.3.1.3 Homophonic Cipher

This method is quite similar to the monoalphabetic cipher; here, one plain text alphabet can be replaced by a fixed number of alphabets [6]. For example, A can be swapped by D,H,P,R, and B can be replaced by E,I,Q,S, and so on.

Mapping- ABCDEFGHIJKLMNOPQRSTUVWXYZ
DZSFXEHCVITPGAQLKJRUOWMYBN
9 7 3 50 8
2

Plain text- HELLO
Cipher text- C7 PPQ

11.3.1.4 Polyalphabetic Cipher

This technique uses multiple one-character keys, the first key encrypts the first plaintext character, the second key encrypts the second plaintext character; these ciphers effort the same way as monoalphabetic ciphers but rotate through several maps. This makes character occurrence analysis tougher [6].

Plain text- ABCDEFGHIJKLMNOPQRSTUVWXYZ
Cipher text #1- BDFHJLNPRTVXZACEGIKMOQSUWY
Cipher text #2- ZYXWVUTSRQPONMLKJIHGFEDBA
Using #1 for every first letter and #2 for every second letter
Plain text- HELLO
Cipher text- PVXPC

11.3.1.5 Playfair Cipher

Playfair also called Playfair square designed by Charles Wheatstone in 1854. It was used in the First World War, that is, from 1914 to 1918 by the British army. This is based on a 5 × 5 matrix of letters constructed using a keyword [13].

Playfair Example

P	L	A	Y	F
I/J	R	E	X	M
B	C	D	G	H

K	N	O	Q	S
T	U	V	W	Z

As per following rules the plaintext is encrypted two letters at a time [13].

(1) In the first step, the plaintext message that we want to encrypt is broken into two alphabets.
(2) If both alphabets in the pair appear in the same row, then it replaces them with the immediate right, respectively.
(3) If both alphabets in the pair appear in the same columns in our matrix, replace them with the alphabet immediately below, respectively.

Example- MY NAME IS ATUL
Plaintext- MY NA ME IS AT UL
Ciphertext- XF OL IX MKPU LR

11.3.2 Transposition Technique

In this technique, rearrangement of the position of the plaintext is done in order to get the ciphertext. There are various techniques that come under the transposition method [7].

TABLE 11.1
Comparison of Cryptography Techniques

Property	Substitution	Transposition
Definition	Substitution technique involves the replacement of the letters by other letters and symbols. It is a fundamental method of codifying the plaintext message into the ciphertext.	The transposition cipher does not transform one symbol from one more, rather it changes the area of the symbol. And the identity of the characters remains unchanged but their positions are changed to create the ciphertext. It basically reorganizes the characters of the plaintext.
Aims	Replacement procedure intends to adjust the identity of the element.	While the transposition strategy adjusts the position of the element instead of its identity.
Types	Caesar cipher, Monoalphabetic cipher, Polyalphabetic cipher, Playfair substitution cipher.	Rail fence, Simple columnar transposition technique.
Disadvantage	The last letter of the alphabet, which is generally low recurrence will in general remain toward the end.	Keys extremely near the right key will uncover long segments of clear plaintext.
Alteration	The identity of the character is changed while its position stays unaltered.	The location of the character is changed despite its identity.

TABLE 11.2
Comparison of Substitution and Transposition Techniques

Property	Caesar Cipher	Monoalphabetic Cipher	Polyalphabetic Cipher	Playfair Cipher	Rail Fence	Simple Columnar Transposition
Developed by	Julius Caesar in the 19th century	Blaise de Vigenère in the 16th century	Leon Battista Alberti in around 1467	Charles Wheatstone in 1854	Develop by Greeks	-
Definition	In Caesar cipher, the plaintext is replaced by another letter a fixed distance away.	A is replaced by any letter from A to Z and such type of substitution is called monoalphabetic substitution cipher.	A polyalphabetic cipher uses multiple substitution alphabets. These make frequency analysis harder.	This technique encrypts pairs of letters.	In a rail fence cipher, letters are not changed, but only changed in positioning.	Columnar transposition involves writing the plaintext out in rows and then reading the ciphertext off in columns.
Key type	Substitution	Substitution	Substitution	Substitution	Permutation	Permutation
Attack type	Brute force attack	Known plaintext attack	Cipher text and known plaintext attack	Cipher text only	Brute Force attack	Frequency analysis attack
Key size	Fixed Number	Fixed (26!)	Equal to message length	Fixed (25!)	Depth size is variable	Variable

11.3.2.1 Rail Fence Technique

In this technique, the plaintext is written as an arrangement of diagonal and then read row by row to produce the ciphertext [6].

Plain text- MY NAME IS SATYA

```
M   N   M    I    STA
              \ / \ /   \ / \ / \ / \ /
 Y   A   E   S   AY
```

Cipher text- MNMISTAYAESAY

11.3.2.2 Simple Columnar Technique

In this technique, arrange the plaintext message row by row in a rectangle of a pre-defined size.

Read the message column by column, it is not necessary that the order is 1, 2, 3, it can be a random order like 1, 3, 2, etc.

11.4 DISCUSSION

In this Table 11.2, we defined the substitution and transposition techniques. Substitution techniques are mainly used for replacing one set of letters with another. And transposition techniques are used for changing the position of the characters [11] (Tables 11.1 and 11.2).

11.5 CONCLUSION

This paper has given a detailed study of the cryptography technique. Information security plays a very important and powerful role in the field of networking and the internet. Cryptography is a very essential way to provide security and to protect the secret messages from unauthorized users or hackers. In this paper, we have given a brief introduction of cryptography, its principle, various types of techniques for encryption of data, and a comparison of all techniques. In this paper, our purpose is to provide more secure information which cannot be deciphered by hackers.

REFERENCES

1. David Naccache, Jacques Stern "A New Public-Key Cryptosystem" W. Fumy (Ed.): *Advances in Cryptology: EUROCRYPT '97, LNCS 1233*, pp. 27–36, 1997. Springer-Verlag, Berlin Heidelberg, 1997.
2. Ashwak M. AL-Abiachi "A Competitive Study of Cryptography Techniques over Block Cipher" 2011 UKSim 13th International Conference on Modelling and Simulation 978-0-7695-4376-5/11 $26.00 © 2011 IEEE, Boston, 2011. DOI:10.1109/UKSIM.2011.85

3. Ganesh Chandra "ECC Public Key Cryptosystem for Security Services in Mobile Communication: A Study" *International Journal of Computational Intelligence and Information Security*, Vol. 3, No. 1, pp. 157–173, January 2012.
4. Vinod Kumar Yadav, A. K. Malviya, D. L. Gupta, Satyendra Singh, and Ganesh Chandra "Public Key Cryptosystem Technique Elliptic Curve Cryptography with Generator G for Image Encryption" *International Journal of Computer Applications in Technology*, Vol. 3, No. 1, pp. 298–302, 2012.
5. Sourabh Chandra, Smita Paira "A Comparative Survey of Symmetric and Asymmetric Key Cryptography" International Conference on Electronics, Communication and Computational Engineering (ICECCE), Lucknow. 978-1-4799-5748-4/14/$31.00 © 2014 IEEE.
6. Aditi Verma, Harsha Singh "A Review on Cryptography and its Various Techniques" *International Journal of Advanced Research in Computer Science*, Vol. 5, No. 3, pp. 1445–1462, March–April 2014. ISSN No. 0976-5697.
7. Bidisha Mandal, Sourabh Chandra "A Comparative and Analytical Study on Symmetric Key Cryptography" International Conference on Electronics, Communication and Computational Engineering (ICECCE), Paris, 2014. 978-1-4799-5748-4/14/$31.00 © 2014 IEEE.
8. Laiphrakpam Dolendro Singh "Image Encryption using Elliptic Curve Cryptography" Eleventh International Multi-Conference on Information Processing-2015 (IMCIP-2015), New York, 2015.
9. Priyadarshini Patila, Prashant Narayankar "A Comprehensive Evaluation of Cryptographic Algorithms: DES, 3DES, AES, RSA and Blowfish" International Conference on Information Security & Privacy (ICISP2015), 11–12 December 2015, Nagpur, India, 2015. Published by Elsevier B.V. This is an open access article under the CC BY-NC-ND license
10. Payal Patel, Rajan Patel "Integrated ECC and Blowfish for Smartphone Security" International Conference on Information Security & Privacy (ICISP2015), 11–12 December 2015, Nagpur, India, 2015.
11. Preeti Poonia, Praveen Kantha "Comparative Study of Various Substitution and Transposition Encryption Techniques" *International Journal of Computer Applications*, Vol. 145, No. 10, July 2016. (0975 – 8887).
12. Sarika Y. Bonde *Analysis of Encryption Algorithms (RSA, SRNN and 2 Key Pair) for Information Security*. IEEE, 2017. 978-1-5386-4008-1/17/$31.00 ©2017 IEEE.
13. K. Sahu, F. A. Alzahrani, R. K. Srivastava, and R. Kumar. "Evaluating the impact of prediction techniques: Software reliability perspective," *Computers Materials and Continua*, vol. 67, no. 2, pp. 1471–1488, 2021.

12 A Critical Analysis of Cyber Threats and Their Global Impact

Syed Adnan Afaq, Mohd. Shahid Husain, Almustapha Bello, and Halima Sadia

CONTENTS

12.1 Introduction 201
12.2 Classifications of E-crimes 205
12.3 Beginning and Growth of E-crimes 207
12.4 Advances in Cyber Threats and Computer Crime 209
12.4.1 System Exploitation 209
12.4.2 Phishing 210
12.4.3 A Man-in-the-Middle Attack 211
12.4.4 DoS Attack 211
12.4.5 Wi-Fi Exploitation 211
12.4.6 Ransomware 211
12.5 Indian Cyber Situation 212
12.6 Cyberattacks in India of Late 212
12.7 Global Organizations Fighting against Cybercrime 213
12.7.1 Careless Behaviors 216
Understanding and optim 216
Phishing Emails 217
12.8 Conclusion 218
References 220

12.1 INTRODUCTION

In the field of information technology, cybersecurity is essential. Data and information security has become one of the most significant concerns in the modern day. Several governments and corporations are pursuing a variety of steps to manage and reduce cybercrime. Despite various precautions and approaches, many individuals continue to be concerned about cybersecurity. The difficulties that cybersecurity tackles in the present day are the topic of this research. It also includes one of the most up-to-date facts on cybersecurity strategies, principles, and tendencies that are redefining the field of cybersecurity [1]. With the growth of digital technology,

DOI: 10.1201/9781003323426-12

an entirely new forum for criminal behavior has emerged. As a result of technical advancements, cybercrime refers to crimes involving the use of computer networks for unlawful purposes such as security breaches, fraud, economic fraud, and adult material. Hackers, also known as cybercriminals, frequently use systems to gain valuable information including credentials, credit card details, and other identifying details for malicious or manipulative reasons. By precisely evaluating possible criminals and attacks, cybersecurity teams gain a good understanding of who cybercriminals are, what techniques they employ, and what measures may take to fight and avoid potential cybercrimes. As cybercrime evolves, businesses must continue to train their staff and assist them in raising awareness of information technology (IT) security issues [2].

In the modern era, cyber society has become a common and essential origin of information sharing as well as other business activities such as marketing, purchasing, financial transactions, promotions, and services in the present age. This rapid increase in cyberspace use has led to an enormous rise in cybercriminal behavior. The extensive usage of Web applications in virtually every aspect of life is the fundamental driver of this development. These Web apps contain design flaws that cybercriminals use to gain unauthorized access to networks. With a single click, someone may transmit any kind of data, whether it is an email, a video/audio file; however, has he really thought about whether safely his file is being delivered or received to the other individual without even any disclosure. Cybersecurity is the answer. The internet is the quickest technology in today's world. In today's technological world, many technologies are transforming the face of humanity. However, we are often unable to secure our confidential information as efficiently as we would like because of these new technologies, and as a result, cybercrime is going up. As over 60% of all financial activities are now handled online, this industry requires a high degree of protection to assure that those operations are effective and efficient. Cybersecurity encompasses not just the security of privacy in the IT industry but also other areas such as cyberspace [3].

In our society, economics and basic infrastructure, computer networks, and information technology solutions have grown increasingly important. Cyberattacks are becoming more fascinating and potentially dangerous as our reliance and requirements on information technology grow. Cyber threat rates are approximately US$114 billion each year, as per the record of Symantec cybercrime. When the time being spent by organizations struggling to rebuild from cyberattacks is taken into account, the overall cost of cyberattacks is $385 billion. The majority of individuals who have been affected by cyberattacks continue to rise. As per a Symantec report on 20,000 people from 24 various countries, 69% had experienced a cyberattack at a certain stage of life. As per Symantec, 14 persons are affected by a cyberattack every second, amounting to over one billion each day [3]. The internet is a set of connected networks that links a large set of computers around the whole world by using different appropriate internet protocols. The internet has evolved into one of the most important components of contemporary life. As a result of the information technology revolution, the internet serves two primary purposes. On the one hand, it has contributed to the advancement of good principles throughout the world. On the other hand, technology has caused a bunch of problems that undermine society order and has generated a new wave of crime all across the world [4]. The internet and computers connect individuals all over the world, thanks to web communication and information technology trends toward digitalization. People become addicted to innovative technologies when their weakness is exploited. Cybercrime is defined as those with a computer or other hardware to commit an offense. It refers to criminal activity conducted over computer networks in breach of norms, regulations, and laws. Cybercrime includes detecting theft, harm, transaction fraud, phishing, and software privacy. Violent and nonviolent cybercrime are the two types of cybercrime that exist. The majority of cybercrimes are nonviolent in nature, as they occur in the absence of any physical touch connection. Cyber trespass, cyber theft, and cyber fraud are examples of nonviolent cybercrimes. The internet is an online medium that allows individuals of diverse backgrounds to create profiles and communicate with other users on compatible sites. These are destinations for person-to-person conversation, and various social networking sites and applications like Facebook, Twitter, Instagram, and other similar sites have become so common among all internet users that they have started to share each and every detail of their personal life on these platforms [5].

As a consequence, academics and governments are increasingly concerned regarding cybersecurity. Cybersecurity is defined as tools, strategies, regulations, security checks, protection regulations, prevention and mitigation methods, exercises, instruction, industry standards, and security verification can all be used to protect users' resources in cyberspace. Cybersecurity, which entails securing information by detecting, avoiding, and responding to cyberattacks, has been a topic of global attention and importance in recent years.

Various organizations using defensive methods for protecting their cyberspace are inefficient to defend their cyber environments from ever-increasing security risks. Therefore, over the last decade, it has gained the attention of academics and practitioners as one of the most pressing scientific concerns.

As social networking sites become a part of daily life and the number of users grows, they can share opinions and insights with unknown (stranger) friends and interact well with recognized friends. It can be broadly characterized as internet-based social sites that allow networks of connections to communicate, collaborate, and share material. The most common users connect with existing networks, make and increase friends, create an online presence for their users, view content, find information, create and customize profiles, and so on. It is based on how users utilize sites like Facebook and other social networking sites in general [3].

There are numerous definitions of cybercrime and cyberterrorism available today. There has recently been some debate as to whether the phrases cybercrime and cyberterrorism are synonymous. Some authors, on the other hand, endorse the concept that the two names are synonyms, while others disagree and provide two alternative definitions. In a broad sense, the most significant definitions are those that are based on the United States' formulation terminology.

However, in order to make a clear distinction between the terms cybercrime and cyberterrorism in the following paragraph, distinctions must be addressed. It is imperative to define the word cyberterrorism, and policymakers are pushing for it. The main goal of cyberterrorism is to infiltrate a system in a specific institution to create violence and harm (financial damage, property damage) in order to destabilize and weaken the target country's security.

Cyberterrorists can attack on well-defined targets that are important strategic points for certain countries, but it does not rule out the possibility of a broader occurrence to accomplish a precise objective. As an example, a cyberterrorist's objective could be an electrical plant that supplies electricity to citizens living in the surrounding area. Cyberterrorists can be effective in a wide range of situations by committing this type of attack with few resources. If the energy supply chain is disrupted as a result of this type of attack, it has an impact on citizens' daily routines for meeting basic demands [6].

Another type of cyberterrorism is the hacking of a hospital's computer system and modifying medical prescriptions, causing harm to the patient by giving the wrong drug. As a result, anyone could become a victim of this terrorist attack.

Any unlawful action that enables a computer as its principal means of offense and theft is referred to as cybercrime. The US Department of Justice has broadened the definition of cybercrime to include any illegal behavior that involves the storing of evidence on a computer.

Computer-based crimes, such as network intrusions and the propagation of computer viruses, are classified as cybercrime, as are system-based versions of commonly committed crimes, such as identity theft, data theft, extortion, trolling, and violence, which have become a large concern for governments and individuals. Cybercrime is described as a crime committed with the use of a computer and the internet to steal someone's identity, sell contraband, stalk individuals, or interfere by using malevolent software. Cybercrime, also known as e-crime, refers to crimes dedicated against individual person or groups with the intent of intentionally harming the victim's reputation, causing physical or mental harm, or causing monetary or information loss, either directly or indirectly, through the use of the internet and electronic devices [7]. Any criminal activity involving computers and networks is classified as cybercrime. Cybercrime also covers typical crimes that are carried out on the internet. Hate crimes, phishing, spammers, and card account fraud are all examples of cybercrimes done with the help of computer software and the internet. As the digital age grows increasingly dependent on the internet, a new cyber threat emerges. Everyday components of our life are gradually being incorporated technologically, putting them at risk. From kids in school to members of the company's board of directors, the internet is widely used.

However, the benefits of this digital life are outweighed by the risks posed by the species identified as hackers in our world. They're placing their knowledge and abilities to good use in order to benefit from mankind's online activity. Linking the company's official server to the person's social networking account is part of the process. As a result, nearly everyday, there must be some type of engineer or security analyst in society to defend against these types of threats and to increase awareness about network security. From the perspective of cyberattackers and security experts, there are many advancements in this arena to protect these types of threats. E-crimes affect the community in many ways.

- A loss in online commerce and consumer trust in the digital economy.
- The risk of essential infrastructure being harmed, which could disrupt water supply, health care, national communications, energy distribution, financial services, and transportation.
- Personal financial resources are depleted, resulting in emotional pain.
- Damage to company properties.
- The rate of re-establishing credit accounts, financial records, and individualities used for administration establishments and industries.
- Costs to businesses in enhancing cybersecurity measures.
- Costs to law enforcement authorities in terms of time and resources spent on other criminal activity.

12.2 CLASSIFICATIONS OF E-CRIMES

Cybercrime is a broad word that refers to criminal action in which a computer or computer network, including computers themselves, serves as a tool, target, or place for criminal behavior. The term is used to describe a wide range of crimes, such as kidnapping of kids as a result of chatroom fraud. It also encompasses the use of such computer systems to facilitate criminal conduct. The US Department of Justice classified cybercrime as when a computer is utilized against a target or when the medium of computer use is marginal to other crimes, in directive to style, investigations are easier.

Through the misuse of computer networks, there are numerous sorts of criminal activities involving tech-savvy individuals on the internet. These categories cover illicit operations that could be carried out in other ways, but criminals choose to use computers as a tool.

Computer crime: Is the unauthorized obtaining of data and information through the use of direct electronic operations can compromise security. The types of cybercrime are depicted in Figure 12.1.

High-tech crime: Refers to a wide range of criminal acts involving computers that are carried out unlawfully and are in violation of country or federal laws. These crimes are committed through phishing, financial fraud, virus, bullying, digital, and data theft.

White collar crime: Is a crime committed or attempted by an individual or group of persons who have respect in society with the purpose of fraud and stealing money.

Cyberterrorism: A deliberate and politically driven attack on information, computer networks, computer programs, and data that results in harm against civilian areas. Cyberterrorism might attack the banking industry, military sites, power stations, and air traffic control centers [6].

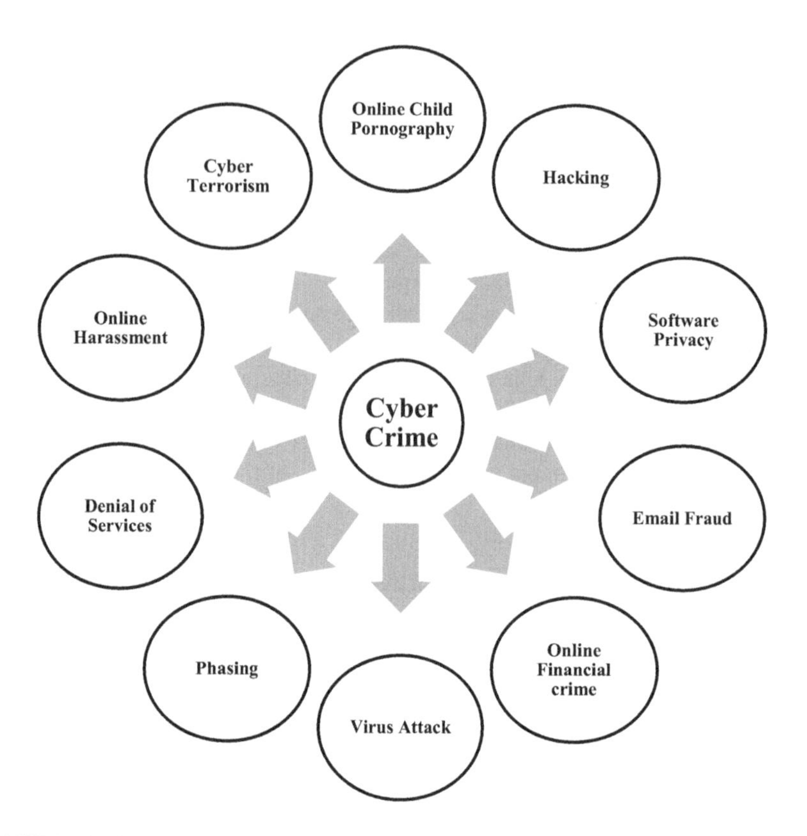

FIGURE 12.1 Types of cybercrime

1. Online Financial Crimes

Due to the prevalence of organized crime in specific types of criminal activity, the criminal site appears to be additionally diversified. By the use of cyberattacks as a weapon by organized criminal groups, drug traffickers and money launderers have gotten a lot of attention. Virtual currencies like Bitcoin, Litecoin, Bitcoin Cash, Monero, and Ethereum are used in this form of crime.

2. Cyberterrorism

Terrorist groups are enthralled by the concept of cyberterrorism and cyber warfare, and have stated that if they can, they will commit vast strikes. Despite the fact that several terrorist groups lack the skills to plan sophisticated cyberterrorist attacks, they have sought to gain the technological capabilities needed to transmit out critical maneuvers.

3. Child Pornography

Child pornography is defined as pornography depicting a child engaging in actual or simulated explicit sexual acts, or any depiction of a child for the purpose of sexual assault. Law enforcement agencies and international organizations all across the world regard the sale and distribution of child pornography to be highly serious crimes.

The expansion of social media and the internet has led to a rise in the frequency of child pornography-related offenses. The kind of person who commits these crimes usually has no prior criminal record, particularly in regard to childhood sexual exploitation.

4. Online Harassment

Because of the widespread use of the internet in everyday life, online harassment has become a huge danger to many people's safety and well-being. Threatening emails, threats of physical violence, and posting material on the internet are all illustrations of bullying online. Stalking could also occur on the internet, when someone anonymously transmits threats via a digital system such as a phone or computer. Online harassment, online stalking, computer hacking, online trolls, the publication of unlawful, obscene, or offensive content, as well as harassment and threats, are all examples of online harassment.

Unlike physical harassment, cyberbullying typically targets the younger generation and can lead to threatening, disgrace, and humiliation.

5. Computer as a Target

Any criminal action using computers, networking systems, or networks is classified as cybercrime. This is done by attacking computers with malicious programs, which then propagate through a network, comprising networks, servers, and other computers. A criminal act in which the main aim is a computer or network is known as a computer as a target. Computers are considered targets for distributive denial of service (DDoS) attacks and viruses.

6. Computer as a Tool

When individuals are the primary targets, computer crime is classified as a tool crime. A low degree of technical ability is necessary for these types of cybercrimes. Identity theft, phishing scams, and spam are all examples of computer-based cyberattacks.

12.3 BEGINNING AND GROWTH OF E-CRIMES

- Computer crimes were primarily committed by dissatisfied individuals and dishonest personnel in the early decades of modern information technology (IT).
- Up until the 1980s, physical damage to computer systems was a major concern (Sterling, 1992).
- Criminals frequently used authorized access to bypass security mechanisms, modifying data for financial gain or destroying data for retribution.

- Long-distance telecommunication disruption was used for entertainment and resource stealing in the 1960s. As telecommunications technology spread throughout the IT sector, criminals learned how to hack into systems and networks for fun.
- To disrupt home computers, programmers began developing computer viruses, including self-replicating viruses, in the 1980s.
- The internet has enabled criminals to gain unauthorized access to an expanding number of networks around the world; they use this access for sabotage, political activity, and financial gain.
- Financial crime involving computer penetration and destabilization intensified as the 1990s progressed.
- Malware evolved in the 1990s, taking advantage of new weaknesses as operating systems improved, only to be replaced by new attack vectors.
- From the mid-1990s onwards, illegal email programs exploded, resulting in a flood of unwanted commercial and fraudulent emails.
- Cyberattackers are increasingly turning to social media to recruit individuals to assist them in their shady deals all around the world.

The real start of e-crime was in 1960, when attacks on telecommunication infrastructure in the United States resulted in the loss of long-distance phone connections. In 1971, the rogue program Creeper, which spread through early bulletin board networks, increased wire fraud through communication in the United States. In the same year, a man named Draper created a blue box that enabled free long-distance calls for ten minutes [1].

The first instance of phishing emails was discovered in 1976 when that was sent via the Advanced Research Projects Agency Network (ARPANET). The ARPANET kernel paved the way for the present internet (Ping, 2011). Ian Murphy was the first person to be sentenced to an e-crime in 1981. Murphy broke into American Telephone & Telegraphs (AT&T) and changed the billing clock so that individuals could enjoy cheaper prices during normal business hours. In 1982, the first virus on an Apple computer was discovered in the early decades of modern information technology.

Unauthorized circumstances led to the formation of the oldest virus, known as "Pakistani Brain," which infected the systems of International Business Machines (IBM) Corporation in 1986. Kevin Mitnick was sentenced to prison in 1988 for spying on emails for a multinational company.

The first ARPANET worm arrived on government computer networks in the same year and expanded out of control, shutting down colleges and government organizations as it expanded over 6000 networked systems.

RT Moris, a student of Cornell University who was dismissed and sentenced to three years of suspension and a $10,000 fine, accomplished this.

Criminal organizations started developing malicious software programs, with self-executing and replicating programs, to disrupt personal computers [8].

As the use of the internet has grown, criminals have begun to employ malicious software to achieve their objectives. By the mid-1990s, e-crime had progressed to the point where it was utilizing software systems for computer breakthroughs and email-based frauds.

The first virus, named "Dark Avenger," was released in 1992. (Melissa, a well-known virus, first appeared in the late 1990s. The other well-known virus, "Chen Ing-Hau" (CIH), was also released to internet users all over the world [9].

The technological advancements of e-crime increased significantly at the turn of the millennium. A denial-of-service (DoS) attempt was performed toward corrupt websites including Yahoo, eBay, CNN, Amazon, and others in the year 2000. "I LOVE YOU" malware, which was popular at the time, was propagated by emailing all of the recipients' contacts.

The most well-known e-crime occurred in 2001 when Microsoft's Web sites were banned for two days after being attacked and corrupted by a new domain name server (DNS). In addition, over the millennium, several new worms were identified. The L10n worm, Code Red, Sadmind, Nimda memory-only, Klez. H, many MyDoom worm variants, and Storm Worm are among them.

One of the most serious cyberattacks is the Structured Query Language (SQL) injection attack, which is initiated using internet browsers and leaves a variety of access doors open for attackers to exploit and get access to private information contained in website server databases.

12.4 ADVANCES IN CYBER THREATS AND COMPUTER CRIME

12.4.1 System Exploitation

This type of cyberattack involves exploiting a system's flaws after they have been identified. Microsoft for desktop computers and Android for the majority of smartphones may be utilized. The introduction of Trojan horses with a platform extension will be the starting point for the majority of these attacks. The executable virus will then be sent to the victim by any means available, including an email attachment or even a link to the piece of malware. The target will be affected if they execute the program or click on the malicious link.

The attacker will establish a session with the victim system. Antiviruses or applications like Windows Defender had previously discovered Trojans.

Types of Exploitation

Windows Exploitation:
 a. Veil-Evasion exploitation
 b. sAINT exploitation
Android Exploitation:
 a. Evil-Droid exploitation

Veil-Evasion Exploitation
It is a software platform for making Metasploit payloads that are resistant to antivirus software. To keep the Malware hidden from antivirus software, it employs cryptographic methods. The Trojan's or Payload's output would be the same as the Windows program with the.exe extension. Up to this point, the upgraded Windows 10 2018 version has been unable to detect this malicious program as a Trojan.

sAINT
It's a spyware maker for Windows systems that produces a JAR file as a result. When the file is opened on Windows, it appears to be a normal JAR file, but it contains malicious code that allows it to spy on the system. Antivirus applications and Windows Defender do not identify the sAINT output file on a properly updated computer. Following the execution of the file, the webcam snaps after a set time period. All files are emailed to the attacker's chosen email address.

Features of sAINT

1. Keylogger
2. Take screenshot
3. Webcam capture
4. Persistence

Evil-Droid
Evil-Droid is an Android breach framework that designs, generates, and embeds apk payloads. It takes advantage of Android phones. They are undetectable by Android security up to the most recent version. Evil-Droid has a unique feature in that it can be linked to any Android application file. After successfully exploiting smartphones, the attacker will be able to obtain the following things:

Features of Evil-Droid

a. It provides Live Webcam Stream facility.
b. It can download various files from any device.
c. Keylogger can capture the key activities.
d. It can capture screenshots.
e. Provide the facility to record conversation.
f. Dump and retrieve contacts.
g. Dump and retrieve SMS.

12.4.2 Phishing

Phishing is the most common method used by hackers since it is simple to execute and get outcomes with slight effort. Fake e-mails and phishing links are designed to appear to come from a legitimate source. The hackers first develop a site based on the individual's interests. When the victim inputs their credentials into the login box after receiving the phishing page, the credential is sent to the attacker automatically. Earlier days, after designing a phishing website, the website was detected as a phishing website by the majority of browsers. However, because of developments in this field and research, cybercriminals have established a means to get around this and generate a page that is HTTPS and appears to be real.

Socialfish
It is a platform that's used to make phishing sites. It works in conjunction with internet attacks. It generates a webpage that comes with a pre-template, like a Facebook

Login Page or a LinkedIn Page. Phishing pages created with this platform are undetected by any browser, making them impossible to identify as phishing or fake links.

12.4.3 A Man-in-the-Middle Attack

This is a type of cyberattack in which a hostile attacker disrupts a link between different parties, falsifies both stakeholders, and obtains the information that the two or more parties were attempting to convey. In a summary, it records traffic and packets between each sender and receiver in the network. To accomplish phishing or eavesdropping, an attacker can introduce malicious code.

12.4.4 DoS Attack

When we input a URL into a computer browser, we make a request to the site's computer server to display a page. At any given time, each server can indeed handle a fixed number of requests. If an attacker rushes the server with requests, it will be impossible to manage your request. It was forced to shut down and restrict access to genuine users due to the huge amount of requests

12.4.5 Wi-Fi Exploitation

We often use Wi-Fi with any credential and encryption method, such as WPA2, WPA, or Wi-Fi Protected Setup (WPS). Initially, hackers used brute force, DoS attack, or several WPS pins on a WPS-configured router to exploit Wi-Fi. However, because Wi-Fi encryption techniques and firmware must be updated, exploiting Wi-Fi with this type of attack takes a long time and is dangerous. Recently, a security researcher discovered a way to exploit Wi-Fi and access the Wi-Fi password without having to try many passwords or using brute force cracking.

Phishing by Wi-Fi
The hacker uses this method to send a sequence of DE verification packets to the routers, deauthenticating all clients on the Wi-Fi connection. At the same time, the attacker will create a bogus network interface of the same identity as the previously connected internet access. Whenever the client attempts to access it, the client is sent to the router's Bogus Automatic Update page, where the client must enter the router credentials to proceed. Without using brute crack, the hacker will obtain the credentials in a text format after properly submitting them.

Wi-Fi phisher is a cybersecurity tool that utilizes Wi-Fi auto-association threats to trick wireless users into connecting without their awareness to an attacker-controlled access point.

12.4.6 Ransomware

It's a virus or harmful software that encodes all of the records/knowledge on a computer, rendering it inaccessible and unusable. Cybercriminals threaten the victim with the decryption of their data and demand payment in bitcoins as a ransom.

Additionally, if an amount is paid, there is no assurance that decryption keys for safe data retrieval will be provided.

WannaeCry

"WannaCry" is the most current ransomware to wreak widespread damage in the digital world, and it is a different form of virus from previous ransomware. This virus infects by exploiting the vulnerability in the way Windows handles the Server Message Block (SMB) communication.

The following is how it works:

On affected Windows systems, WannaCry encrypts the hard drive.

- A worm and a ransomware program are the two main components.
- It spreads laterally among machines on the same local area network (LAN).
- Malicious email attachments are often used to distribute it.
- ETERNAL BLUE is the name of the exploit.
- The first ransom was US$300; however, it was later increased to $600 in Bitcoin [10]. The top 20 counties impacted by cybercrime are depicted in Figure 12.2.

12.5 INDIAN CYBER SITUATION

India is third in the world regarding internet subscribers, after the United States and China, with a percent per annum of 44% in the same period and 2017.

The United States is among the top 10 spam-sending countries in the world, as per reports. India is one of the five most important nations affected by cybercrime, according to a study issued on October 22 by internet security company "Symantec Corp."

12.6 CYBERATTACKS IN INDIA OF LATE

HEIST AT THE UNION BANK OF INDIA (JULY 2016): Hackers used a malicious email sent to an employee to get entry to the credentials needed to conduct a cash transfer, robbing Union Bank of India of $171 million. The bank was able to recover nearly all of the funds due to its quick reaction.

WANNACRY RANSOMWARE: In India, the universal malware threats acquired their peal through ransomware-seeking hackers encrypting thousands of computers. The attack also affected the systems of the Andhra Pradesh police and West Bengal's state utilities.

ZOMATO DATA THEFT (MAY 2017): An "ethical" hacker requested that the food tech company admit its security weaknesses, so he stole data from 17 million customers, including names, email addresses, and hashed passwords, and sold it on the Dark Web.

RANSOMWARE, PETYA: The malware incident had a worldwide effect, notably in India, where cargo processing operations at a port owned either by Danish Corporation Port were impacted [11].

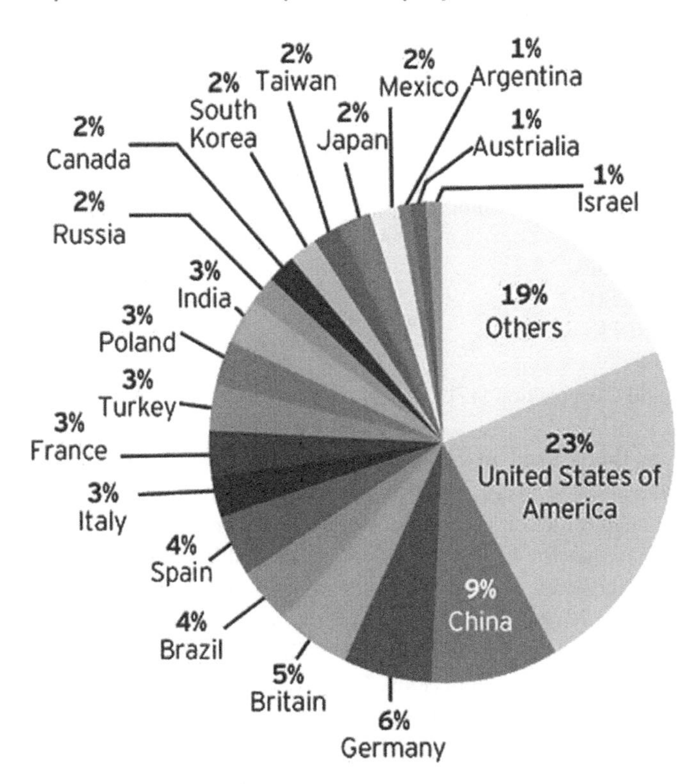

FIGURE 12.2 Countries impacted by cybercrime

12.7 GLOBAL ORGANIZATIONS FIGHTING AGAINST CYBERCRIME

1. *International Criminal Police Organization (INTERPOL):* INTERPOL's main goal is to allow investigators from its 190 member countries to work together to combat international crime, notably cyberattacks and crimes against minors.
2. *Europol European Cybercrime Centre*
   The European Cybercrime Centre is a division of the European Union's Police Office that organizes bend police to combat computer fraud and serves as a technical resource center on the subject.
3. Council of Europe
   This gateway provides details on the Council of Europe's ongoing ways to tackle cybercrime, such as the status of the Paris Convention on Cybercrime and the effort of the Resolution Recommended Group.
4. International Telecommunication Unions
   In addition, the International Telecommunication Unions and UNICEF have collaborated to create Kid Online Protection Guide.

5. United Nations Office of Drugs and Crimes
   This office is in charge of the United Nations (UN)'s ways to tackle illicit trafficking, particularly cybercrime. The UN Office of Drugs and Crime established an Active Inter-Governmental Advisory Committee in 2011 to study cyberattacks and the worldwide community's response. The Inter-Governmental Expert Group's meetings are summarized on this website, together with links to meeting recordings and supporting documents.
   The following are some examples of non-governmental organizations:
   - APWG
   - Spamhaus
   - eNACS0
   - INH0PE
   - IWF
   - Rand Corporation [12]

Tools and methods used in cyber issues at the time of Covid-19: One of the deadliest pandemics in history is currently sweeping the planet. The Covid-19 epidemic had a huge effect on the whole world and ground various countries to a halt already. Not only has the Covid-19 epidemic had health and financial consequences for enterprises, individuals, governments, and administration, but it has also evolved into a weapon for hackers and cybercriminals to utilize in cyberattacks. Because the environment is ideal for cyber thieves to attack, cybersecurity becomes even more vital during these critical periods. The World Health Organization (WHO) witnessed a substantial rise in the number of cyberattacks directed at its staff, as well as email frauds targeting the general public, shortly after the Covid-19 epidemic began. According to WHO, 450 WHO email accounts and passwords were stolen online in April 2020 [13]. Scammers posing as WHO have been sending emails to the general populace, requesting money for a bogus Covid-19 Solidary Response Fund rather than the real Covid-19 Solidary Response Fund. The Covid cybersecurity graph is depicted in Figure 12.3 and various malicious cyberattacks and threats are defined in Figure 12.4.

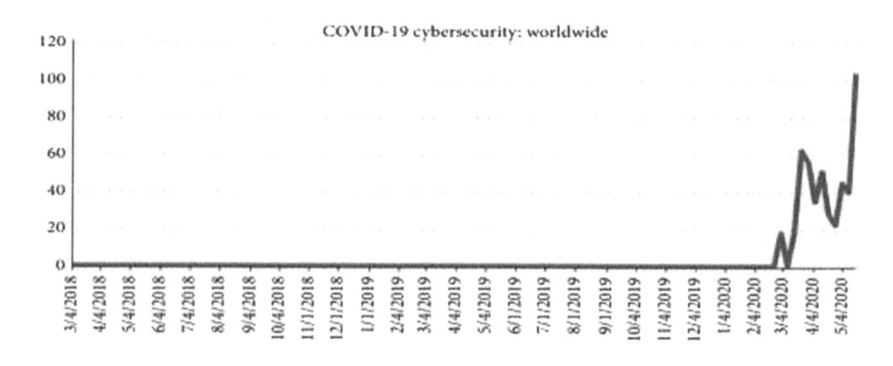

FIGURE 12.3 Covid-19 worldwide graph

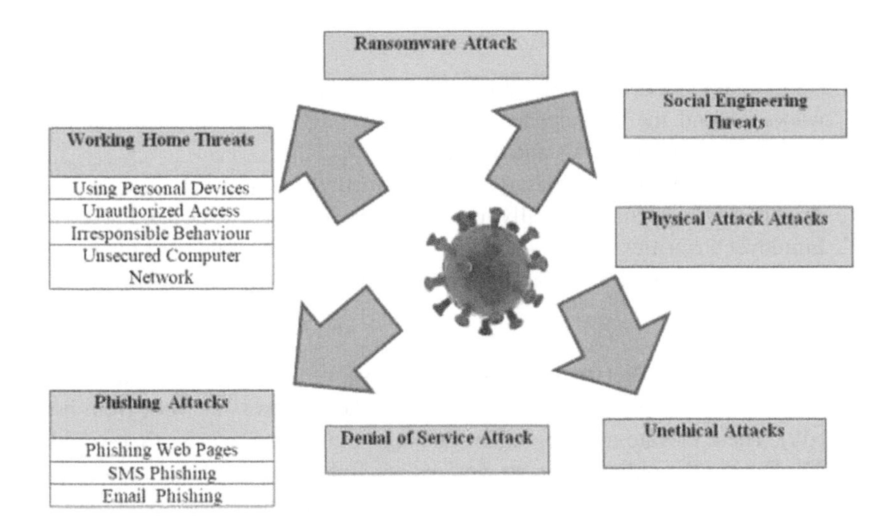

FIGURE 12.4 Various malicious cyberattacks/cyber threats

Several attackers have effectively leveraged the pandemic situation to build ransomware and spoofing attempts against innocent persons and companies, according to WebARX Security. The attackers have created a slew of new phony and hazardous websites that utilize phishing to steal sensitive information from users. Phishing attacks have increased by 350% since the outbreak began, according to research. Malware has been placed on smartphone platforms meant to follow Covid-19's activity, giving the apps the capacity to collect sensitive data from users [4].

Numerous Categories of Malevolent Threats

Working from Home Malicious Cyber Threats: Many people were compelled to work from home for the first time due to the Covid-19 pandemic. Working from home exposes you to additional cybersecurity risks, such as deliberate cybercrime. Improper access to data saved on a computer or a smartphone phone can have serious repercussions on someone's private, psychological, economic, and business life.

Using Personal Devices: Employees who work from home typically utilize their personal gadgets since they are more comfortable with them. Personal computers or laptops, on the other hand, are quite likely to exist. Therefore, the following threats may exist.

(1) Inadequate Performance: High-performance workstations with more processing components and memory storage capacity than personal computers, laptops, or mobile phones were regularly used by industries.
(2) The Use of Untrustworthy Programs: They expose data to a variety of risks that might result in data loss or incorrect operations and processes, increasing the chance of malware infection.
(3) Inadequate Backup Plans: It's conceivable that some information will be deleted, or at the very worst, the most recent changes.

12.7.1 Careless Behaviors

A. Unreliable Connections: Some employees worked from public Wi-Fi networks beyond their residences, which are thought to be an ideal access point for computer threats and data theft.
B. Surveillance Without Permission: Untrustworthy persons, such as an angry neighbor or a spy, may illegally monitor employees who work from home.
C. Employee Priorities: Employees have distinct priorities while working from home, and extraordinary family care requirements have an influence on staff availability.

Unsecured Computer Network: To reduce the risk of transmission of the coronavirus, some firms have encouraged or compelled their workers to work from home (Covid-19). These elevate new apprehensions regarding cybersecurity that would be managed. During the outbreak, most workers are linked to their local network, but are not as secure as their corporate network, putting them in danger. All home networks, as well as machines, frequently lack security features found in corporate networks, such as the ones listed below.

1. Antivirus Programs: Malevolent software will be perceived by the antivirus package. A successful antivirus system will always detect whether a file contains suspicious actions, preventing data loss or theft.
2. Firewalls: It is a network circulation controller device. Firewalls can allow all network packets while blocking just suspect ones, or they can refuse all packets while allowing only those considered suitable.

Following are some cases which are faced during an outbreak

Understanding and optim

The Covid-19 epidemic forced Germany to adopt some action, including restricting the restaurant business. Therefore, Germans have become increasingly reliant on still-operational delivery systems. Liefrando, for example, sources meals from over 15,000 restaurants. The German food delivery company "Takeaway.com" has been targeted by cybercriminals in a ransomware campaign (Liefrando.de). Liefrando was supposed to offer consumer refunds if food orders were received but could not be processed. As a result, some businesses may be forced to compensate for cybercriminals or invest in sophisticated threat security.

Phishing Attacks

During Covid-19, the hacker sends out a variety of emails and SMS messages claiming to have a "cure" or seeking donations. These emails and SMS, like other phishing scams, utilize real-world situations to trick users to click. Scam communications (or phishes) can be difficult to spot and are designed to make individuals react without thinking.

Phishing by SMS, phishing through mail, and phishing scams are the three main forms of phishing tactics.

The majority of phishing efforts are made via email; however, the National Cyber Security Centre (NCSC) has discovered some phishing attempts made using text messaging. The SMS phishing pitch has always featured cash incentives, such as grants and reimbursements (such as a tax relaxation).

The UK government concept, for instance, is used in SMS messages to collect usernames, addresses, names, and bank data from individuals. The phishing website is linked directly in the SMS texts from "COVID" and "UKGOV." Financial troubles are likely to continue to be used by malicious cyber actors in their phishing operations. New government-aid programs in response to Covid-19, in particular, will very certainly be exploited as phishing targets.

Phishing Emails

Computer scammers send emails purporting to have a "cure" for the illness, offer money prizes, or encourage the target to give because of the newest coronavirus condition (Covid-19). Such communications, like some of the other phishing techniques, use real-world concerns to persuade you to click the given malicious link are described in Figure 12.5. Coronavirus information, new reported cases, breakouts, and rescue services are all instances of phishing emails. These emails could include a request to visit a Link where hostile cyber actors can steal information such as usernames and credentials, credit card details, and other private information. This email was sent with the purpose of duping the recipient into accessing a website that collects Personal Identification Knowledge under the guise of offering travel advice to countries with Covid-19-verified cases. When hackers get information, they often create bank accounts or bank cards in the names of the victims, then use the victim's funds to buy valuable stuff or exchange currency using unsuccessful cryptocurrency like Bitcoin [14].

The scam texts, sometimes known as "phishes," are designed to get people to act without thinking and can be tough to notice. Cybercriminals regularly design

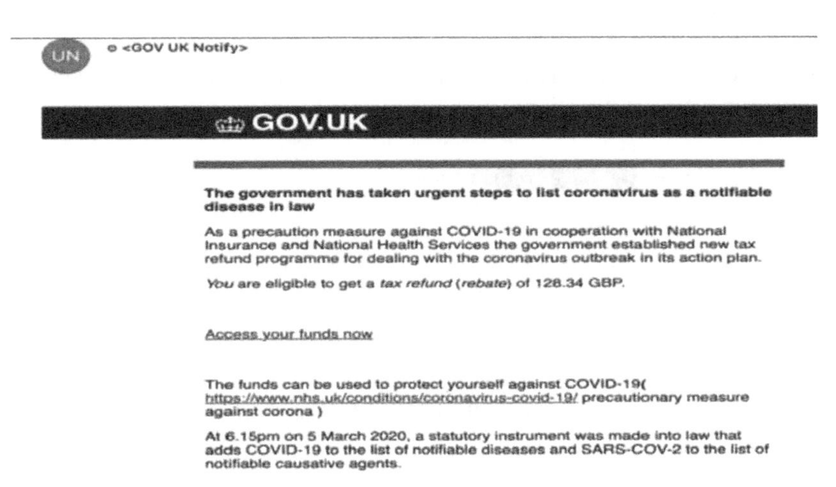

UN <GOV UK Notify>

GOV.UK

The government has taken urgent steps to list coronavirus as a notifiable disease in law

As a precaution measure against COVID-19 in cooperation with National Insurance and National Health Services the government established new tax refund programme for dealing with the coronavirus outbreak in its action plan.

You are eligible to get a *tax refund* (*rebate*) of 128.34 GBP.

Access your funds now

The funds can be used to protect yourself against COVID-19(https://www.nhs.uk/conditions/coronavirus-covid-19/ precautionary measure against corona)

At 6.15pm on 5 March 2020, a statutory instrument was made into law that adds COVID-19 to the list of notifiable diseases and SARS-COV-2 to the list of notifiable causative agents.

FIGURE 12.5 Scams including phishing mail

a variety of schemes geared at a growing number of people who are disadvantaged. The ACSC has been made aware of an international fraud in which people are asked to help the "Coronavirus Relief Fund" as casual employees or volunteers. It is recommended that applicants accept donations for Covid-19's social projects. In actuality, people who are unwittingly drawn into this scam become money mules for cybercrime gangs, converting criminal proceeds into untraceable cryptocurrency.

The Advanced Centre for Computing and Communication (ACCC) warned Australians on March 20, 2020, about a phishing email that asked them to fill out an attached form in order to get help of $2500 for Covid-19. The email attachment comprises a macro that downloads Malware to your computer system. If people receive such phishing emails, they should not open the attachments and instead uninstall the document. The view of scam mail is depicted in Figure 12.6.

Tables 12.1 and 12.2 demonstrate that organizations with fewer workers had a greater incidence of receiving fake emails, contrary to the common perception that breaches only impact major corporations [15].

Covid-19 appeared at the start of 2020 in this present complex cybersecurity landscape, becoming a facilitator for digitalization and boosting human adaptability to new types of employment. Covid-19-related phishing and ransomware attempts rose significantly with its debut, as well as the urgency of information and news sharing. The graph shows the trends of cyberattacks during Covid-19 are depicted in Figure 12.7.

12.8 CONCLUSION

This study reviews cybersecurity and cyber threats. This study also emphasizes threats that occurred during the Covid-19 epidemic. It also explores the different types of cybersecurity that appeared before the pandemic. Computer security is a wide issue that is becoming increasingly important as the world gets more linked and critical activities are conducted via networks.

With each passing New Year, cybercriminals and information security continue to diverge in different directions.

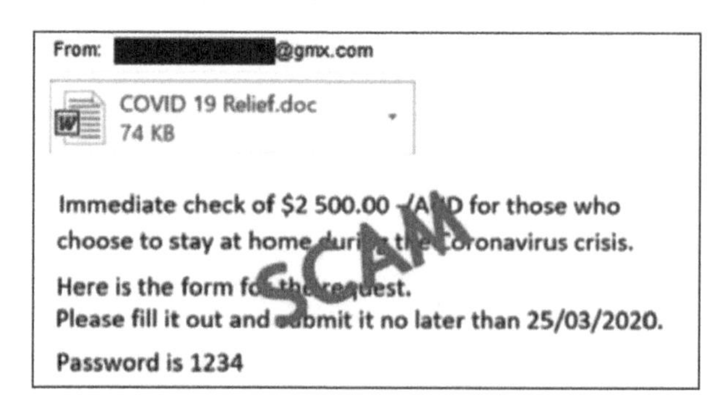

FIGURE 12.6 Scam mail

TABLE 12.1
Malicious Email Rate by Company Size on an Annual Basis

Organizational size	Rate of Malicious Mail
1 to 250	324
251 to 500	357
501 to 1000	393
1001 to 1500	826
1501 to 2500	442
Above 2501	558

TABLE 12.2
Number of Malevolent Emails Based on Organization Size

Organizational Size	Users Who Have Been Affected
1 to 250	7
251 to 500	8
501 to 1000	3
1001 to 1500	5
1501 to 2500	6
Above 2501	12

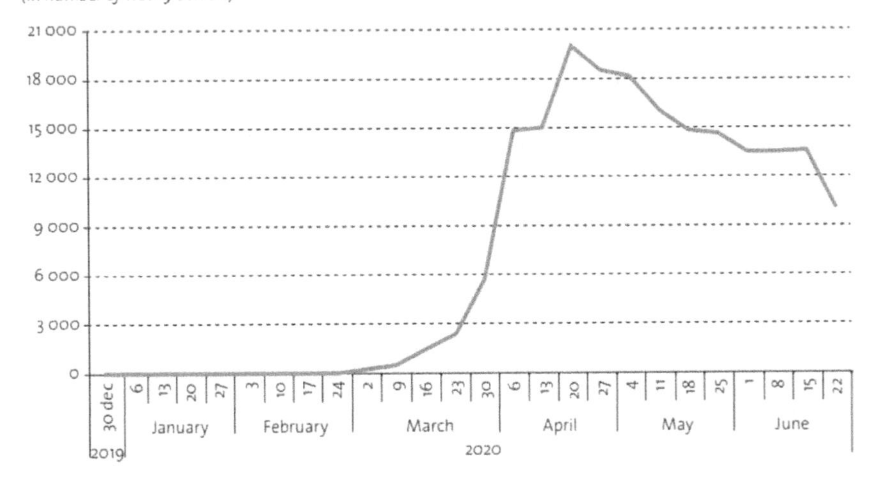

FIGURE 12.7 Cyberattacks since December 2019

This article, which goes beyond standard cybersecurity attacks, covers working from home risk, social engineering threats, extortion threats, phishing, and various types of malicious attacks.

REFERENCES

1. Alansari, M. M., Aljazzaf, Z. M., & Sarfraz, M. (2019). On Cyber Crimes and Cyber Security. In M. Sarfraz (Ed.), Developments in Information Security and Cybernetic Wars, pp. 1–41. IGI Global, Hershey, PA, USA. doi:10.4018/978-1-5225-8304-2.ch001.
2. J. Jang-Jaccard and S. Nepal, "A Survey of Emerging Threats in Cybersecurity," *J. Comput. Syst. Sci.*, vol. 80, no. 5, pp. 973–993, 2014, doi: 10.1016/j.jcss.2014.02.005.
3. M. Ganesan and P. Mayilvahanan, "Cyber Crime Analysis in Social Media Using Data Mining Technique," *Int. J. Pure Appl. Math.*, vol. 116, no. 22, pp. 413–424, 2017.
4. Ramon, M. C. and Zajac, D. A. "Cybersecurity Literature Review and Efforts Report," *Prep. NCHRP Proj.* pp. 3–127, 2018
5. Jahankhani, Hamid, Ameer Al-Nemrat, and Amin Hosseinian-Far. "Chapter 12 - Cybercrime Classification and Characteristics." edited by B. Akhgar, A. Staniforth, and F. B. T.-C. C. and C. T. I. H. vol. 5, pp. 149–64, Syngress, Bosco.
6. J. Achkoski and M. Dojchinovski, "Cyber Terrorism and Cyber Crime: Threats for Cyber Security," Proc. First Annu. Int. Sci. Conf., Global Security and Challenges of the 21st Century - MIT University – Skopje, 2012,[Online]. Available: http://eprints.ugd.edu.mk/6502/2/__ugd.edu.mk_private_UserFiles_biljana.kosturanova_Desktop_Trudovi_Jugoslav Achkoski_Scientific Papers_elektronska verzija_Cyber Terrorism and Cyber Crime – Threats for Cyber Security_rev_JA.pdf.
7. M. Gercke, "Cybercrime Understanding Cybercrime," *Understanding Cybercrime: Phenomena, Challenges and Legal Response*, ITU, p. 366, 2012, doi: 10.1088/1367-2630/11/1/013005.
8. A. Maqsood and M. Rizwan, "Security, Trust and Privacy In Cyber (Stpc Cyber)," *Int. J. Sci. Res. Publ.*, vol. 9, no. 2, p. p8682, 2019, doi: 10.29322/ijsrp.9.02.2019.p8682.
9. A. R. P. Tushar and P Parikh, "Cyber security: Study on Attack, Threat, Vulnerability," *Int. J. Res. Mod. Eng. Emerg. Technol.*, vol. 5, no. 6, pp. 1–7, 2017.
10. R. R. Yadav, "Advances in Cyber Security," *Int. J. Eng. Res.*, vol. V7, no. 03, pp. 117–120, 2018, doi: 10.17577/ijertv7is030091.
11. M. Chaturvedi, "Cyber Security Infrastructure in India: A Study," *Emerg. Technol.*, no. April 2014, pp. 70–84, 2008, [Online]. Available: https://www.csi-sigegov.org/emerging_pdf/9_70-84.pdf.
12. C. Academy and C. Studies, "World Internet Development Report 2017," *World Internet Dev. Rep.*, vol. 2017, pp. 89–117, 2019, doi: 10.1007/978-3-662-57524-6.
13. H. S. Lallie et al., "Cyber Security in the Age of COVID-19: A Timeline and Analysis of Cyber-Crime and Cyber-Attacks during the Pandemic," pp. 1–20, 2020, [Online]. Available: http://arxiv.org/abs/2006.11929.
14. R. A. Ramadan, B. W. Aboshosha, J. S. Alshudukhi, A. J. Alzahrani, A. El-Sayed, and M. M. Dessouky, "Cybersecurity and Countermeasures at the Time of Pandemic," *J. Adv. Transp.*, vol. 2021, no. 2003, 2021, doi: 10.1155/2021/6627264.
15. D. P. Fidler, "Cybersecurity in the Time of COVID-19," *Counc. Foreign Relations*, vol. 2020, pp. 7–9, 2020 [Online]. Available: https://www.cfr.org/blog/cybersecurity-time-covid-19.

13 A Cybersecurity Perspective of Machine Learning Algorithms

Adil Hussain Seh, Hagos Yirgaw, Masood Ahmad, Mohd Faizan, Nitish Pathak, Majid Zaman, and Alka Agrawal

CONTENTS

13.1 Introduction 221
13.2 Supervised ML Approaches 223
13.2.1 Classification-Based Supervised ML Techniques 224
13.2.2 Regression-Based Supervised ML Techniques 228
13.3 Unsupervised ML Approaches 231
13.3.1 Clustering-Based Unsupervised ML Techniques 231
13.3.2 Association-Based Unsupervised ML Techniques 233
13.4 Cybersecurity Perspective 235
13.5 Conclusion 239
References 239

13.1 INTRODUCTION

One of the prominent changes in the present world is the succession of information and communication technology. Over the past few decades, technological revolution has greatly been influencing the whole world and has also been changing people's ways of thinking. One of the well-known technologies in this domain is artificial intelligence (AI). AI as a human brain-simulated technology is defined by various eminent authors as: "Systems that think like humans," "Systems that act like humans," "Systems that think rationally," and "Systems that act rationally" [1]. The term AI was first coined by J. McCarthy in the presence of M. Minsky and Arthur Samuel in 1956 in a workshop [1, 2]. After that, AI accelerates its domain in a vibrant way, and present day it is known as a dominant and revolutionary technology throughout the world. The roots of the AI are commonly attached to the following disciplines such as Philosophy, Logic, Computation, Cognitive science, Neuroscience, and Evolution

DOI: 10.1201/9781003323426-13

[3]. And the well-known branches or subfields of AI are machine learning (ML), evolutionary computation, computer vision, natural language processing, robotics, and planning [3]. Machine learning as a sub-domain of AI was first proposed by Arthur Samuel in 1959. After that, ML gained rapid significance in various fields of life, and today it is recognized as one of the growing technologies that can address issues such as future event prediction, disease diagnosis, market analysis, email filtering, intrusion detection, image and speech recognition, and so on.

Machine learning provides the ability to make programs learn from past (historical) data. Then apply the learning behavior to make predictions for future events and activities with less human intervention and explicit programming. With every correct decision, the computer program improves its performance measure. In a more formal way, ML is defined as "A computer program is said to learn from experience *E* with respect to some class of tasks *T* and performance measure *P*, if its performance at tasks in *T*, as measured by *P*, improves with experience *E*" [4]. Here, the main focus is on these three things: a set of tasks represented by *T*, estimation of performance denoted by *P*, and *E* representing the source of experience for the program. Suppose we have a problem:

- To identify suspicious user access against a digital system, which we represent by *T*.
- Percentage of correctly identified suspicious user accesses against the system that we denote by *P* (performance measure).
- Training itself on a historical data set with given class labels that we mention by *E* (training experience).

All the three events go through in a sequence to make an ML model effectively workable with respect to its results. First, a proposed ML-based model gets trained on some historical data (gaining experience *E*), then the model is tested against both suspicious user accesses (positive events) and normal user accesses (negative events), which represents class *T*. This finally results in performance measure (*P*), which is calculated on true-positive and false-negative, true-negative and false-positive events identified by the proposed model on class *T*.

Machine learning involves different algorithms to devise ML-based models for future event predictions with the help of past data. ML algorithms have a great ability to learn from the structure as well as unstructured data and help automated systems to solve real-life problems in different areas such as marketing, education, healthcare, transport, defense, and many more. Moreover, ML has got significant importance in the field of cybersecurity [5]. Machine learning techniques namely supervised techniques, unsupervised techniques, semi-supervised techniques, and reinforcement techniques are the dominant classes of ML algorithms. Each class has its own significance and strength to address various real-life problems. However, the mainly focusing classes are supervised and unsupervised techniques. *Supervised ML* as the name implies works in a supervised environment. In it, labeled historical data is used to train and test the devised models. Supervised ML algorithms are implemented to build models that map the given inputs to outputs on the basis of existing knowledge. It infers the output class for an input object according to the knowledge perceived from labeled examples of training data. In supervised ML, models are completely subjected to labeled data and the efficiency and accuracy of models are directly proportional to the quality of data. *Unsupervised ML* is a contrastive study against the supervised ML. In unsupervised ML, algorithms are implemented to build models to make classification of given data irrespective of its class labels. Data used in it are completely unlabeled and models are exclusively autonomous to compact internal representation of the given data according to their common characteristics. Models analyze data are significant insights from this data to classify the future based on these insights. Further, two more classes of ML are semi-supervised learning and reinforcement learning (RL). Semi-supervised learning integrates both supervised and unsupervised learning to devise more compact and effective models. This relies on both the labeled and unlabeled data. While RL encompasses the area of ML that particularly concerns action–reward problems, an agent is supposed to take possible suitable action in a specified environment to achieve the reward. Here, every correct decision of the agent toward the goal increases the chances of getting the reward, whereas each incorrect decision decreases them. The agent learns from its experiences gained through every possible decision or step. Thus, there is the concept of learning from labeled data in RL as such in supervised learning.

However, in this chapter, our focus is to make a descriptive study on supervised and unsupervised ML algorithms from a cybersecurity perspective. Cybersecurity is a serious issue in the present era and has very disastrous outcomes after being breached especially in areas like healthcare [6]. Where disclosure of protected data puts calamitous effects on both the organizations and concerned stakeholders, it includes loss of reputation of reputed business organizations and various kinds of threats to individual customer properties, sometimes encompasses life threats too. Anomaly detection, fishing page identification, software vulnerability diagnosis, Malware identification, and denial-of-service (DoS) attacks are the main cybersecurity issues and challenges that needs immediate attention of the research community. Henceforth, ML as a proactive approach to address cybersecurity issues will examine the threats and respond to intrusions and security incidents swiftly in an instinctive way. This study will emphasize prominent machine learning algorithms and their characteristics to address these cybersecurity issues as a proactive security mechanism.

The rest of the chapter will be assembled in the following fashion: Section 13.2 describes the supervised machine learning approaches. It further encompasses the classification and regression algorithms in two different subsections; Section 13.3 enlists the unsupervised machine learning approaches, while Section 13.4 defines the cybersecurity perspective of ML approaches; and Section 13.5 provides the discussion of the study.

13.2 SUPERVISED ML APPROACHES

Supervised learning is among the prominent approaches of ML that is widely practiced in prediction and detection systems. As implied from supervised learning, there exists a supervision mechanism during the model training. Purely labeled data is involved to train, test, and validate the designed model(s) that implements a supervised machine learning algorithm(s) to become workable [7]. Here, labeled data exhibits a mapping function from the input variable (X) to the output variable (Y). On the basis of predefined data attributes, a machine learning model gets trained and tested. Later, the accuracy of the model has to be measured on the separate subset of the same data. For that, various accuracy measuring scales namely true-positive and false-negative, true-negative and false-positive, precision, and recall are used by researchers and experts to measure the accuracy of supervised machine learning models.

On the basis of the data dependency, here we make a conclusive statement about the supervised machine learning approaches that they are entirely dependent on the available training data. Thus, training data should comprehensively simulate all the scenarios of the environment for which it is connotated with a sufficient number of attributes. So the accuracy and efficiency of the supervised model are directly proportional to the accuracy, reliability, and completeness of the data. On the basis of data characteristics, supervised machine learning has been further divided into two prominent subclasses namely *regression and classification.*

13.2.1 Classification-Based Supervised ML Techniques

Classification is one of the notable approaches to supervised learning and allows us to make problem decisions based on discrete values as depicted in Figure 13.1. It includes a class of algorithms that deals with the categorical or discrete values

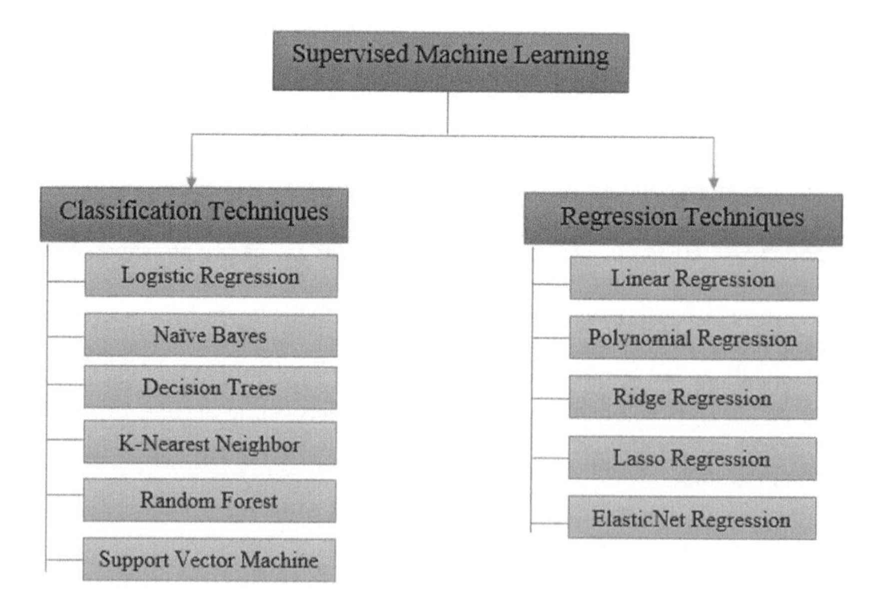

FIGURE 13.1 Types of supervised machine learning.

[8]. Here, the class label for each tuple (it can be a single or multiattribute record) is categorical and it can be binary or ternary and so on. It works more effectively in the areas where available data depicts discrete characteristics and outcomes of predicted values are also demanded in the discrete form. The most commonly used classification algorithms in machine learning are discussed in the following subsection.

i. *Logistic Regression:* Logistic regression belongs to the supervised class of ML techniques. However, among the regression techniques, it is the only the regression techniques that belong to the classification of algorithmic class. It is because of the dichotomous nature of the target variable, which implies that the dependent variable can target only two possible classes [9]. In other words, we can say that the class label can possess only two binary values. One to represent the positive output value and the other to represent the negative output value. The core of this algorithm is the logistic function, which is commonly known as the sigmoid function. This function has an S-shaped representation in the plot and maps any real number input value into the value that lies between 0 and 1 and predicts probabilities. Figure 13.2 provides the view of it. Later the probabilities are converted into binary values (0 or 1) to make predictions as logistic regression is a classification algorithm. The hypothetical expectation of logistic regression and equation is represented as follows:

$$0 \leq h_\theta(X) \leq 1$$

$$F(x) = \left(1/\left(1 + e^{-\text{value}}\right)\right)$$

where *e* represents the natural logarithmic base and the value represents the input value.

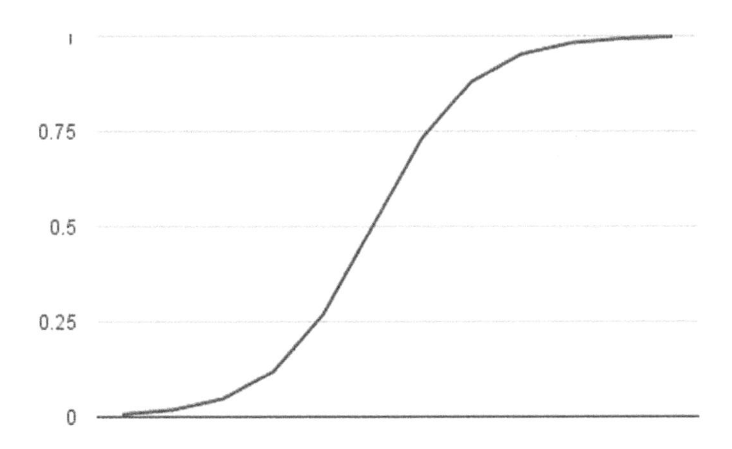

FIGURE 13.2 Logistic regression curve.

However, the logistic regression algorithm can also take multinomial or ordinal form but only in discrete values. In the multinomial mode, it can take more than two values for a target variable without any quantitative significance and in an unordered form. However, in the ordinal mode, it can also take more than two values for the target variable but with quantitative significance.

ii. *Naïve Bayes:* It is among the popular algorithms in the machine learning domain that is premised on the Bayes theorem. It is also a probability-based classifier that predicts the class of an unknown data set. It is one of the fast-working ML algorithms that is easy to implement. It assumes the presence of a specific feature independently with respect to any other features in a class [10]. Here, each feature of the class contributes independently and equally to its total probability. The mathematical formula to calculate the output (class) label for a tuple is as:

$$P(A/B) = P(B/A) \cdot P(A) / P(B)$$

where P represents the probability;
A and B represent the two events;
$P(A/B)$ defines the posterior probability;
$P(B/A)$ defines the likelihood probability;
$P(A)$ defines the prior probability;
$P(B)$ defines the marginal probability.

iii. *Decision Trees:* Decision tree algorithms are one of the powerful and popular algorithms in the classification of problem domains. Decision tree algorithm makes a tree-like structure, which consists of internal nodes, branches, and external nodes (decision nodes) [11]. Internal nodes of the tree define a test on an attribute, whereas the branch of an internet node denotes the outcome of the test, and leaf nodes represent the class label. Decision tree algorithms follow the rule-based approach to find out the result for a given data tuple (vector).

The working procedure of a decision tree is that the tree is initially trained on some labeled historical data set as it is the precondition for all supervised ML approaches. Then with the help of the constructed decision tree, ideal class labels have to be found for each new data tuple. The well-known decision tree algorithms are ID3, CH4, CART, CHAID, and MARS [12].

iv. *K-Nearest Neighbor:* It is known as an efficient lazy learning algorithm and has successful real-life applicability [13]. It is practiced when people have small information or we can say that they have no information regarding the data distribution, primarily we should straightforwardly prefer K-nearest-neighbor classification. The need for finding equations of prediction comprise independent variables, which are used in the individuals to groups classification. When the trustworthy parademic evaluation of probable compactness is not known or impossible to calculate, it leads to the development of K-nearest-neighbor classification. Some of the precise features of the K-nearest-neighbor classification were freaked out later in the year 1967. For example, it was proven that the K-nearest-neighbor classification mismanagement is belted over twice the error rate of Bayes for k=1 and n (Cover & Hart, 1967).

K-nearest-neighbor (KNN) classifier is easy to implement and is robust to the noisy training data. It provides effective results if the training data is larger in volume. However, the computation cost and calculation of k-value remain high and complex, respectively. The Figure 13.3 depicts the working presentation of the KNN algorithm.

v. *Random Forest:* Random forest as the name implies is a collection of individual decision trees that work as an ensemble [14]. It is a fast-running, computationally efficient approach that can handle enormous data sets. In this, the hyperparameters are set differently for each decision tree and each decision tree is trained on a different subset of data. Each individual decision tree in it comes out with a prediction and the prediction of maximum voted decision trees becomes the prediction of the model. It improves the accuracy of the overall model and reduces the problem of overfitting. The following are some advantages of an RF algorithm.
   - Training time for RF is comparatively less.
   - It deals efficiently with the larger data sets with a high accuracy rate.
   - It works well even if some data is missing and retains accuracy.

vi. *Support Vector Machine*: Support vector machine (SVM) is a well-known supervised learning approach for solving classification as well as regression-based problems. However, it is commonly used in ML to solve classification problems [15]. The goal of the SVM method is to identify the best line or decision boundary for classifying n-dimensional space into classes so that subsequent data points may be easily placed in the correct category. The best choice boundary is known as a hyperplane. SVM is used to choose the acute points/vectors that help form the hyperplane. The acute examples are called support vectors, and the technique is termed support vector machine. Figure 13.4 illustrates the usage of a hyperplane to categorize two separate categories:

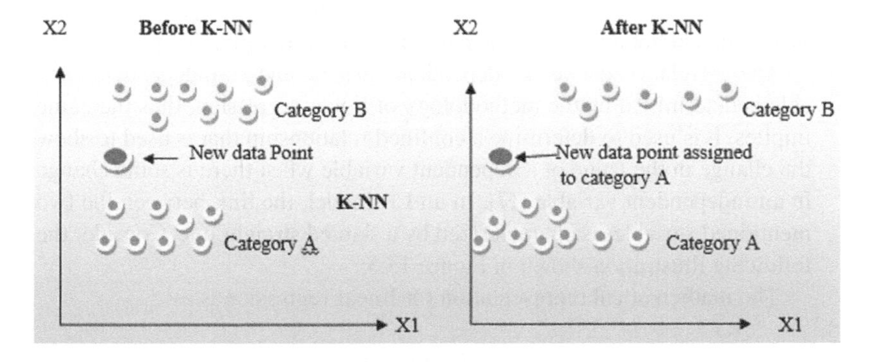

FIGURE 13.3 Working depiction of KNN classifier.

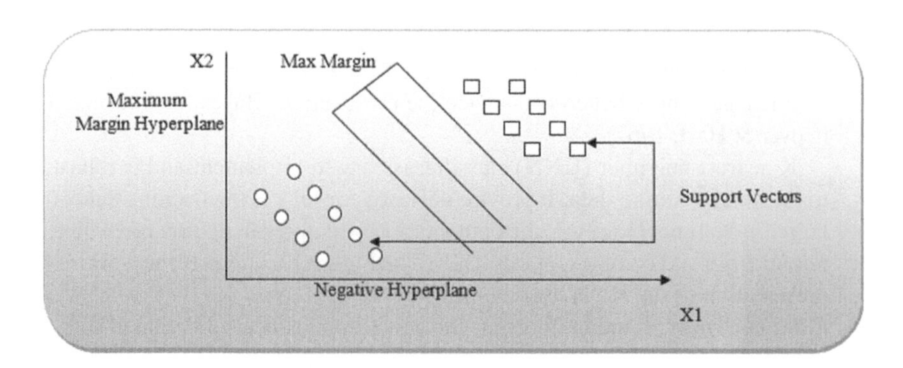

FIGURE 13.4 SVM hyperplane.

13.2.2 Regression-Based Supervised ML Techniques

Regression analysis is another supervised machine learning approach, which incorporates a set of machine learning algorithms that provides flexibility to deal with continuous data. Regression-based algorithms allow us to design mathematically functional models that map an input variable (x) to the output variable (y) having a continuous set of values [16]. Later, the designed equation (mathematical function) can be used to predict the class label (y) on the basis of new values of the predictor variables (x). It represents the relationship between two or more variables and makes an estimation of the impact of independent variables on dependent variables. Regression analysis is more efficient and effective in those areas where the data is generated in continuous form and the effectiveness of the results is directly proportional to the continuity of data. Weather forecasting and market trend analysis are the real-time examples of regression analysis. The most commonly used regression algorithms in machine learning are discussed in the following subsection.

i. *Linear Regression (LR):* The most common and broadly used method in machine learning is referred to as linear regression. It is a statistical method of operating and anticipating investigation. By using linear regression, we can anticipate sales, salary, age, product price, and other repeated/actual or mathematical notations with the help of linear regression.

   The correlation between a dependent variable and an independent variable is determined by the methodology of linear regression, thus the name implies. It is used to determine a confined relationship that is used to show the change in the value of a dependent variable when there is some change in an independent variable [17]. In an LR model, the link between the two mentioned variables is characterized by a slanted straight line. Consider the following illustration shown in Figure 13.5:

   The mathematical representation for linear regression is as:

$$y = a0 + a1x + \varepsilon$$

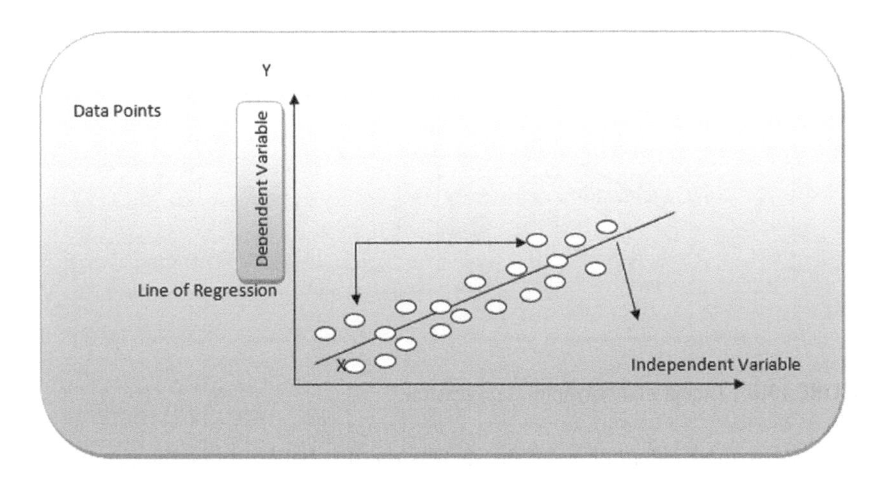

FIGURE 13.5 Linear regression illustration.

where
- Y stands for the "dependent variable,"
- X stands for the "independent variable,"
- $a0$ is the line's intercept,
- a1 = coefficient of linear regression,
- E = unintentional error.

ii. *Polynomial Regression:* As a regression approach, it uses an nth degree polynomial to represent the connection among a dependent variable (y) and independent variable (x) [18]. The equation for polynomial regression is as follows:

$$Y = b0 + b1X1 + b2X13 + \ldots + bnX1n$$

In machine learning, it's also known as the specific case of multiple linear regression. This is because we turn the multiple linear regression equation into polynomial regression by adding certain polynomial terms. It's a precise prototype that has been tweaked a little to boost efficiency. The training data set for this type of regression is nonlinear in character. To fit the intricate and nonlinear behavior and data sets, it utilizes a precise model of linear regression. "In polynomial regression, the original features are transformed into polynomial features of the desired degree (2, 3, … n) and then modeled using a linear model,". Figure 13.6 shows the comparative presentation of linear and polynomial models.

iii. *Ridge Regression:* Ridge regression is a methodology for studying multicollinear data in various regression models. When there is multicollinearity, minimum squares assessments are disinterested, but because their variances are huge, they may be far off the appropriate value [19]. Ridge regression minimizes typical errors by the addition of some degree of unfairness to the

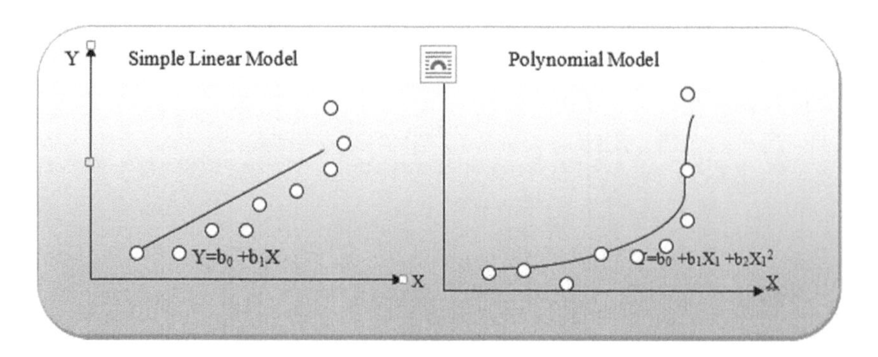

FIGURE 13.6 Linear and polynomial regression.

approximate calculations of the regression. It is believed that the net effect will be to provide more credible estimates. Principal component regression is another biased regression technique accessible in NCSS. The more popular of the two methods is ridge regression. Assume our regression equation is expressed in a matrix form as follows using the standard notation:

$$Y = XB + e$$

where Y denotes the dependent variable, X denotes the independent variables, B denotes the coefficients of regression that are to be calculated, and e denotes residual errors.

iv. *Lasso Regression:* It is a reduction-based method of an LR approach. The values of the data are reduced toward a mid-point, as we calculate the mean in mathematical terms, while reduction [20]. Simple, sparse modeling is cheered by this approach (i.e., models with minimum parameters). This class of regression is quite optimal for models having maximum multicollinearity or when you wish to automate elements during the selection process of the model, like the selection of variables and the removal of parameters.

L1 regularization is used in lasso regression, and it increases a fine which is equal to the supreme value of the magnitude of the coefficients. This form of polarization can lead to sparse models with less coefficients; a few coefficients may lead to zero, and the said model may be removed. Greater penalties provide coefficient characters that are adjacent to zero, which is great for making elementary models. L2 regularization, under other conditions, does not lead to the deletion of coefficients or sparse models (e.g., ridge regression). As a result of this, the lasso is significantly easier to understand than the ridge. Quadratic programming challenges, such as lasso solutions, are best tackled with software (like Matlab). The algorithm's purpose is to minimize:

$$\sum_{i=1}^{n}\left(Y\,i-\sum_{j}XijBj\right)2+\sum_{j=1}^{p}\left|\beta j\right|$$

This is equivalent to reducing the sum of squares with restriction |Bj s (= summation notation). Some of the β s are reduced to zero, which leads us to an easier-to-understand regression model.

v. *Elastic Net Regression:* Elastic net regression arose from the criticism of lasso, whose variable selection can be overly reliant on data, making it ambiguous. The perfect of the two worlds can be had by combining the penalties of ridge regression and lasso [21]. The goal of elastic net is to reduce the following loss function:

$$\text{Lenct}(\beta)=\sum_{i-1}^{n}(Yi-Xi\beta)2+\lambda\left(\frac{1-\alpha}{2}\sum_{j-1}^{m}\beta j2+\alpha\sum_{j-1}^{m}|\beta j|\right)$$

where α represents the mixing parameter between the two, i.e, ridge (α = 0) and lasso (α = 1).

A weighted combination of L1 and L2 regularization is used in the elastic net algorithm. As you can view, the similar function is used for lasso and ridge regression, with the difference being the L1 wt argument. This argument decides how much weight is given to the partial slopes' L1-norm. If L1 wt = 1.0 and the regularization is pure L2 (ridge), then the regularization is pure L1 (LASSO).

13.3 UNSUPERVISED ML APPROACHES

Unsupervised ML is a contrastive study against the supervised ML. In unsupervised ML, algorithms are implemented to build models to make the classification of given data irrespective of its class labels [7]. Data used in it are completely unlabeled and models are exclusively autonomous to compact internal representation of the given data according to their common characteristics [4]. Data are analyzed by models and significant insights are found from this data to classify the future on the basis of these insights. Unsupervised models distribute the given data into clusters or associations on the basis of some common characteristics or data dependencies. Data items that share common characteristics are put into the same cluster, and predictions of new data elements are made on the same phenomenon. Primarily, unsupervised ML learning is divided into two broader classes: Clustering and Association are depicted in Figure 13.7. Both the classes of unsupervised learning are described in the following subsections with their concerned algorithms.

13.3.1 Clustering-Based Unsupervised ML Techniques

Clustering as a subclass of unsupervised machine learning deals with the unlabeled data and has no supervision phase like in supervised algorithms [22]. The core concept of clustering is to distribute data points on some similar characteristics. Different clusters of input data sets are made by the clustering algorithms and each cluster consists of similar data points having some shared common characteristics. However, dissimilarities are shown with other cluster data points. In the following, famous clustering algorithms have been discussed.

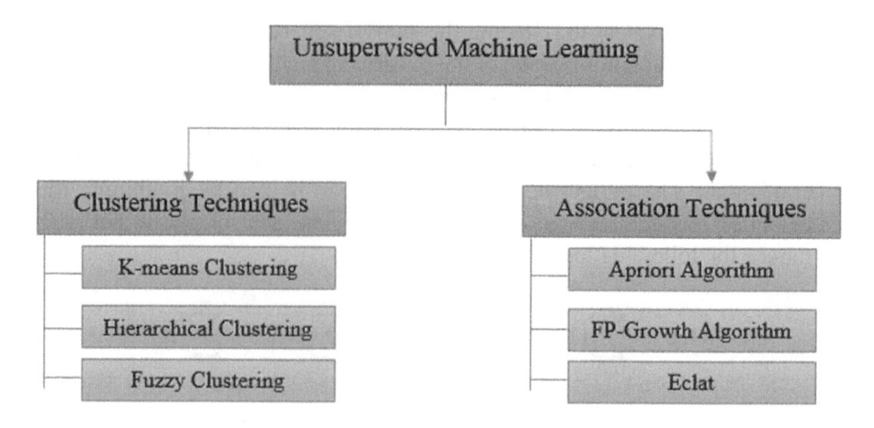

FIGURE 13.7 Unsupervised ML approaches.

i. *K-Means Clustering:* Autonomous K-means clustering divides the unlabeled data set into various groups (clusters) [22]. Here, K determines a predefined number of clusters that must be available at the time of the said process of clustering; let's take an example, if the value of K is 2, it indicates two groups have to be formed, similarly if $K = 3$, then three groups have to be formed, and so on. "It's an iterative technique that splits an unlabeled data set into K clusters, with every data set belonging to only a single group with similar qualities." It allows us to bundle data into separate groups and provides a simple technique to determine the sections of groups in an unlabeled data set beyond any training. It's a centroid-based method, which means that each group has its own area center. This technique's main objective is to reduce the sum of distances between data points and the groups to which they belong. The K-means clustering technique is generally used to achieve two objectives: The ideal value for K-center points or centroids is determined iteratively, and every data point is allocated to the K-center that is closest to it.

ii. *Hierarchical Clustering:* It is another individual machine learning approach for grouping unlabeled data sets into clusters [22]. The hierarchy of clusters is developed in a tree form in this technique, and this tree-shaped architecture is known as the dendrogram. Although the results in two approaches i.e., K-means clustering and hierarchical clustering may appear to be comparable at times, their methods differ. As opposed to the K-means algorithm, there is no need to predispose the number of clusters.

There are two ways to hierarchical clustering:

Agglomerative: is a bottom side-up strategy where the method/process starts with all data points as sole clusters and combines them until only a single cluster remains.

Divisive: As a top-down method, the divisive algorithm is the polar opposite of the agglomerative algorithm.

iii. *Fuzzy Clustering:* A data object can belong to more than one group or cluster in fuzzy clustering, which is a sort of soft technique. Each data set has a set of membership coefficients that are based on the degree of cluster membership. This sort of clustering is exemplified by the fuzzy *C*-means method, which is also known as the fuzzy *k*-means algorithm [23]. The working procedure of this algorithm is as follows:

- Pick a few clusters to work with.
- For being in the clusters, assign coefficients to each data point at random.
- Repeat until the algorithm converges (i.e., the change in coefficients between iterations is less than ε, the set sensitivity threshold):
- Calculate the centroid of each object.
- Calculate the coefficients of being in the clusters for each data point.

13.3.2 Association-Based Unsupervised ML Techniques

Association rule mining is a class of unsupervised learning that as the name implies it finds relationships among data items that show some kind of dependency on one another or one data item has some kind of relation with other items [22]. It is very famous in the marketing and production industry as it finds associations among different data items that help business organizations manage these items accordingly to make them more profitable. However, it has very little usage in the cybersecurity domain. But still, we discuss this subclass of unsupervised ML to provide some basic insights to the readers of the book chapter. The most famous association-based algorithms have been discussed below:

i. *Apriori Algorithm:* This algorithm generates association rules by using persistent item sets, and it is created to work with transaction databases. It determines how energetically or weakly two things are associated using these association rules [24]. To efficiently calculate the item set relationships, this approach uses a breadth-first search and a Hash tree. In the year 1994, R. Agrawal and Srikant presented this algorithm. It is mostly used for market basket analysis, which aids in the discovery of products that can be purchased together. It can also be utilized to discover drug reactions in patients in the healthcare field.

   Step 1: Determines the transactional database's support for item sets and chooses the lowest level of backing and determination.

   Step 2: Chooses all elements in the transaction that have a greater support value than the lesser or chosen support value.

   Step 3: Identifies all of these subsets' rules with a greater determination value than the point of departure or minimal determination.

   Step 4: Arrange the rules in descending sequence of lift.

ii. *FP Growth Algorithm:* The Frequent Pattern (FP) growth algorithm is an advanced version as compared to the Apriori algorithm. Beyond the generation of candidates, this method is used to locate frequent itemsets in a transaction database [24]. The fact that FP growth is a divide-and-conquer strategy is one of the reasons behind its efficiency. We also recognize that a skilled method must have been using data structures and advanced programming techniques. These techniques or data structures include a tree, linked list, and the Depth First Search (DFS) concept.

The FP growth algorithm has the following advantages:

- 1. It is faster than the Apriori algorithm.
- 2. There is no selection of candidates.
- 3. There are only two runs across the data set.

The FP growth technique has the following drawbacks:

- 1. The FP tree may not fit in memory.
- 2. The cost of constructing an FP tree is high.

iii. *Eclat:* Equivalence class clustering and bottom-up lattice transversal algorithm are abbreviated as Eclat algorithm. It's a method for locating frequently occurring item sets in a transaction or database [25]. This method is among the most effective approaches for learning association rules. In a database, the Eclat algorithm is used to build frequent itemsets as depicted in Figure 13.8.

For finding common item sets, the Eclat algorithm uses a depth-first search, whereas the Apriori method uses a breadth-first search. It acts in place of data in a vertical pattern, as opposed to the horizontal pattern used by the Apriori method. The Eclat method is faster than the Apriori algorithm because of its vertical layout. As a result, the Eclat algorithm is a faster and more scalable variant of the Association Rule Learning method.

Advantages

- The Eclat method utilizes less memory than the Apriori algorithm since it uses a Depth-First Search approach.
- When compared to the Apriori method, the Eclat algorithm is obviously faster.
- The Eclat method does not require repeated data scanning in order to determine individual support values.
- Unlike Apriori, which scans the original data set, the Eclat algorithm scans the recently created dataset.

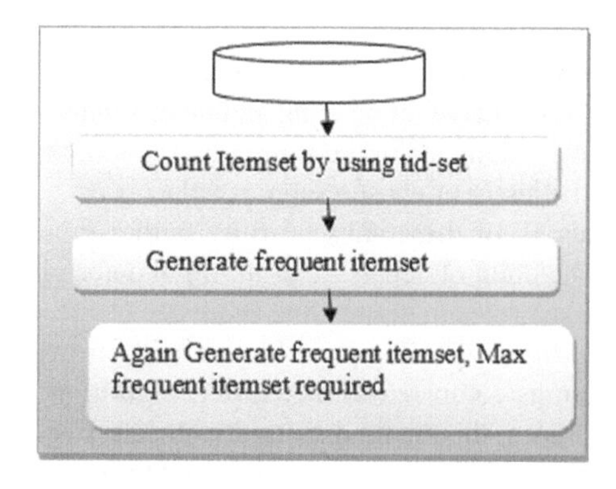

FIGURE 13.8 Working process of Eclat.

13.4 CYBERSECURITY PERSPECTIVE

Machine learning as an emerging technology provides great flexibility to make insights into big data, which helps researchers to analyze the existing huge amounts of data and find interesting patterns from it. The insights examined from historical data through machine learning provide enormous benefits to modern industries and business organizations. And one of the interesting characteristics of machine learning techniques is to provide proactive security mechanisms in the cybersecurity domain. ML as a proactive approach to address cybersecurity issues will examine the threats and respond to intrusions and security incidents swiftly in an instinctive way. Here, in this section of the book chapter, we will discuss the role and significance of machine learning algorithms from a cybersecurity perspective. The various cybersecurity fields where machine learning plays a vital role are discussed as under:

i. *Spam Detection:* Spam as a technical term is mostly concerned with electronic mails and is known for some other names such as junk mail or unsolicited bulk mail. It is an unwanted and unwelcome digital content that is used by spammers through different messaging systems [26]. Mostly, it comes in the form of unwanted and unnecessary mails through the internet. Most of the time, spams are used for commercial purposes and are just unpleasant in nature. However, sometimes spam messages can be catastrophic for the system and system user. In this scenario, the intention of spammers is to send malicious codes, execute phishing scams, and earn money. The denial of services like attacks can also be executed with the help of spam messages.

   Different machine learning approaches namely naïve Bayes algorithm, Term frequency inverse document frequency (TF-IDF), support vector machine, and boosting technique have been practiced by cybersecurity experts and researchers to detect spam on the network. It is commonly known as spam filtering and is performed by textual content of the documents. The ML-based models need a huge amount of training data first to identify spams. It also needs huge processing power to detect spam content [27]. Thus, huge processing power and large training data are the main limitations of these spam-filtering techniques.

ii. *Phishing Identification:* Cyber intrusions are very common at the present time and are showing a rapid growth with time. And phishing is one of the common and interesting social engineering attacks used by intruders to steal confidential data often including credit card details and login credentials. In phishing, cybercriminals use the concept of spoofing which helps them to pretend as a legitimate and known source for the victim [28]. Mostly they use it to spoof the websites of reputed organizations so that the victims can easily trust and share their confidential data. In addition to credential stealing, phishing attacks are nowadays also used to spread Malware for system cookie stealing and keystroke capturing.

Phishing identification comes under the domain of the supervised classification problem. Thus, it needs the labeled data to train the ML-based model which consists of both the phishing data pages and non-phishing (legitimate) data pages. Mostly, support vector machines, decision trees, and naïve Bayes are practiced by researchers for phishing detection. The main features used in phishing detection problem are URL-based features (which incorporate the URL length, the number of characters in URL, URL subdomains in number, and the number of domain names which are misspelled, etc.); domain-based features (which incorporate domain name, present domain age, domain availability, and register-name availability); page-based features (which incorporates page rank, page references in average by each visit, duration of each visit, and estimated domain visits); and content-based features (which incorporates meta tags, document titles, body content, and hidden text).

iii. *Malware Identification:* Malware as a collective suit of various malicious software mainly comprises viruses, spyware, key logger, and ransomware. Malware is a code designed by cyberattackers, with the intention to make severe damage to the victim system or to acquire illegitimate network access. Generally, it is a coded file that is spread by cyberattackers through different messaging systems such as email and needs the victim's action to make the execution of Malware. Different types of Malware programs are designed for different purposes by intruders and are mostly financially compromised. Both individuals and organizations have been threatened by different types of Malware programs since the 1970s in the form of creeper virus.

Although there exist different Malware identification and detection approaches namely static, dynamic, and hybrid analysis. Machine learning algorithms have been practiced by researchers for both the Malware detection and Malware classification into different classes or families. To detect well-known malwares, whose signatures are clearly known can be detected through signature-based machine learning techniques. These techniques work for static Malware analysis and use features such as the signature of source code, application programming interface (API) calls, function calls, and so on. However, they fail to detect the malwares having dynamic nature and show frequent changes in its characteristics. Statistical mining is another approach to detect Malware softwares, and in this scenario features like usage of CPU, battery, memory, and network traffic. ML-based algorithms such as decision trees, naïve Bayes, SVM, and neural network, and logistic regression have been practiced by researchers to detect Malware applications. *K*-means clustering and hidden Markov model are also used to identify whether an application is Malware or goodware.

iv. *Detection of DoS and distributive denial-of-service (DDoS) attacks:* The three main components of security or cybersecurity are confidentiality, integrity, and availability (CIA). Commonly known as CIA triad and are considered the basic components for the security of any system or network. Among the three, one of the vital components is availability. Availability literally defines the character that is to be used or obtained; but in information security, it ensures that whenever information and other resources are needed by authentic users, there should be timely and reliable access to them (Forouzan, 2007). To make interrupts in the way of the system and system resource access for its users, cyber intruders use the DoS and DDoS attacks. DoS and DDoS attacks are used to make the online system resources unavailable to its users by flooding a server with traffic. The key difference between DoS and DDoS is that DoS shows a one-to-one relationship between the cyberattacker system and the victim system, whereas DDoS shows a many-to-one relationship. This means that in DDoS, several systems are used in a distributed environment to attack a single system or server. The different types of DoS and DDoS attacks are teardrop attack, flooding attack, IP fragmentation attack, protocol attack, and application based attack.

The two common approaches of machine learning to detect these attacks are signature-based and anomaly-based intrusion detection. Signature-based intrusion detection works only when there are some known signatures or patterns already stored in the database. The designed ML model examines the incoming traffic on the network. It then compares with stored patterns to determine whether the incoming packets are from a normal user or from the intruders [29]. But the main issue with signature-based techniques is that they cannot detect zero-day attacks. To overcome the shortcoming of signature-based techniques, anomaly-based techniques have been practiced by researchers to detect the dynamic nature of DoS and DDoS attacks on the network. These techniques use packet header information, size of packet, rate of the packet, and connections to a host to identify different DoS and DDoS attacks on a network. However, these techniques have a high false-positive rate. Support vector machine, KNN, naïve Bayes classifier, and *K*-means clustering are commonly used ML-based algorithms to detect these attacks on the network.

v. *Biometric Recognition:* Rapidly growing rate of data breaches and cyberattacks causes serious issues for both the individuals and organizations. Provides tough challenges for both the security experts and researchers to design and develop more robust and strong authentication mechanisms. In this line, one of the effective and efficient authentication mechanisms designed by researchers is the biometric-based authentication mechanism. In biometric technology, different parts of the human body are used to identify the real entity. It utilizes physiological or behavioral data of a human to make verification of an entity as a legal or illegal user access. In the physiological domain, the physical body parts are included such as the eyes retina, fingerprint, face structure, and in the behavioral domain pattern of typing and signatures, voice tunes, etc., are included. Each biometric system consists of two phases: the first phase is the enrollment phase in which a person makes his/her registration on a biometric system by providing his/her necessary data that is stored in the database; the second phase is the verification phase in which the claimed identity is verified by the biometric system by comparing the current entity data with the stored data.

Machine learning algorithms play a vital role in biometric technology to improve the efficiency of biometric systems [27]. To create automatic matching such as one-to-one and one-to-many in biometric systems, ML-based algorithms work efficiently in it. SVM, Artificial neural network (ANN), genetic algorithms, and probability-based classifiers provide effective performance in fingerprint biometrics; whereas PCA and LDA show good performance in Iris biometric identification. And deep neural networks, SVM, kernel PCA and LDA show better performance in face biometric recognition. The commonly used attributes in biometric recognition are distance between eyes, Fourier transform, core points, Discrete cosine transform (DCT), wavelet transform, principal components, and ridge ending.

vi. *Detection of Software Vulnerabilities:* Revolution in Information and communications technology (ICT) and concept of digitization greatly increases the demand for software. Thus, the causes of a software crisis are mainly when it is concerned about the quality of software. Software vulnerabilities are mostly the outcome of quality compromise. Software vulnerability defines some kind of deficiency in the software product that makes it prone to attacks and threats. The inadequacy in software code can allow intruders to make unauthorized access to the system and system resources. These vulnerabilities can be because of the flaws in the software design or in the source code. These flaws can take the system into an abnormal state that can lead to system crash, invalid output, or unexpected system behavior. The most commonly found vulnerabilities are buffer overflows, misuse of operators, type-conversation errors, privileged and file permission issues, SQL injection, cross-cite script, access control flaws, and structure padding.

Software vulnerability identification is a process to examine the software product for any kind of vulnerability that can lead to software security compromise. ML-based algorithms play a significant role to detect software vulnerabilities. ML-based algorithms have been practiced by researchers to model the syntax and semantics of code, make code analysis and inferences for code patterns, assisting in the process of code auditing and understanding. ML algorithms are broadly divided into two categories with regard to the detection of vulnerabilities. First is the anomaly-based detection approaches and second is the pattern recognition approaches [27]. The attributes used in first one are usage patterns of API, missing checks, insufficiency in validation of inputs, and problems in access controls and in the second approach (pattern recognition) attributes that have been used are system call API, and syntax trees, etc. The most commonly used ML techniques to detect software vulnerabilities are K-NN, logistic regression, random forest, ANN, and BLSTM.

13.5 CONCLUSION

Cybersecurity ensures the real-time protection of information, information systems, and networks from intruders. It is depicted from various prominent security and privacy reports that cybersecurity breaches have revealed a rapid elevation in the last decade. To address these cyber security issues, organizations have spent huge amounts and researchers have made various efforts to overcome these intrusions. Different approaches and techniques have been practiced by experts and researchers to provide reliable and robust security mechanisms. One of the prominent among them is machine learning, which plays a vital role in the cybersecurity domain. ML has a proactive character to address that cybersecurity issues will examine the threats and respond to intrusions and security incidents swiftly in an instinctive way. Thus, it is more beneficial in the cybersecurity field for detecting and classifying various kinds of cyberattacks. Especially supervised and unsupervised machine learning techniques possess great ability to address different cybersecurity issues. In this book chapter, we have discussed different supervised and unsupervised machine learning algorithms. Supervised techniques work with labeled data whereas unsupervised techniques work with unlabeled data. Supervised techniques are further classified into classification and regression techniques, where classification-based techniques deal with discrete data and regression-based techniques deal with continuous data. Further, unsupervised techniques are divided into clustering-based techniques and association-based techniques. Clustering-based techniques divide the data into clusters or groups on the basis of similarities identified in the data, whereas association-based techniques find patterns or interesting associations among the variables of data on the basis of dependencies among data items. Finally, in this chapter, we have discussed various cybersecurity issues and the role of ML to address these cybersecurity issues. Mostly, KNN, SVM, ANN, decision trees, *K*-means clustering, and naïve Bayes algorithms have been practiced by researchers to address the cybersecurity issues. Hence, there is a need to practice other ML algorithms in the cybersecurity domain to examine their efficiency.

REFERENCES

1. K. Sahu, F. A. Alzahrani, R. K. Srivastava, and R. Kumar. "Evaluating the impact of prediction techniques: Software reliability perspective," *Computers Materials and Continua*, vol. 67, no. 2, pp. 1471–1488, 2021.
2. M. Haenlein and A. Kaplan, "A brief history of artificial intelligence: On the past, present, and future of artificial intelligence," *California Management Review*, vol. 61, no. 4, pp. 5–14, 2019.
3. R. Kumar, S. A. Khan, and R. A. Khan. "Durability challenges in software engineering," *Crosstalk-The Journal of Defense Software Engineering*, pp. 29–31, 2016.
4. T. M. Mitchell, *Machine Learning*. McGraw-Hill, 1997.
5. G. Apruzzese, M. Colajanni, L. Ferretti, A. Guido, and M. Marchetti, "On the effectiveness of machine and deep learning for cyber security," in 2018 10th International Conference on Cyber Conflict (CyCon), New Delhi, 2018, pp. 371–390.
6. A. H. Seh et al., "Healthcare data breaches: Insights and implications," in *Healthcare*, 2020, vol. 8, no. 2, p. 133.
7. E. Alpaydin, *Introduction to Machine Learning*. MIT Press, 2020.

8. S. B. Kotsiantis, I. Zaharakis, and P. Pintelas, "Supervised machine learning: A review of classification techniques," *Emerging Artificial Intelligence Applications in Computer Engineering*, vol. 160, no. 1, pp. 3–24, 2007.
9. A. Field, "Logistic regression," *Discovering Statistics Using SPSS*, vol. 264, p. 315, 2009.
10. G. I. Webb, "Naïve Bayes.," *Encyclopedia of Machine Learning*, vol. 15, pp. 713–714, 2010.
11. C. Kingsford and S. L. Salzberg, "What are decision trees?," *Nature Biotechnology*, vol. 26, no. 9, pp. 1011–1013, 2008.
12. S. B. Kotsiantis, "Decision trees: a recent overview," *Artificial Intelligence Review*, vol. 39, no. 4, pp. 261–283, 2013.
13. Z. Deng, X. Zhu, D. Cheng, M. Zong, and S. Zhang, "Efficient kNN classification algorithm for big data," *Neurocomputing*, vol. 195, pp. 143–148, 2016.
14. G. Biau and E. Scornet, "A random forest guided tour," *Test*, vol. 25, no. 2, pp. 197–227, 2016.
15. S. V. M. Vishwanathan and M. N. Murty, "SSVM: A simple SVM algorithm," in Proceedings of the 2002 International Joint Conference on Neural Networks. IJCNN'02 (Cat. No. 02CH37290), New Delhi, 2002, vol. 3, pp. 2393–2398.
16. I. Uysal and H. A. Güvenir, "An overview of regression techniques for knowledge discovery," *Knowledge Engineering Review*, vol. 14, no. 4, pp. 319–340, 1999.
17. D. C. Montgomery, E. A. Peck, and G. G. Vining, *Introduction to Linear Regression Analysis*. John Wiley & Sons, 2021.
18. E. Ostertagová, "Modelling using polynomial regression," *Procedia Engineering*, vol. 48, pp. 500–506, 2012.
19. G. C. McDonald, "Ridge regression," *Wiley Interdisciplinary Reviews: Computational Statistics*, vol. 1, no. 1, pp. 93–100, 2009.
20. J. Ranstam and J. A. Cook, "LASSO regression," *Journal of British Surgery*, vol. 105, no. 10, pp. 1348–1348, 2018.
21. C. Hans, "Elastic net regression modeling with the orthant normal prior," *Journal of the American Statistical Association*, vol. 106, no. 496, pp. 1383–1393, Dec. 2011, doi: 10.1198/jasa.2011.tm09241.
22. R. Gentleman and V. J. Carey, "Unsupervised machine learning," in *Bioconductor Case Studies*. Springer, 2008, pp. 137–157.
23. "Data clustering algorithms: Fuzzy c-means clustering algorithm." https://sites.google.com/site/dataclusteringalgorithms/fuzzy-c-means-clustering-algorithm (accessed Jun. 29, 2021).
24. T. A. Kumbhare and S. V. Chobe, "An overview of association rule mining algorithms," *International Journal of Computer Science and Information Technologies*, vol. 5, no. 1, pp. 927–930, 2014.
25. X. Yu and H. Wang, "Improvement of eclat algorithm based on support in frequent itemset mining," *Journal of Computers*, vol. 9, no. 9, pp. 2116–2123, 2014.
26. A. Attaallah Attaallah, H. Alsuhabi, S. Shukla, R. Kumar, B. K. Gupta, and R. A. Khan. "Analyzing the big data security through a unified decision-making approach," *Intelligent Automation and Soft Computing*, vol. 32, no. 2, pp. 1071–1088, 2022.
27. T. Thomas, A. P. Vijayaraghavan, and S. Emmanuel, *Machine Learning Approaches in Cyber Security Analytics*. Springer, 2020.
28. A. Handa, A. Sharma, and S. K. Shukla, "Machine learning in cybersecurity: A review," *Wiley Interdisciplinary Reviews: Data Mining and Knowledge Discovery*, vol. 9, no. 4, p. e1306, 2019.
29. A. H. Almulihi, F. Alassery, A. I. Khan, S. Shukla, B. K. Gupta, and R. Kumar "Analyzing the implications of healthcare data breaches through computational technique," *Intelligent Automation and Soft Computing*, pp. 1763–1779, 2022.

14 Statistical Trend in Cyber Attacks and Security Measures

Shirisha Kakarla, Deekonda Narsinga Rao, Geeta Kakarla, and Srilatha Gorla

CONTENTS

14.1 Introduction 242
14.1.1 Health Expenditure Indicators and Allocations 242
14.1.2 Role Players in the Medical Sector 244
14.1.3 Essential Considerations in the Healthcare Domain in Resource-Poor Contexts 244
14.1.3.1 Data Sharing among the Medical Sector Players 247
14.1.3.2 Rising Usage of Smart Devices among the Key Players of All Sectors 247
14.1.3.3 Economic and Educational Advancement of the Users Using Smartphones/Tablets 247
14.1.3.4 Need of Lightweight Computing for Maintaining the Data Secrecy of the Patients' Sensitive Data among the Resource-Poor Nations due to Weak Infrastructures 248
14.1.3.5 Adaptable Data-Sharing Protocols and Standards in Resource-Poor Setups 248
14.2 Role of Fundamental Elements for Enhancing Healthcare Quality in Resource-Poor Settings 248
14.2.1 Adaptive Systems 249
14.2.2 Participatory Approach 249
14.2.3 Accountability 249
14.2.4 Evidence and Audit-Based Intervention 249
14.2.5 Innovative Assessment and Evaluation 250
14.3 Data Security Threats and Countermeasures 250
14.3.1 Classification of Security Threats Breaching the Data Privacy 251
14.3.2 Healthcare Data Protection Laws 252

DOI: 10.1201/9781003323426-14

14.3.3 Healthcare Sector Security Practices 252
14.3.3.1 Database Standards Compliance and Monitoring 253
14.3.3.2 Datasets Classification and Assessment of User Rights Management 253
14.3.3.3 Data Masking 253
14.3.3.4 Data Confidentiality Using Encryption 254
14.4 Concluding Remarks 256
Contributions 257
References 257

14.1 INTRODUCTION

With the advent and mass-scale production of computing machinery and interconnected installations, the economy has surged in almost all the sectors, namely primary, secondary, and tertiary, especially in developed countries. Be the primary sector involving the farming, fishing, and mining, or the secondary sector involving the product manufacturing, or the tertiary which includes banking, education, retails, healthcare, hotels and recreation, media and communications, information technology and information technology enabled services (IT and ITeS), civic amenities supply, the financial aspect has been affected largely in a positive manner. In the resource-rich nations, the computing infrastructure, and the production machines have increased the economy manifold in the secondary and the tertiary sectors. Besides, the supplementary sectors of the economy are quaternary and its sub-section quinary [1]. These are largely associated with the services offered in the tertiary sector. The quaternary relates to intellectual services provided in the ecosystem that drive technological advancement. These majorly include governmental bodies, scientific research organizations, education systems, and cultural habitats. Quinary is the smallest group which serves as the crucial role player in decision-making of the economy and includes the top officials/bureaucratic representatives of the universities, government, scientific bodies, media agencies, cultural societies, healthcare, and the non-profit public services like police and fire departments. In the developed countries, the developments in one sector will have a proportional impact on its subsidiaries.

However, in the developing or the underdeveloped nations, the inadequate supplies of resources remain a major bottleneck in the overall financial system. The resources can be classified basically as three types: capital, human, and natural. In order to meet the demand–supply chain in any establishment in an ecosystem, the resources play a crucial role. To tap the potential of the reserves available in a nation, the capital investments, and the human resources must be closely associated for thriving in the economy.

14.1.1 Health Expenditure Indicators and Allocations

Out of all sectors impacted in the resource-poor settings, healthcare is usually observed to be the worst-hit. In Table 14.1, the level of capital investments in health

TABLE 14.1
Capital Health Expenditure (% of GDP) in Resource-Poor and Few Developing Nations

Country Name/Year	2004	2008	2009	2010	2011	2012	2013	2014	2015	2016
Rwanda	..	..	..	..	..	..	..	..	..	..
Haiti	0.35	0.23	0.23	0.33	0.59	0.54	0.87	0.36	0.28	0.24
Zimbabwe	..	..	..	0.79	..	..	..	..	0.01	..
Mali	0.53	0.20	0.20	0.20	0.18	0.21	0.30	0.03	0.00	0.00
Uganda	..	..	..	0.20	0.57	0.13	0.16	0.18	0.17	..
Ethiopia	0.01	0.00	0.00	0.01	0.01	0.02	0.01	0.49	..	..
Nepal	1.09	0.99	0.86	0.28	0.35	0.33	0.46	0.31	0.31	0.43
India	0.14	0.17	0.17	0.18	0.19	0.20	0.28	0.26	0.24	0.45
Brazil	..	0.08	0.10	0.14	0.13	0.48	0.10	0.08	0.07	0.07
China	0.26	0.29	0.41	0.41	0.39	0.41	0.44	0.49	0.56	0.62

Data Source: Health Nutrition and Population Statistics Database, World Bank Organization.

infrastructure involving buildings, machinery, IT, and other medical essentials is expressed as a percentage of gross domestic product (GDP) in some of the resource-poor nations and few developing countries.

In Table 14.2, the current health expenditure level is expressed as a percentage of GDP for the above-selected resource-poor nations and developing countries. Current health expenditures' estimates contain healthcare commodities and services that are consumed year-wise [2, 3]. This indicator is exclusive of the above-mentioned health infrastructure elements.

According to the Global Report of 2019 [4], Figure 14.1 demonstrates the key groups with their respective share of funding for the healthcare facilities among the low-income resource-poor nations. With the Government spending near to 25% till the recent times out of the total investments and expenditure, as shown in Figure 14.2, the major share of spending on healthcare in the low-income nations is borne by the patient and the supportive family entity.

Amidst the low spending by the Governments in low-income and resource-poor countries on the healthcare environment, providing effective and economical medical care to the patients remains primarily the goal rather than the privacy of the patients' data in the healthcare setup.

14.1.2 Role Players in the Medical Sector

The physical health information (PHI) in the healthcare domain remains in the form of the physical records, electronic records, or audio records. The components of the PHI may be health records, pharmacy bills, diagnostic test reports, family medical history, and other personal information like medical insurance policy details, residential address, and professional information.

The medical sector is encompassing the four major role-players: Patient, Care Team, Organizational Infrastructure, and Regulatory Environment [5, 6], as depicted conceptually in Figure 14.3. With the patient at the helm of the healthcare ecosystem, the front liners are the healthcare professionals and the team involving doctors, diagnostics, nurses, care providers, and family. The next layer to provide the interactivity among the patient and the healthcare providers is the organizational infrastructure like hospitals, clinics, nursing homes, and so on. The regulations and policy frameworks are provided by the regulatory environments governed by the public and private regulating agencies, insurance firms, third-party verifiers, research organizations, and so on.

14.1.3 Essential Considerations in the Healthcare Domain in Resource-Poor Contexts

With resource-poor environmental setup and the rising standards on the global stage to provide the confidentiality to the patients' data irrespective of the economic level

TABLE 14.2
Current Health Expenditure (% of GDP) in Resource-Poor and Few Developing Nations

Country Name/ Year	**2004**	**2008**	**2009**	**2010**	**2011**	**2012**	**2013**	**2014**	**2015**	**2016**
Rwanda	9.30	9.05	9.10	9.37	8.48	8.45	7.54	6.90	6.52	6.76
Haiti	5.85	6.02	6.17	8.15	10.23	9.65	7.24	7.80	5.37	5.39
Zimbabwe	..	..	..	11.83	8.01	6.87	6.59	8.67	9.26	9.41
Mali	5.21	4.81	4.70	4.43	4.10	4.40	5.25	4.48	4.11	3.82
Uganda	10.29	10.40	9.74	10.52	8.92	7.89	7.22	6.70	6.33	6.17
Ethiopia	4.30	4.28	4.41	5.47	4.47	4.54	4.08	4.03	3.98	3.97
Nepal	4.62	4.35	4.45	4.97	5.08	5.17	5.32	5.77	6.22	6.29
India	3.96	3.51	3.49	3.27	3.25	3.33	3.75	3.62	3.60	3.66
Brazil	8.06	8.12	8.45	9.72	9.78	10.04	10.24	10.82	11.46	11.77
China	4.26	3.88	4.32	4.21	4.33	4.55	4.71	4.77	4.89	4.98

Data Source: Health Nutrition and Population Statistics Database, World Bank Organization.

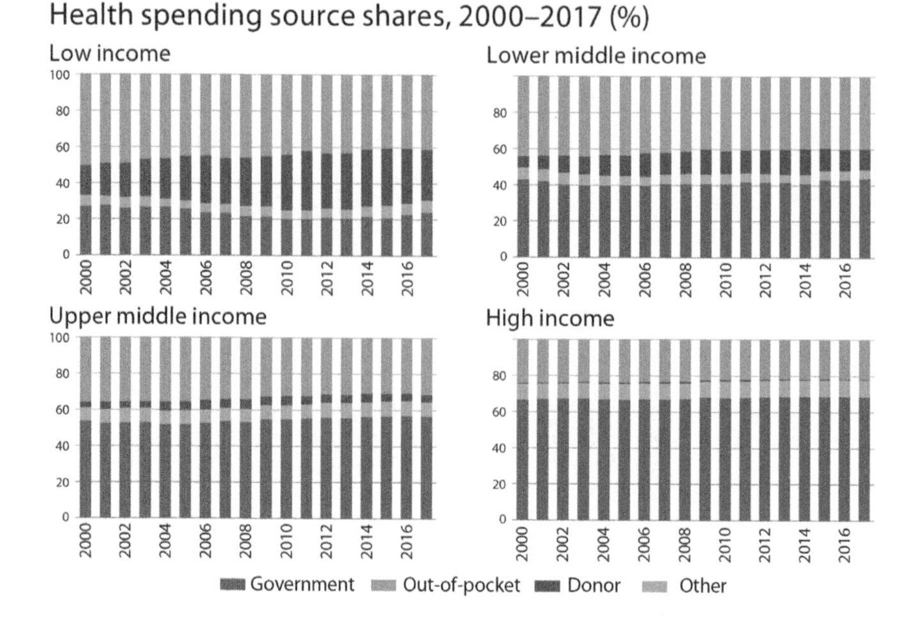

FIGURE 14.1 Image source: Global Health Expenditure Database, World Health Organization, 2019.

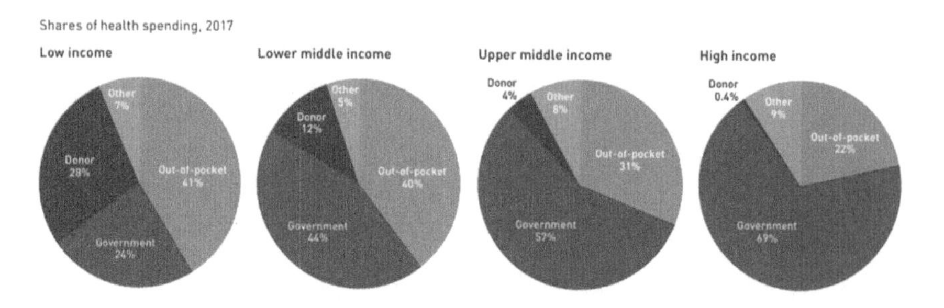

FIGURE 14.2 Image source: Global Spending on Health: A World in Transition, World Health Organization, 2019.

of the nation, the challenges faced by the medical sector are seen in the multiple parameters:

- Data sharing among the medical sector players.
- Rising usage of smart devices among the key players of all sectors.
- Economic and educational advancement of the users using smart phones/tablets.
- Need for low-cost computing for maintaining the data secrecy of the patients' sensitive data among the resource-poor nations due to weak infrastructures.
- Adaptable data-sharing protocols and standards in resource-poor setups.

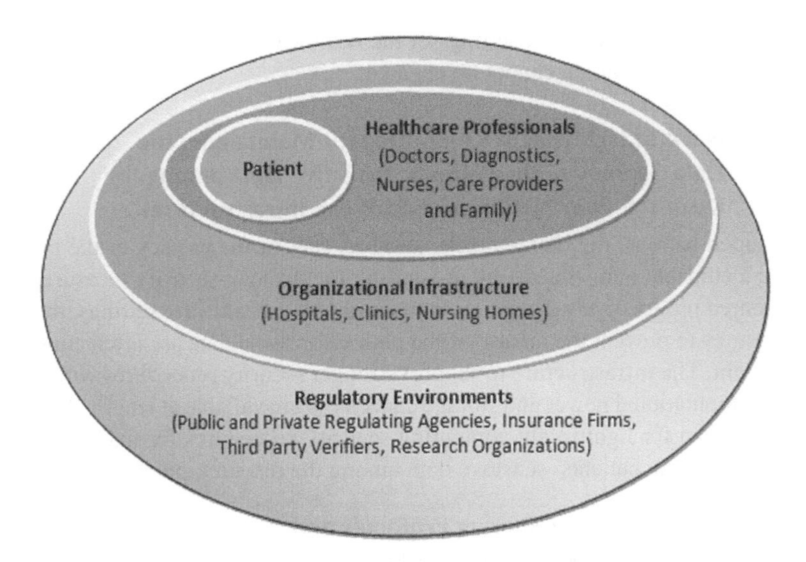

FIGURE 14.3 Conceptual representation of the major role-players in the healthcare system.

14.1.3.1 Data Sharing among the Medical Sector Players

In the course of providing critical care, the patients' data including his existential information is shared among the healthcare professionals, diagnostic agencies, insurance firms, governmental and regulatory bodies, and third-party verification teams. The confidential information of the patient is everywhere leaving a wide scope for susceptibility.

14.1.3.2 Rising Usage of Smart Devices among the Key Players of All Sectors

With the prices of smartphones getting more and more affordable, the usage of smart phone technology among all strata people in almost all sectors of any nation [3] is also rising proportionally. This provides the scope for different application vendors to design and develop software solutions in the healthcare domain to be used conveniently by all the users.

14.1.3.3 Economic and Educational Advancement of the Users Using Smartphones/Tablets

The growing ease of using smartphones and tablets, irrespective of the economic class and the educational level of the users, has triggered a wide range of applications for a purpose within a domain. More and more people are embracing smartphone technology with a wider choice of applications at their disposal for a task [7]. Many of the applications available for use are also not completely secure and may contain malicious code, thereby risking the user's private data. In the healthcare-related apps, the users, especially the patients, can interact with the medical team, healthcare workers, and other insurance providers by forwarding their health, financial, and other personal information. In high-income nations, specific security standards, and measures are strictly enforced while handling the users' private information.

However, in the resource-poor settings of the healthcare sector, the rampant misuse of confidential information cannot be negated.

14.1.3.4 Need of Lightweight Computing for Maintaining the Data Secrecy of the Patients' Sensitive Data among the Resource-Poor Nations due to Weak Infrastructures

In developed nations, the rising attacks and breaches of the privacy of the patients' data are a culpable punishment and are calling for stronger security measures to be implemented progressively. However, in resource-poor healthcare settings, the security measures to prevent the misuse of the patients' private data are often considered insignificant. The infrastructure to support stronger security procedures with respect to their computational power and storage capacity is unavailable at length. This gives rise to the need for lightweight computing security procedures for maintaining the data secrecy of the patients' sensitive data among the resource-poor nations.

14.1.3.5 Adaptable Data-Sharing Protocols and Standards in Resource-Poor Setups

The data stored about the patients' details and other users involved in the healthcare setup may be huge in terms of a few hundreds of MB in the data storage units. The storage devices can be augmented relatively easier as compared to augmenting the processing units in the already established infrastructures. The weak infrastructures available in resource-poor healthcare environments demand for security settings which take less computation albeit offer strong resistance to modern attacks and threats with the need to focus on technologies and procedures that are cost-effective and sustainable.

The chapter is organized as follows. In Section 14.1, a detailed study of the expenditure indicators, players, and essential considerations in the healthcare sector is made. In Section 14.2, the fundamental elements for enhancing healthcare quality are described with their roles which are considered for resource-poor settings. In Section 14.3, data security threats, their classification, along with their countermeasures in terms of the protection laws in place and the security practices that are followed are elaborated. The concluding remarks are detailed in Section 14.4.

14.2 ROLE OF FUNDAMENTAL ELEMENTS FOR ENHANCING HEALTHCARE QUALITY IN RESOURCE-POOR SETTINGS

For any nation, a healthier population is attributed to the improvement in healthcare quality [8]. The developed and few developing countries have considered healthcare as one of the key components of the budgetary allocations. Otherwise, the healthcare sector in low-resource settings remains neglected of health strategies and indicators to measure the healthcare quality [9]. Although different indicators to measure the quality of healthcare globally remain the same, they are not essentially considered of importance in the resource-poor settings or the extent to which these mechanisms are practiced may be different [10]. The investments without governmental healthcare policies, the protocols and standards of the healthcare, adherence to the norms and violation acts, privacy rules of the vulnerable data, and the extent of data sharing among the key players of the system would have little role in strengthening the health systems of any nation [11]. As recommended by Nambiar et al. [8], the five fundamental elements considered and described here for the improvement of the parameters to maximize the potential of quality improvements are Adaptive Systems, Participatory Approach, Accountability, Evidence and Audit-based Interventions, and Innovative Assessment and Evaluation. The smaller level considerations of each of these elements which have to be designed with an adequate understanding of relevant contextual parametric inputs, their different conceptual hierarchy and inter-dependencies that control the healthcare deliverance, and the health outcomes are enlisted below. The tangible measures are also to be put in place to improve the degree of desired deliverables.

14.2.1 Adaptive Systems

The healthcare systems remain the complex, adaptive, and dynamic systems consisting of the players: patient, care team, organizational infrastructure, and regulatory environment. The interrelationships among the players, the different levels of the healthcare structure ranging from the neighborhood medicare (microlevel), health facility center (mesolevel) to tertiary referral system (macrolevel) and material resources, supervision, training, and management bodies must be adequately assessed.

14.2.2 Participatory Approach

The participatory contributions by all the players involved in the healthcare system are really important to bring about quality improvement. The collective inputs of the population framework representatives, research efforts, and novel designs with smooth adaptability, considering the socioeconomic and sociocultural contexts of the specific region, into the legacy healthcare system are necessary for the progress to be effective.

14.2.3 Accountability

The efforts of the people involved in the healthcare framework right from the community health centers to the tertiary hospitals for improving healthcare qualitatively must be assessed. The assessment metrics chosen must be accountable based on the data available in the healthcare ecosystem. This provides the mechanism to improve the attainment of the qualitative objectives and/or recalibrate the chosen metrics.

14.2.4 Evidence and Audit-Based Intervention

The data pertaining to the representative inputs, collective efforts, research findings, decision-making, and the policies imposed in the pursuit of improving the healthcare system must all be evidenced appropriately and audited. The strategies adopted to conclude the decisions in social contexts must be harmonizing the data, users of the data, and the global standards to the maximum extent and must be meeting compliance set by the regulatory and authoritative bodies of the region.

14.2.5 Innovative Assessment and Evaluation

All the research findings, designs, and outcomes should be supported by the assessment and the evaluation of the prototyped contextual ecosystem. The innovative interventions should be rigorously determined to conclude the ways and means of the definitive circumstances under which they work with no means of violating the norms.

Unless the fundamental elements and their interdependencies are addressed [12], especially in resource-poor settings, the interventions of modernization and policy frameworks may plausibly further cause inconvenience in using equipment and result in low outcomes in the healthcare system. Inadequately trained personnel, under-used expensive equipment, under-staffed healthcare system, socioeconomic traits of the staff, and the absence of risk monitoring and mitigating plans in healthcare systems may further downgrade the health services.

14.3 DATA SECURITY THREATS AND COUNTERMEASURES

Every year, incidents concerning the breaches of digital healthcare data are reported, globally. The personal health information of the owners is subjected to mishandling and misuse, either purposefully or accidentally, and remains at risk of losing its confidentiality. The statistics projected in references [13–16] about the healthcare data breaches, and an ever-rising upward trend is observed in this decade. The data breaches reported in the healthcare sector were categorized as willful disclosure of the data by an insider, accidentally leaking the data by the patient himself through apps/applications accessed via electronic gadgets, data collectors like the hospital, pathological laboratories, blood donation camps, and so on, sending sensitive information to the wrong person, credential-stealing malware crept into the patients' data information system, data accessed from lost laptops/devices, third-party vendors for data management, and phishing attacks.

The Healthcare Cybersecurity Statistics, as given in [17], sum up that there is a rise of more than 400% in healthcare data frauds. Almost one-fourth of the health employees are not trained adequately in the cybersecurity area [18], amounting to 90% of the cyberattacks due to email phishing. Less than half of the employees are unaware of the security measures that are adopted in their organization [19, 20] and stay non-contributory in maintaining the security of the organization's sensitive data as well as inefficiency in balancing the security budget. According to healthcare data breaches in 2018 [19], the biggest causes were incidents related to hacking/IT which accounted for 43.29%, unauthorized access/disclosures with a share of 39.18%, together accounting for 82.47% of all data breaches reported in 2018.

14.3.1 Classification of Security Threats Breaching the Data Privacy

Keeping in view the purpose and manner of the data breaches reported worldwide, the security threats [21] are categorized generically as follows:

Denial of Service (DoS): Very widely prevalent and impacting menace that leads to data inaccessibility which occurs when the server crashes for one or the other reason. If persisted unaddressed for a longer time, then this will eventually cause loss of data and the related services due to the non-availability of applications for storing/accessing the patients' data.

Ransomware: Caught unaware of the phishing emails containing a malicious attachment, or viewing the content containing the malware or clicking the malicious link, ransomware is triggered by the legitimate user of the data thereby infecting the victim's machines and rendering them inaccessible until a ransom is paid. Ransomware attacks [L] in the healthcare system cripple the critical processes and make them completely inoperable. In many cases, the services are accounted for using pen and paper, thus making the medical processes tedious.

Mismanaged Sensitive Data and Storage Media Leaks: For the regular checks and to counter the DoS attacks, the data copies are multiplied and maintained. However, these copies are left unprotected and vulnerable to breaches and exploitations.

SQL Injection Attack and No-SQL Injection Attack: Considered as the rampant, SQL and NoSQL attacks take in the unauthorized access requests to the server for the sensitive data, thus causing the data leakage of the confidential data.

Privilege Escalation: With the default access given for all the controls of the data units, the users of the information system may lead to intentional/unintentional usage and bring in the inconsistencies. The unwarranted updation and elimination of the legitimate data and adding of the malicious ones would disrupt the data integrity. Legitimate Privilege Abuse is the subtype of privilege escalation involving the misuse of the privileges offered for illegal purposes.

Infiltration through Common Cybersecurity Attacks: The other popular threats and attacks staged to disrupt the operational modules handling the PHI are brute force attacks, phishing, Sybil attack, selective forwarding, internal attacks, sinkhole threats, eavesdropping, and so on.

Often the root cause of the data breaches is human negligence and due to the lack of expertise and availability of skilled staff to handle sensitive data and security procedures, which contribute to 30% of the overall data breaches [21]. The unawareness of the security policies, procedures to be enforced, and conducting and handling the incident response processes remain the main challenges faced by the data handler or the data owner, which threaten the confidentiality of the sensitive information.

14.3.2 Healthcare Data Protection Laws

Worldwide, in many developed countries, the respective governments have made laws to curb the menace of data breaches, especially the healthcare domain. In the United States of America, according to the Health Insurance Portability and Accountability Act (HIPAA) of 1996, breaching the privacy of the patients' medical information and mishandling is a punitive offense. Similarly, the Information Privacy Act has been in force in Australia since 2009 to control the sensitive data information leakage. In Canada, the Personal Health Information Protection Act came into force in 2004, whereas in New Zealand, the Health Information Privacy Code has been in force since the year 1994. In India, Digital Information Security in Healthcare Act (DISHA) came into existence in 2019 and is meant to institute privacy and security measures for electronic health data and regulate its storage and exchange.

The developed nations have been proactively engaged in support of privacy preservation of the patients' sensitive data from illegal sharing and misuse for financial gains [20] for a long time. The developing nations are studying the consequences of the improper handling of sensitive data and drafting the bills and making the laws in the same line. The underdeveloped countries with resource-poor settings are struggling to upkeep the primary concerns of the healthcare facility and health services.

With the better security procedures and policies enforced like adopting strong and stringent encrypting standards, many of the easily preventable breaches can be reduced. In the resource-rich context of healthcare systems, the latest encryption standards are enforced to optimally use the high processing power of the equipment installed. In resource-poor settings, the presence of the low processing units proves to be a bottleneck for using stringent security features. In order to address this scenario, the resource-poor settings of the healthcare sector need to be supported by the encrypting procedures which demand low processing speeds of the installed equipment and still does not compromise the security of the patients' sensitive and private data.

14.3.3 Healthcare Sector Security Practices

According to the study made and published in Reference [21], the primary set of guidelines set forth for addressing the insecurity challenges in the healthcare industry can be aptly tailor-made for the type and size of the healthcare ecosystem. These recommendations are threefold:

1. The security practices should be cost-effective on the installations and minimize the security risks for a range of healthcare establishments.
2. The security procedures and methodologies should be voluntarily adopted in the best interests of the patients and secondary stakeholders.
3. The continuous security-auditing mechanism must be in place scrutinizing the threats and trigger the incident response processes in the eventuality to counter security breach in cognizance of the stakeholders.

Although a variety of security controls and procedures are available, selecting the most appropriate one for the health system under consideration remains the intimidating task. The major practices following the recommendations that are adopted and deployed for ensuring the security presently can be classified [22] as follows:

- Database Standards Compliance and Monitoring (DSCM)
- Datasets Classification and Assessment of User Rights Management
- Data masking
- Data confidentiality using encryption

14.3.3.1 Database Standards Compliance and Monitoring

PHI stored in the electronic form in databases is termed electronic health records (EHRs) and requires necessary security procedures in place. Few of the compliant security standards are maintaining regular back-ups of the datasets, storing the information enciphered form by using strong encryption techniques, usage of antivirus, using authorized access with strong passwords, incorporating multifactor authentication, updating regularly the operating systems with patches, securing the channel with secure socket layer (SSL) so as to avoid the security glitches for the dataset at rest or in motion.

Apart from the security procedures adopted to maintain the privacy of the patients' sensitive and identifiable information staged in the databases, the other key component of the dataset security is checking its compliance with the standards and monitoring the activities performed on the data sets. The primary task is to monitor the database activity in the real time. As a part of the DSCM, all the legitimate and malicious accesses to the data maintained in the patients' dataset are logged and incident responses are triggered in the advent of unauthorized or spurious activity that may otherwise harm the data integrity.

14.3.3.2 Datasets Classification and Assessment of User Rights Management

The datasets storing the sensitive information of the patients must be differentiated from the false/altered datasets that have crept into the system. It involves designing the user rights management module for the real datasets for granting access to only the users performing the genuine and required tasks associated with the data stored and discarding the malicious and unwarranted access. A good design puts in place the initially adopted compliance standards to thwart the breaches and invoke the incident response processes.

14.3.3.3 Data Masking

The main set of tasks of the security is to maintain the confidentiality and integrity of the sensitive and identifiable patients' data present in the datasets of the healthcare system. Data masking is one viable solution using which the susceptible information of the patients is garbled either partially or fully to conceal the identity using secret techniques. The same masked data is shared across the public or the private networks by authorized users. On receiving, the contents are unmasked by the reverse procedures to perform the legitimate data processing tasks. Further, the data is re-masked using the secret procedures and stored in the storage units, thus restoring the secrecy of the information.

14.3.3.4 Data Confidentiality Using Encryption

The earliest means to transform the secret data into the insensible one leaving little room and incentive for the attackers to steal the information is encryption. With the emerging threats and customized stealthier attacks on the healthcare applications and systems, the healthcare security system needs to be periodically assessed to measure the counter effects. The state-of-the-art encryption standards must be integrated into the system to combat cyber offenses [23]. At the same time, deployment of robust cryptosystems with complex processing instructions proportionally demands for high-configured processing equipment and abundant storage capacity. The developed nations can afford to meet the demands with the budgetary allocations to build cybersecure healthcare systems. However, in resource-poor countries, the procedures to thwart the ever-rising attacks and secure patients' sensitive data need to be deployed in the resource-strapped environment of healthcare.

The traditional encrypting procedures of the information stores in the dataset are grouped as column-level encryption [24], field-level encryption [25], encrypting files system [26, 27], application-level encryption [28], encryption of data at rest [24], encrypting data in motion [29], and transparent database encryption (TDE) [30]. In Figure 14.4, the pictorial representation of the potential layers for encryption with varying degrees of development complexity in relevance to the protection offered is shown.

TDE is also widely addressed as external database encryption (EDE) which encrypts the whole dataset staged in the storage media and proves to be a rigorous one [22] as it offers protection against unauthorized access. Additionally, the applications installed upon need not be changed. However, TDE is vulnerable to insider threats and difficult to encrypt the high-dimensional data like healthcare systems, seismographic applications, consumer–financial information datasets, stock market

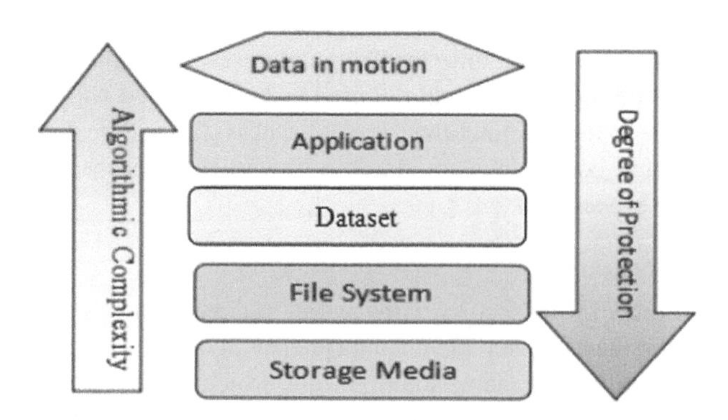

FIGURE 14.4 Potential encryption stages influencing algorithmic complexity and protection offered.

applications, and likewise. Compression of the enciphered data is the other issue that works the other way round, occasionally. TDE is further subclassified into column-level encryption and the field-level encryption techniques. The former one use dissimilar keys for encrypting the distinctive feature of a dataset and avoid the data leakage by preventing the rainbow tables generation. Field-level encryption, also termed as probabilistic or the randomized encryption, enciphers the sensitive fields in such a way that the operations can be performed on the encrypted data resulting in outcomes similar to the same operations done on the raw data without revealing the confidentiality of the sensitive data. However, it lacks the support of the data filters.

For datasets with high degree and cardinality, other popular techniques are file system encryption and the application-level encryption. With the masquerading, the file system encryption can easily be compromised for personal gains as the private key is locally stored. Although the application encryption is the apparent choice, designing an appropriate and robust encryption procedure is very time-consuming, given the consideration that it is of worth. Moreover, as the underlying dataset is subjected to several applications, storing, indexing, and retrieval degrade the overall system performance. The other stage for encrypting the sensitive data is during the transmission. The sensitive fields/columns of the records retrieved by the application may first be encrypted before sent over the communication channel. In the conventional encryption techniques available, enhancement in the key size would exponentially increase the computational cost on the sender and receiver systems.

In the present scenario of large-scale mobile technology ranging from the low-end to high-end devices and ever-changing configuration, meeting the high computational demands of the latest cryptographic procedures at both the ends in all connection sessions across the population framework of the application is very likely in the developed nations. Majority of the modern cryptographic systems used in the healthcare applications designed and deployed in the developed nations in the recent times are either the block or the stream ciphers and keep the algorithmic operational tasks public, whereas the key remains obscure, unlike the classical ciphers that maintain the algorithmic obscurity and manipulate traditional characters. The primitive operations used in these cryptographic methods are enlisted in Table 14.3.

The healthcare applications in resource-poor settings lack high-configured installations and are devoid of frequent upgrades in the technology. However, with the need of presence of mind, the governments in underdeveloped states install and deploy the systems which could meet the minimalistic demands stated by the premier medical councils of the homeland and the world. Thus, the applications tend to maintain the data of the patients enrolled in a medical care system. The administration team of the healthcare application are trained and bestowed with the responsibility of adding more storage units to handle the growing patients' data and also add new additional attributes per patient if needed. The security measures to restore the confidentiality of the sensitive data of the patients from misuse or maligning remain the answerable issue under resource-poor settings. The typical cryptographic methods used are unsuitable in case of data space with high degree and cardinality in low-configured healthcare systems because of their intricate modular arithmetic multiplications in the binary formats expending high amount of computing resources and memory

TABLE 14.3
Comparison among the Stream and Block Ciphers

Cipher Types → Features ↓	Stream Cipher	Block Cipher
Input data type	Single binary bit	Block(s) of bits
Output data type	Single binary Bit	Block(s) of bits
Length of the key	Byte	64 bits
Cipher sub-types	Asynchronous and synchronous	Block ciphers with 128-bit keys, block ciphers with 64-bit keys
Popular ciphers	Asynchronous cipher, synchronous cipher, A5/1 cipher [36]	AES-128 [29], DES-64 [30], Triple DES-64 (3DES) [31]
Operations included	Ex-or, Arithmetic addition	Modular arithmetic additions, modular arithmetic multiplications, transpositions, round functions, padding, substitution
Demand on resources	Presence of fewer processor instructions makes the cipher lighter and faster on resources	Performing multiple processor instructions makes the cipher demand a higher set of resources in terms of storage and processor speeds
Features in	Applications with lesser computational resources: Internet of Things (IoT) devices, GSM mobile handsets for voice encryption	Internet communication
Degree of security	Less secure	More secure

space resulting in low throughput. Additionally, key enhancement is also infeasible as this would worsen the response times in legacy healthcare systems. An efficient lightweight cryptosystem would essentially be put in place to safeguard the data especially in the resource-poor healthcare sector, which is of computationally low cost and provides strong resistance in countering the popular attacks.

14.4 CONCLUDING REMARKS

In this chapter, a discussion is made on the advances in cyber threats and computer crimes prevailing in the healthcare industry. With the budgetary allocation and the efforts channeled towards opposing the attacks, there exists a need to view this situation leading to many unforeseen problems. Especially, in the state of condition where there is a deficit of the high-configured computing installations, and stringent and fool-proof measures should be adopted that would not exceed the available processing capacity and storage requirements.

With these constraints prevailing in the resource-poor settings for the healthcare applications, a novel cryptographic set of procedures' developments is desired that is devoid of bulkier modular arithmetic multiplications and still prove to be the robust encryption technique for efficiently securing the high voluminous and dimensioned patients' data space containing confidential information. In the next section, the novel block cipher is described in detail which takes the key bunch matrix and an additional matrix for enhanced security.

CONTRIBUTIONS

The ideas presented in this manuscript are based on discussions of all authors. Shirisha and Geeta made the analysis of the attacks and the variations in the healthcare budget allocations. Shirisha and Narsinga Rao wrote the first draft of this manuscript. Srilatha provided the latest trend in cyberattacks in the healthcare sector. All authors reviewed and improved the manuscript. All authors read and approved the final manuscript.

REFERENCES

1. Employment Projections, U.S. Bureau of Labor Statistics. 2019. *Employment by Major Industry Sector.* Available online at https://www.bls.gov/emp/tables/employment-by-major-industry-sector.htm
2. Chung H, Mayes J, White A. 2016. *How Smartphone Technology Is Changing Healthcare In Developing Countries.* Newcastle University. Retrieved from https://www.ghjournal.org/how-smartphone-technology-is-hanging- healthcare-in-developing-countries/.
3. Statista. 2016. *Smartphone Users Worldwide 2014–2019.* Retrieved from http://www.statista.com/statistics/330695/number-of-smartphone-users-worldwide/.
4. Global Report. 2019. *Global Spending on Health: A World in Transition.* released by WHO.
5. Bevan, Helderman, Jan-Kees et al. 2010. Changing choices in health care: implications for equity, efficiency and cost. *Health Economics, Policy and Law*, 5(3): 251–267.
6. Hernandez P et al., 2009. Measuring expenditure on the health workforce: concepts, data sources and methods. *Handbook on Monitoring and Evaluation of Human Resources for Health.* World Health Organization.
7. Miakotko L 2017. The impact of smartphones and mobile devices on human health and life. Retrieved from http://www.nyu.edu/classes/keefer/waoe/miakotkol.pdf.
8. Nambiar B, Hargreaves DS, Morroni C et al. 2017. Improving health-care quality in resource-poor settings. Retrieved from https://www.who.int/bulletin/volumes/95/1/16-170803/en/#R1.
9. Horton R. 2014. The third revolution in global health. *Lancet*, 383(9929): 1620. Retrieved from https://doi.org/10.1016/S0140-6736(14)60769-8, (Offline).
10. Leatherman S, Ferris TG, Berwick D et al. 2010. The role of quality improvement in strengthening health systems in developing countries. *International Journal Quality Health Care*, 22(4):237–243. Retrieved from http://dx.doi.org/10.1093/intqhc/mzq028 pmid: 20543209.
11. Crisp N. 2010. *Turning the World Upside Down: The Search for Global Health in the 21st Century. Commonwealth Health Minster's Update.* Royal Society of Medicine Press Ltd., 1–210.
12. Fulop N, Robert G, 2015. *Context for Successful Quality Improvement.* The Health Foundation, 1–116.
13. HIPAA Journal. 2020. Healthcare data breach statistics. Retrieved from: https://www.hipaajournal.com/healthcare-data-breach-statistics.

14. Health IT Security, Xtelligent Healthcare Media. 2020. The 10 biggest HEalthcare Data Breaches of 2019, so far. Accessed on https://healthitsecurity.com/news/the-10-biggest-healthcare-data-breaches-of-2019-so-far (2020).
15. Ragan, S. 2016. *Ransomware Takes Hollywood Hospital Offline, $3.6M Demanded by Attackers*. CSO. Retrieved from https://www.csoonline.com/article/3033160/security/ransomware-takes-hollywood-hospital-offline-36m-demand ed-by-attackers.html.
16. Donovan, F. 2018. *Healthcare Data Breach Costs Remain Highest Among Industries*. Health IT Security. Retrieved from https://healthitsecurity.com/news/ healthcare-data-breach-costs-remain-highest-among-industries.
17. Morgan S. 2020. Healthcare industry to spend $65 billion on cybersecurity from 2017 to 2021. *Cybercrime Magazine*. Retrieved from https://cybersecurityventures.com/healthcare-industry-to-spend-65-billion-on-cybersecurity-from-2017-to-2021/.
18. Health IT Security. 2020. *24% of US Health Employees Never Received Cybersecurity Training*. Retrieved from https://healthitsecurity.com/news/24-of-us-health-employees-never-received-cybersecurity-training (2020/04/23).
19. HIPAA Journal (Online). 2020. Analysis of 2018 healthcare data breaches. Retrieved from https://www.hipaajournal.com/analysis-of-healthcare-data-breaches/.
20. Allor P, 2017. *Cost of Data Breach Study: Global Overview*. Ponemon Institute.
21. The Primary Publication of the Cybersecurity Act of 2015. 2018. *Health Industry Cyber Security Practices: Managing Threats and Protecting Patients (HICP)*. Section 405(d) Task Group. Retrieved from https://www.phe.gov/Preparedness/planning/405d/Documents/HICP-Main-508.pdf (2018).
22. Kakarla, S. Dr. . 2019. Securing large datasets involving fast-performing key bunch matrix block cipher. *Healthcare Data Analytics and Management, Advances in Ubiquitous Sensing Applications for Healthcare*. Elsevier Publications, Paperback ISBN: 9780128153680, https://doi.org/10.1016/C2017-0-03245-7, Vol 2, 111–132.
23. Suciu D. 2012. Technical perspective: SQL on an encrypted database. *Association for Computing Machinery*. Commun. ACM.
24. Database Encryption in SQL Server 2008 Enterprise Edition. 2015. Technet.microsoft.com. Retrieved.
25. Spooner DL, Gudes E. 1984. A unifying approach to the design of a secure database operating system. *IEEE Transactions on Software Engineering*, 10(3): 310–319.
26. Application Encryption from Thales e-Security. 2015. www.thales-esecurity.com.
27. SANS Institute InfoSec Whitepaper. 2007. *Regulations and Standards: Where Encryption Applies*.
28. Babu R. 2019. *Transparent Data Encryption with Azure SQL Database*. SQL Shack.
29. Federal Information Processing Standards Publication 197. 2001. *United States National Institute of Standards and Technology (NIST)*. Advanced encryption standard (AES).
30. Diffie W, Hellman ME 1977. Exhaustive cryptanalysis of the NBS data encryption standard. *Computer*, 10(6): 74–84.
31. De Cannière C 2005. Triple-DES. In: van Tilborg H.C.A. (eds) *Encyclopedia of Cryptography and Security*. Springer.

Index

A

Accountability, 241, 249, 252
Accuracy, 87, 88, 90, 91, 224
Acknowledgement (ACK), 1, 3, 5, 8, 9, 12, 18, 20
Actions, 216
Adaptability, 218
Adaptable Data sharing protocols, 241, 246, 248
Adaptive Systems, 241, 249
Administration, 205, 214
Advanced encryption standard (AES), 112, 114
Advancement, 202, 205, 209
Advantage, 208, 210, 218
Agglomerative, 232
ALARM, 6
Algorithms, 221–228, 231, 233, 236
Alleles, 28, 37, 39
Analysis, 201
Analyze of the data, 122
AngularJS, 128
Annual, 219
Annum, 212
Anomaly-based, 12, 19
Anomaly detection, 223
Antivirus, 209, 210, 216
Appears, 206, 210
Application-level encryption, 254, 255
Apriori Algorithm, 233, 234
Architectural model, 102
Area Under Curve (AUC), 87
Artificial intelligence (AI), 221, 222
Association, 231, 233, 234, 239, 240
Asymmetric Key Cryptography, 192–194, 200
Authentication, 168, 172
Authorized, 207
Automatic, 211
Avalanche effect, 99, 101, 110–112, 117
Awareness, 204, 211

B

Background and Driving Forces, 121
Backgrounds, 203
Banking, 205
Behavior, 202, 204, 205, 216
Bernoulli NB, 90
Beyond, 216
Big Data, 121–123, 130
Binary, 225
Bio-medical Sensors, 52, 53, 55
Biometric Recognition, 237, 238
Block Cipher, 99, 101, 102, 110, 112–115, 117
Blowfish, 112, 114
Body Control Unit (BCU), 55
Bogus, 211, 214
Box-plot, 83
Breaches, 218
Breakthroughs, 208
Brute force attack, 115
Brute-force technique, *see* Brute force attack
Bullying, 205, 207
Bunch, 203
Bypass, 207

C

Caesar cipher, 191, 195, 196
Campaign, 216
Capacity, 215
Capital, human, and natural, 242
Cascading Style Sheets (CSS), 128
Casual, 218
Categories, 205, 215
Caused, 203
Cell nucleus, 27
Changed, 208
Change in Key_Enc, 111
Change in Plaintext, 111
Change in Sup_Key, 112
Charitable, 2
Chosen-ciphertext, *see* chosen-ciphertext attack
Chosen-ciphertext attack, 115
Chosen-plaintext, *see* chosen-plaintext attack
Chosen-plaintext attack, 115
Chromosomes, 25–28, 32, 34–36, 40, 43
Ciphertext-only attack, *see* Brute force attack
Classification, 221, 223–227, 231, 236, 239
Classification of Cyber Crime, 175, 176
Classification of Cyber Terrorism, 173–185
Classification of Cyber Threats, 174
Cloud Computing, 63–68, 72
Cloudera, and MS Azure, 126
Cloud Service Model, 68
Clustering, 231–233, 237, 239, 240
Clustering Based, 221, 231, 239
Collaboration, 14, 23
Column level encryption, 254, 255
Comma Separated Values (CSV), 102
Commit, 203, 206
COMMIT, 5, 15, 21

Community Cloud, 65, 66
Community's, 214
Comparative Analysis, 173, 179, 184–187
Comparative metrics, 101
Comparison of Analytical Tools in BDA, 126
Complex, 218
Comprising, 207
Computer program, 222
Computing resources, 100
Concern, 207, 217
Conclusion & Future Scope, 173, 187
CONFIDENT, 6, 15
Confidentiality, 164, 168
Confusion, 105, 106, 112
Confusion Matrix, 87, 90, 91
Connected, 202, 211 contemporary, 202
Conversation, 203, 210
Cooperation Techniques, 1, 3, 14–19
Cooperative communication, 1, 2
Corporation, 208, 212
Correlation, 228
Correlation Coefficient (ρ), 84, 85
Cost Comparison among Amazon, 126
Council, 213
Counter Measures, 241, 248, 250
Countermeasures against Cyber-crime and Cyber Terrorism, 173, 183
Credential, 202, 210–212, 217
Crediting Mechanism, 1, 3, 4, 15, 16
Crime, 202–205
Crossover, 30, 31, 35, 36
Cryptanalysis, 99, 102, 115
Cryptocurrency, 217, 218
Cryptographic, 9, 10, 18
Cryptographic system, 100
Cryptography Techniques, 191–195, 197
Currency, 217
Current health expenditure, 244, 245
Current health expenditure level, 244, *see* Current health expenditure
Cyberattack, 202–207, 209, 211–214
Cyber-criminals, 202, 210, 211, 214–218
Cyber Law, 177, 180, 181, 187, 189
Cybernetics, 161, 162, 166–171
Cyber security.163, 165, 167, 171
Cyber Terrorism, 180, 183, 189
Cyber-terrorist, 204, 206
Cyber Threat, 173, 180, 181, 188, 189

D

Damage, 204, 205, 207
Danger, 207, 216
Databases, 209
Database Standards Compliance and Monitoring, 242, 253
Data Breach, 161, 163, 164
Data Confidentiality using Encryption, 242, 243, 254
Data Dissemination, 131–134, 138, 140–144, 146, 152, 153, 157, 158
Data encryption standard (DES), 112
Data Masking, 242, 253
Data Security threats, 241, 248, 250
Datasets Classification and Assessment, 242, 253
Data sharing, 246, 247, 249
Date Physical Health Information (PHI), 53–55, 57, 58, 60
Deciphered, *see* decryption
Decision Trees, 226, 227, 236, 239, 240
Decoding, 5
Decryption, 101, 211, 212
Decryption key bunch vector, 109
Dedicated short-range communication (DSRC), 53
Defend, 203, 204
Definitions, 203
Denial of Service, 251
Denial of services attacks, 235
Destabilization, 208
Destroying, 207
Detecting, 203
Detection of DoS and DDoS attacks, 236
Detection of software vulnerabilities, 238
Developments, 210
Diffusion, 106, 109, 112
Digital Information Security, 252
Discrete Markov Chain, 91
Disruption, 208
Diverge, 218
Divisive, 232
Dot product, 101
Dynamic Mobility, 131, 132, 143
Dynamic Source Routing (DSR), 10, 13, 14, 19

E

Eavesdropping, 211
EBCDIC, 105, 107, 108, 111
ECB mode, 112
Eclat, 234, 240
Economic and educational advancement, 241, 246, 247
Efficient lightweight cryptosystem, 256
Effort, 213, 217
Email filtering, 222
Emergency routing protocol VehiHealth, 54, 61
Emerges, 204
Enciphered, *see* encryption
Encipherment, 102, 109
Encoding, 37, 47
Encompasses, 202, 205

Encrypting data in motion, 254
Encrypting files system, 254
Encryption, 102, 115
Encryption key bunch square matrix, 108
Encryption model, 101
Encryption of data at rest, 254
End-to-End, 1, 8, 9, 18
Evidence and Audit based Interventions, 241, 249
Expanded, 208
Explores, 218
Extortion, 204

F

F1 Score, 87, 89–91
Facility, 210
Fake, 210, 211, 218
False negatives (FN), 87, 88
False positives (FP), 87, 88
Falsifies, 211
Fascinating, 202
Fast block cipher, 112, 113
Fast Dataset Block Cipher
Features, 210, 216
Feedback, 162, 167
Field-level encryption, 255
Fishing page identification, 223
Fitness function, 30, 46
Flying Ad Hoc Networks (FANET), 54
FP-Growth Algorithm, 233, 234
Framework, 210
Fraudulent, 208
Funds, 212, 217
Fuzzy k-means algorithm, 233

G

Gadgets, 215
Game, 10–12, 17
Gaussian NB, 90
Gene, 29
Gene pool, 28
Generation, 207
Genotype, 25, 28, 29, 43
Geometric flight plan, 144
Government, 201, 203, 204, 208, 214, 217
Groups, 204, 206

H

Hackers, 204, 210–212, 217
Hadoop, 123
Hadoop Distributed File System (HDFS), 123, 124
Harassment, 207
Health-care, 257
Healthcare Act, 252
Healthcare Cybersecurity Statistics, 250
Healthcare data, 163, 166, 167
Health-care dataset, 107, 117
Healthcare Security, 163, 165, 167, 171
Healthcare systems, 161, 162, 167, 169, 170
Health Expenditure Indicators, 241, 242
Health Information Privacy Code, 252
Health Insurance Portability and Accountability, 252
Health Monitoring System, 51, 54, 55, 60, 61
Heterogeneity, 74
Hidden Semi-Markov Models (HSMMs), 80
Hierarchical clustering, 232
High Availability, 79
High cardinality
Homophonic cipher, 191, 196
Hybrid Cloud, 65–67
Hybrid WBSN and VANET frameworks, 59

I

Identity, 206, 207, 211
Impact, 201, 204, 218
Important, 202, 204, 212, 218
Incentive, 1, 3, 4, 7, 15
Increase, 202–204
Individuals, 202–205, 208, 214, 217
Infiltration through common cyber security attacks, 251
Information, 226, 237, 239, 240
Information Privacy Act, 252
Initial population, 28, 29, 32, 43, 44, 46
Innovative assessment and evaluation, 241, 249, 250
Input variable, 224, 228
Intelligence Transport System (ITS), 53
Internet, 202–205, 211, 212
Internet of things, 63, 64, 69–77, 79
Inter-Quartile Range (IQR), 85
Intrusion detection, 222, 237
IoE, 25
IoE based genetic algorithm, 25–27, 29, 33, 34, 46
IoT, 25, 42, 43, 47
Iteration, 106, 107, 113, 117

J

JSON, 127, 128
Jupyter Notebook, 87
Justice, 204, 205

K

Kernel, 208, 238
Keylogger, 210
Key vector, 99, 101–105, 116, 117

Kidnapping, 205
K-Means algorithm, 232, 233
K-Means Clustering, 232, 236, 237, 239
K-Nearest Neighbor, 226
Knowledge, 204, 211, 217
Known-plaintext, *see* Known-plaintext attack
Known-plaintext attack, 115, 117

L

Labeled data, 223, 224, 236, 239
Lack, 206, 216
Life, 202–204, 207, 215
Light weight computing, 241, 248
Lightweight cryptosystems, 100
Light-weight security, 101
Likelihood, 89
Linear, 228–230, 240
Linear Regression, 228–229, 240
Link, 202, 209–211, 214, 217
Logistic Regression, 225–226, 237–238, 240
Long-distance, 208

M

MAC, 12, 17, 21
Machine learning, 79, 221–229, 231–233, 235–240
Machines, 208, 212, 216
Malevolent, 204, 215, 216, 219
Malicious, 2, 3, 6, 12–14, 22, 23
Malware, 164
Malware Identification, 223, 236
Mapping function, 224
Map Reduce, 123, 125, 126
Markov Property, 80
Mean time to failure (MTTF), 81
Medical emergencies, 51
Medical sector, 241, 244, 246, 247
Methodology, 166
Methodology, 121–123, 129
Micro-services, 79
Mismanaged Sensitive Data and Storage Media Leaks, 251
Modern, 201, 202, 207, 208
Mono Alphabetic ciphers, 191, 196
Multinomial Naïve Bayes, 89–92, 96
Multiplicative inverse, 102, 104, 109, 117
Mutation, 30–31, 33, 34, 39, 40, 44–47

N

Naïve Bayes, 226, 235–237, 239, 240
Naïve Bayes, 79, 81, 89–92, 96
Need, 206, 212
Network, 202–205, 207, 208, 211, 212, 216, 218
Node.js, 128
Node Association, 135, 138, 139, 146, 149, 150
Non-cooperative, 1, 6–8, 10–12, 17–21
Non-linear, 229
Nonviolent, 203
Numerous, 203, 205, 215
Numpy, 87

O

Objective, 204, 208
Occurred, 209, 218
OCEAN, 7, 16
Outbreak, 215, 216
Outlier Detection, 79, 85
Output variable, 224, 228
Overall Equipment Effectiveness (OEE), 80
Overall Production Effectiveness (OPE), 80

P

Packets, 211, 216
Pandas, 87, 90
Participatory Approach, 241, 249
Passwords, 211, 212, 214
Path Planning, 149, 152, 159, 160
PdM, 79–81, 83
Performance, 222, 238
Performance Analysis, 99, 112
Personal Health Information Protection Act, 252
Persons, 202, 205, 215, 216
Perspective, 221, 223, 225, 227, 229, 231, 233, 235, 237, 239
Phantomjs, 128
Phenotype, 25, 28, 29, 43
Phishing, 201, 203–205, 207, 208, 210, 211, 215–218
Phishing Identification, 235, 236
Physical, 203, 204, 207
Physical health information, 244
Platform, 203, 209–211, 215
Playfair cipher, 191, 196, 198
Poly alphabetic cipher, 191, 196
Polynomial Regression, 230, 239, 240
Population, 25, 27–35, 41, 43–46
Pornography, 206, 207
Posterior Probability, 89
Precision, 88–91, 96
Prediction, 222, 223, 225–227, 231
Predictive routing, 132, 136, 155
Primary sector, 242
Prior Probability, 89
Private Cloud, 65–67
Privilege Escalation, 251
Proposed Framework of Cyber Terrorism, 180, 183, 189

Pseudo codes, 99, 101, 105–107
Public Cloud, 65–67
Punishment, 1, 10, 11, 17, 20, 21
Python, 127, 128
Python, 127, 128

Q

Quaternary, 242
Quick, 202, 212
Quinary, 242

R

Radio-frequency Identification, 70
Rail fence technique, 191
Random Forest, 227, 238, 240
Randomness, 102, 106
Range, 204, 205
Ransom, 211, 212
Ransomware, 211, 212, 215, 216, 218
Ransomware, 251
RCare, 54, 61
Recall, 87–91, 96
Receiver Operating Characteristics (ROC), 87
References, 173, 188
Refers, 202–205
Regression, 221, 223–231, 236, 238–240
Related key attack, 101
Remaining Useful Lifetime (RUL), 80
Requirements, 202, 216
Research Implications, 173, 185
Research Limitations, 173, 187
Resource poor countries, 244, 254
Resource poor health care, 248, 256
Resource poor settings, 241, 242, 248, 250, 252, 255–257
Resources, 242, 249, 256, 257
Response, 214, 217
RFID, 54, 61
Ridge Regression, 229–231, 240
Roadside units (RSU), 53, 55, 58
Role players, 241, 244
Route reply (RREP), 6
Route request (RREQ), 6, 11
Routing protocol, 131–136, 139–142, 154–157
Routing protocols, 2, 5, 21, 22
RTS, 12, 19
RTS/CTS, 12

S

Scammers, 214, 217
Scheduling, 131, 132, 138, 142, 148–151, 156, 158
Schemes, 218
Secondary sector, 242
Security, 201–208, 210–212, 214–218
Security and privacy, 72, 76
Selection, 25, 27, 30–35, 43, 44, 46, 47
Self-executing, 208
Selfish, 2, 3, 5–8, 10–14, 19–23
Send Output, 128
Send Output, 128
Server, 204, 207, 209, 211
Sexual, 206, 207
Simple columnar technique, 191
Sklearn, 84, 87, 90
Smart vehicular ad hoc network (SVANET), 54, 60
Social, 203, 204, 207, 208, 218
Software, 203, 204, 208, 209, 211, 216
Software vulnerability diagnosis, 223
SORI, 6, 7, 16, 22
Spam Detection, 235
Spammers, 204
SQL Injection attack and No-SQL Injection Attack, 251
Stalking, 207
State Transition Diagram, 93, 94
Stream and Block Ciphers, 256
Substitution Technique, 191, 195
Summary, 131, 154
Supervised machine learning, 223, 224, 228, 239
Supervised techniques, 222, 239
Supplementary key block, 101
Support Vector Machine (SVM), 227, 235, 237
Support Vector Machine, 227, 235–237
Symmetric key Cryptography, 192–194, 200
System Design, 121, 125
Systems, 202, 205, 207, 208, 210, 212, 216

T

Target 204, 205, 207, 209, 214, 216, 217 technologies, 202, 203
TCP, 9, 18
Technology Used, 121, 128
Technology Used, 121, 128
Telecommunication, 208, 213
Ternary, 225
Terrorism, 203–206
Tertiary, 242, 249
Tertiary sector, 242
Tested, 222, 224
Theoretical analyses, 101
Time to Failure (TTF), 94, 95
Tool and Technology, 121, 123
Total Effective Equipment Performance (TEEP), 80
Training, 222–224, 227, 229, 232, 235
Trajectory Optimization, 131–133, 149–152, 159, 160

Transparent Database Encryption, 254, 255
Transposition technique, 191, 198
Trojan, 209
True negatives (TN), 87, 88, 93, 95
True positives (TP), 87, 88
Trusting, 5
Types of Cryptography, 192

U

UAV placement, 131–134, 138, 139, 145
Unauthorized, 202, 205, 208
Undetectable, 210
Unsuccessful, 217
Unsupervised, 221–223, 231–233, 239, 240
Unsupervised machine learning, 223, 231, 239, 240
Unsupervised techniques, 222, 239

V

Ad-hoc Vickrey, Clarke, and Groves (VCG), 4, 5, 15, 21
Validation of data, 168
Various, 201–203, 210, 214, 215
Vectors, 208
Vehicle Communication (V2V), 53, 54
Vehicle to Infrastructure Communication (V2I), 53
Vehicular Ad-hoc Network (VANET), 51–57, 59–61
Victim, 204, 209–211, 217
Virus, 204, 205, 207–212, 216–218
Voltage Sensor, 81, 84, 85, 88, 89
Volunteers, 218
Vulnerability, 173, 174, 179, 212
Vulnerability Types, 174

W

Web, 202, 203, 209, 212
Web Scraping, 127
Website, 209, 210, 214, 215, 217
Weibull, 79, 95
Weighted, 231
Weighting, 8, 16
W-GeoR, 54, 55, 61
Wifi/Wi-Fi, 51, 201, 211, 216
Wireless Body Sensor Network (WBSN), 51–53, 55–57, 59–61
Wireless communication, 2, 22, 23
Wireless sensor network (WSN), 54
Worldwide, 212, 214

Y

Year, 202, 203, 208, 209, 218
Younger, 207
Your, 211, 218

Z

ZOMATO, 212
Z-Score, 85